Stem Cells

Suchandra Chowdhury • Shyamasree Ghosh

Stem Cells

Biology and Therapeutics

Springer

Suchandra Chowdhury
Department of Zoology
Bethune College
Kolkata, West Bengal, India

Shyamasree Ghosh
SBS
NISER
Bhubaneswar, Odisha, India

ISBN 978-981-16-1637-2 ISBN 978-981-16-1638-9 (eBook)
https://doi.org/10.1007/978-981-16-1638-9

This Springer imprint is published by the registered company Springer Nature Singapore Pte Ltd.
The registered company address is: 152 Beach Road, #21-01/04 Gateway East, Singapore 189721, Singapore

Contents

About the Authors

Suchandra Chowdhury is currently working as an Assistant Professor of Zoology at Bethune College, Kolkata. She has worked in the field of immunology and glycosylation in stem cells, protein chemistry working on haematopoietic malignancy. She did her Ph.D. from CSIR-IICB, Kolkata, India, Jadavpur University, post-graduation and graduation from the University of Calcutta.

Shyamasree Ghosh has worked and published in the domain of glycobiology, cancer biology, nanotechnology, immunology, and computational biology. She completed her graduation and post-graduation studies under Calcutta University and Ph.D. from IICB, Kolkata, India. In her Post-Doctoral Research, she worked in stem cells and nanotechnology. Later she joined as a faculty and held the position of Chair in college affiliated to Bangalore University. Currently, she is serving as Scientific Officer at NISER, Bhubaneswar, Orissa, India. She is a member of many scientific bodies and has received many awards.

Stem Cells

Stem Cells an Overview

1

Contents

Abstract

Stem cells (SC) are the progenitor cells with the ability to self-renew and differentiate into various lineages. They find importance in stem cell therapy, gene therapy, and tissue engineering. While embryonic stem cells (ESCs) are bestowed with the hallmark property of self-renewal and ability to differentiate into different lineages, the mesenchymal stem cells (MSCs) reveal restricted lineage potential as compared to the ESCs. In this chapter, we discuss the different types of stem cells including ESCs, adult stem cells, stem cells from umbilical cord blood, tissue-specific stem cells (TSSCs), induced pluripotent cells, cancer stem cells, and stem cell plasticity.

Keywords

Stem cells · Embryonic stem cells · Adult stem cells · Induced pluripotent stem cells · Cancer stem cell · Stem cell plasticity

Abbreviations

ADSCs	Adipose-derived stem cells
ASC	Adult stem cell
BM	Bone marrow

S. Chowdhury, S. Ghosh, *Stem Cells*, https://doi.org/10.1007/978-981-16-1638-9_1

CarScs	Cardiac Stem cells
CESCs	Corneal epithelial stem cells
CSC	Cancer stem cell
DC	Dendritic cells
DFPCs	Dental follicle progenitor cells
DPPSCs	Dental pulp pluripotent-like stem cells
DPSCs	Dental pulp stem cells
EISC	Epithelial intestine stem cells
EMT	Epithelial-to-mesenchymal transition
EndoSCs	Endometrial and Myometrial stem cells
ESC	Embryonic stem cells
GbSC	Gallbladder stem cells
GIFM	Genetic inducible fate mapping
HAP	Hair follicle-associated-pluripotent stem cells
hESCs	Human ESCs
HpSCs	Hepatic stem cells
HSC	Haematopoietic stem cells
ICM	Inner cell mass
iPSC	Induced pluripotent stem cell
KSC	Keratinocyte stem cells
LESC	Lingual epithelial stem cells
LSC	Limbal stem cells
MaSc	Mammary epithelial stem cells
MEP	Megakaryocytic-erythroid progenitors
MerSc	Merkel cell stem cells
MeSc	Melanocyte stem cells
MFs	Macrophages
MOPD	MF/OC/DC development progenitors
MSCs	Mesenchymal stromal/stem cells
MuSC	Muscle stem cell
MyoSCs	Myometrial stem cells
NSC	Neural stem cells
OCs	Osteoclasts
OEpSCs	Odontogenic epithelial stem cells
OeSC	Oesophageal epithelial stem cells
OESC	Oral epithelial stem cells
PDLSCs	Periodontal ligament stem cells
PrSc	Prostrate Stem cells
PSCs	Pluripotent stem cells
SC	Stem cells
SCAP	Stem cells from apical papilla
SGSC	Salivary gland stem cells
SHED	Stem cells from exfoliating deciduous teeth
SKP	Skin derived precursors
SSCs	Spermatogonial stem cells

SVZ	Subventricular zone
TA	transit-amplifying cells
TESC	Thymus stem cells
TSSCs	Tissue-specific stem cells
UCB	Umbilical cord blood
VSELs	Very small embryonic-like stem cells

1.1 Introduction

The stem cells (SCs) are unspecialized, undifferentiated, progenitor cells of living organisms, with the ability of self-renewal and differentiation into any functional cell of the body including the entire organism [1, 2]. Two basic criteria define a stem cell. Primarily, they have the potential of renewing themselves all through the life, i.e., the division of these cells produces identical daughter cells to maintain the primitive stem cell population. Secondarily, they differentiate into different cell types of diverse lineage [3]. When the stem cells differentiate, division is asymmetric, leading to the production of an identical cell and a daughter cell having properties of a specific cell type, for example, explicit phenotype, morphology, and physiological properties which label it as a cell, specific to a certain tissue (Fig. 1.1) [4, 5].

Plants also have innately differentiated stem cells located in the meristem of plants called as plant stem cells which will be discussed in the subsequent chapter. In this chapter we have discussed stem cells from the human and animal origin.

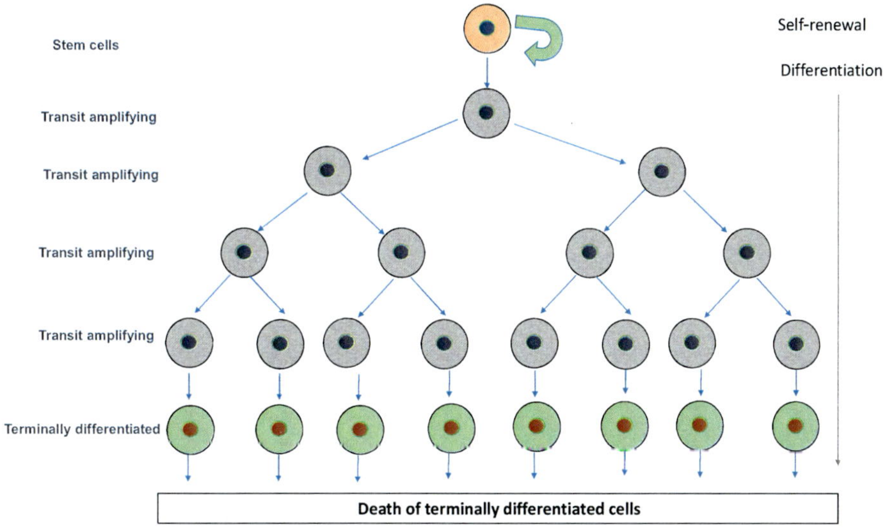

Fig. 1.1 Illustration of the potential of a stem cell. Stem cell(s) divide asymmetrically for self-renewal, forming transit amplifying cells (TA), whose functional capacity evolves concurrently with a decrease in division potential. TA cells later become terminally differentiated (TD) and are programmed to decease in a tissue specific manner

1.2 Concept of "Stem Cell"

The stem cells have captivated biologists for over a century. The derivation of the term "stem cell", dates back to nineteenth century where it originated in the context of two major embryological questions regarding the continuity of the germplasm and the origin of the haematopoietic system. Theodor Boveri and Valentin Ha¨cker used the term stem cell to define cells committed to form the germline. Artur Pappenheim, Alexander Maximow, Ernst Neumann, and others described these cells to be the expected progenitor of the blood system [5].

The concept of "stem cell" was first introduced by Till and McCulloch following their ground-breaking studies of the regeneration of blood system in vivo [6]. The detection of spleen colony forming units in the bone marrow (BM) of mouse was one of the first experiments to study stem cell diversity [7]. The ability of self-renewal of cells to form colonies was established in 1963 [8]. Since then the study has come a far way with discoveries of stem cell from different sources at different sites having variable differential potential.

1.3 Stem Cell "Niche"

A three-dimensional physiological microenvironment, consisting of specialized cells, where stem cells reside, support, and regulate their ability to self-renew and produce progeny is known as stem cell "niche" [9]. Thus the association of tissue cells and extracellular substrates, which supports the survival of stem cell in vivo is known as the "niche" [10]. The concept of a stem cell "niche" enunciating the practical features of a specialized microenvironment on stem cell function in vivo was proposed by Raymond Schofield. According to him, a stem cell "niche" is a site which restricts the entrance of a stem cell into the cell cycle for further differentiation program, integrates signals reflecting tissue and organismal state, imposes stem cell features on daughter cells, and also elicits mechanisms for limiting mutational errors [11].

The survival and development of the committed cells are governed by the interaction with other cell types and the components of the extracellular matrix. One of the most considered receptors is the integrins which hold the cells in place, and their absence leads to mobilization through the "niche" [12–14]. The capability of a cell to remain in an undifferentiated state, proliferate, or differentiate relies on the expression of numerous transcription factors within it which is derived from extrinsic or intrinsic sources or, a combination of the two [4].

The presence of a stem cell "niche" has been verified from the mammalian adult stem cells of the intestinal, neural, epidermal, and hematopoietic systems. Stem cell "niche" or the haematopoietic microenvironment is an elaborately studied system which consists of haematopoietic and non-haematopoietic (both cellular and extra-cellular) components [11, 15, 16]. The stem cell "niche" provides the composite molecular signals that guide the process of haematopoiesis and is also partly accountable for the control of the differentiation and maturation of haematopoietic stem cells (HSCs) [17]. The crypt base of the small intestine, in which intestinal epithelial stem cells reside [18], and the bulge region of hair follicles, where cutaneous epithelial stem cells are found, are some additional examples of stem cell "niche"s [19].

The presence of cancer stem cell (CSC) "niche" also exists, which interacts with the self-renewing population of tumour cells. Evidences suggest that the factors obtained from tumour microenvironment serve to control cancer cells. Disintegrating such aberrant "niche"s may impair the CSC self-renewal, thus suggestively hindering tumour progress, and providing a solemn aspect of cancer therapy [9].

1.4 Stem Cells: Classification

On the basis of their ability to differentiate, stem cells are classified into totipotent, pluripotent, multipotent, oligopotent, tripotent, bipotent, and unipotent. Developmental potency reduces with every step of differentiation [1, 20, 21]. Stem cells are grouped into two types on the basis of their source of origin: the embryonic and the adult stem cells [1, 2]. Tissue-specific or adult stem cells are located in specific "niche"s of nearby organ or the tissue of origin where the microenvironment supports them to remain quiescent, to multiply as undifferentiated cells (self-renewal), and to differentiate into progenitors or terminally differentiated cells which migrate from the "niche" to perform specific roles [20].

1.4.1 Classification of Stem Cells on the Basis of Potency

Totipotent stem cells possess maximum differentiation potential, being capable of dividing and differentiating into the whole organism including embryo and extra-embryonic lineage. The best example of a totipotent cell is the zygote, formed after the fertilization of sperm and ovum. These cells differentiate into any/all of the three germ layers and also can form the placenta. Blastomere from a two-cell embryo has the potential of developing into a whole organism and is also totipotent [20, 22, 23].

After about 4 days of fertilization, inner cell mass (ICM) from the blastocyst becomes pluripotent. The pluripotent stem cells (PSCs) can generate cells of all germ layers but not the extra-embryonic structures, like placenta [1]. The embryonic stem cells (ESCs) can be derived from ICM cells and are pluripotent [24]. The discovery and study of ESCs pioneered the discovery of another kind of pluripotent cells named induced pluripotent stem cell (iPSC) through transcriptional and epigenetic networks [25, 26]. iPSCs are generated from somatic cells through artificial means, and their function is similar to that of the PSCs [1].

Multipotent stem cells possess a narrow range of differentiation potential producing cells of specific lineages. They are considered to be adult stem cells due to their restricted ability of differentiation [2]. HSCs are an example which forms all types of blood cells [27]. Another example is the multipotent stem cells of the brain, which

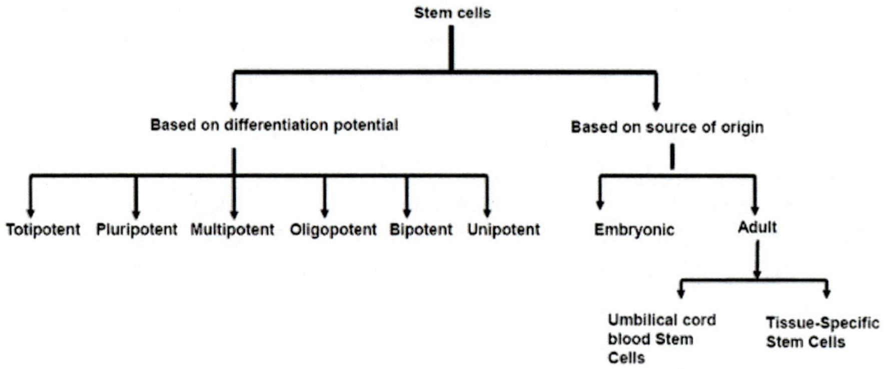

Fig. 1.2 Classification of the stem cells

can generate dissimilar neural cells and glia. The BM is an important source of multipotent stem cells. It harbours a varied population of cells like HSCs, marrow endothelial progenitor cells, stromal cells, and very small embryonic-like stem cells (VSELs) [2]. The neuronal stem cells and epithelial stem cells are few more examples [20].

With another step of differentiation, multipotent stem cells become oligopotent stem cells, which differentiate into several cell types. After differentiation, HSC becomes oligopotent. Their differentiation abilities become restricted to cells of its own lineage. An example is the myeloid stem cell which differentiate into blood cells other than leucocytes. Another example is the oligopotent stem cells (holoclones) distributed throughout the ocular surface including the cornea, having the ability to form individual colonies of corneal and conjunctival cells [1, 28].

Tripotent stem cells differentiate into highly specific cell types like monocytes/ macrophages (MFs), osteoclasts (OCs), and dendritic cells (DCs), differentiating from the oligopotent progenitors [21].

Bipotent stem cells differentiate into two functionally different cell types like the megakaryocytic-erythroid progenitors (MEP), which generate two totally diverse functional outputs: platelets and red blood cells [29]. Other examples include the cardiac/skeletal myogenic progenitors [30].

Unipotent stem cells are categorized by the narrowest differentiation potential and a specific property of repeated divisions. These cells form only one cell type, e.g. dermatocytes [1] and spermatogonial stem cells (SSCs) [20]. The latter feature of these cells makes them a promising candidate for therapeutic use. They determine the homeostasis and development of mammary gland [31], hepatocytes [32], megakaryocytes [33] (Fig. 1.2).

1.4.2 Classification of Stem Cells on the Basis of Source

1.4.2.1 Embryonic Stem Cells

Embryonic stem cells (ESC) are the pluripotent stem cells with the potential of indefinite self-renewal and differentiating into all somatic cells of the embryo. Human ESCs (hESCs) are isolated from inner cell mass (ICM) of blastocysts [34] and the germ cells referred to as embryonic germ cells (EG), derived from spontaneously aborted foetuses [35]. ESCs help to understand the mechanisms of development of dedicated cells and formation of organ structures. The plasticity and self-renewal of ESCs enable generation of infinite number of distinct cell types in vitro, thereby opening new avenues for regenerative medicine [36, 37].

1.4.2.2 Adult Stem Cells

The conservation and repair of adult tissues depend on trivial populations of residential stem cells known as adult or postnatal stem cells (ASCs). These dedicated cells are defined by their skill to maintain themselves over prolong periods of time ("self-renewal") and produce all differentiated cell types of the tissue ("multipotency"). Mostly, ASCs are multipotent to unipotent in nature [2, 38–40]. They reside in almost all adult tissues, like brain, blood, liver, intestine, skeletal muscle, skin, and cornea. Stem cells isolated from these adult tissues can regenerate the respective tissues, helping in healing of ailing and injured tissues, serving as gene-delivery vehicles [41]. The list of adult stem cells is shown in Fig. 1.2.

In addition to the aforesaid defining characteristics, ASCs are dormant within their "niche", dividing irregularly to produce one stem cell copy and a rapidly cycling cell or transit-amplifying cells (TA). TA cells undergo a limited number of cell divisions before they terminally differentiate into the functional cells of that particular tissue. This type of stem cell-driven tissue renewal is particularly obvious in the epithelial lining of intestine, stomach, and skin, which is constantly exposed to a bombardment of environmental mugging [39].

The gold standards for the identification of adult stem cells are genetic inducible fate mapping (GIFM) and transplantation [42]. ASCs can be broadly classified as endodermal, mesodermal, and ectodermal on the basis of their origin of development. The subsequent types of stem cells are endodermal in origin, namely, gastrointestinal tract SCs, pancreatic SCs, hepatic SCs, mammary and prostate gland SCs, testicular and ovarian SCs. The list of SCs having mesodermal origin are the HSCs, cardiac SCs, mesenchymal stromal SCs, and the satellite cells located in the muscle to name a few. The ocular, skin, and neural SCs are of the ectodermal origin [43].

ASCs usually fall under one of the following types, namely, epithelial, mesenchymal, neural, and muscle based on the type of cells. Epithelial stem cells are rarely cycling cells with superior proliferative capacity. These cells are biochemically and morphologically primitive and multipotent in nature [44]. The progeny of the stem cells mobilizes from the stem cell "niche" as TA cells, having limited proliferative potential and becomes postmitotic when such potential is exhausted [45].

Mesenchymal stromal/stem cells (MSCs) which are multipotent with self-renewal ability are located in many human organs and encompass a heterogeneous population of cells. They range from resident SCs in liver, bone marrow, skin, skeletal muscle, heart, and lung and are linked to regeneration of tissue and maintenance of homeostasis. Some of these cells exhibit extensive plasticity and differentiate into cells of different lineage. MSCs from adult human tissues are useful for regeneration of tissue and tissue engineering. Additionally, they also exhibit immunomodulatory and anti-inflammatory features which can regulate tissue repair by modulation of local environment [46].

The stem cells of the nervous system are referred to as the neural stem cells (NSCs). All through ontogenesis, they form the complete nervous system. In adults, few NSCs remain quiescent, exhibiting their roles in plasticity, disease, ageing, and regeneration of the nervous system [47, 48].

Muscle stem cells (MuSc) enable prolonged capacity of regeneration of skeletal muscle. These cells continue in a dormant state in intact muscle, and are stimulated in case of any muscle disorder to produce committed progeny [49].

The ASCs can be broadly divided into the umbilical cord blood (UCB) stem cells and tissue-specific stem cells (TSSCs) based on their sites of residence. We have introduced these cell types here (Fig. 1.3, Table 1.1) and discussed them in detail in the following chapters.

Umbilical Cord Blood Stem Cells

Blood which remains in placenta and in the attached umbilical cord post-child birth is the umbilical cord blood (UCB). Human UCB is a very strong source of stem cells that is widely used for therapeutic reconstitution. These stem cells are the multipotent progenitors, with the ability of self-renewal as can be seen by their replating efficiency in colony assay in vitro where they differentiate to form the multipotential erythroid (BFU-E), CFU-GEMM, and granulocyte-macrophage (CFU-GM) progenitor cells. Progenitor cells collected from human UCB can be preserved for many weeks in long-term liquid culture systems, signifying their production from primitive cells [104, 105]. UCB is a substitute of HSC in patients suffering from haematologic diseases, that can be cured only by allogeneic HSC transplantation [106]. The stem cells from UCB hold potential for treating blood malignancies, many genetic diseases, and immune deficiencies. UCB can also be of therapeutic use in case of a sick sibling or relative [107].

Tissue-Specific Stem Cells (TSSCs)

The tissue-specific stem cells (TSSCs) cells are residents of individual organs (Fig. 1.3) and are responsible for maintaining the homeostasis in them. They also become functional during injury or damage. Most of these are multi- to unipotent in nature. The details of these cells have been given in Table 1.1 and discussed in detail in Chap. 2. Most of these stem cells are used for therapeutic purpose in regenerative medicine [47, 48, 51–103].

1.4.2.3 Induced Pluripotent Cells

Cells which have already differentiated are reprogrammed to a de-differentiated embryonic-like state, by the transfer of nuclear contents into oocytes or by fusing

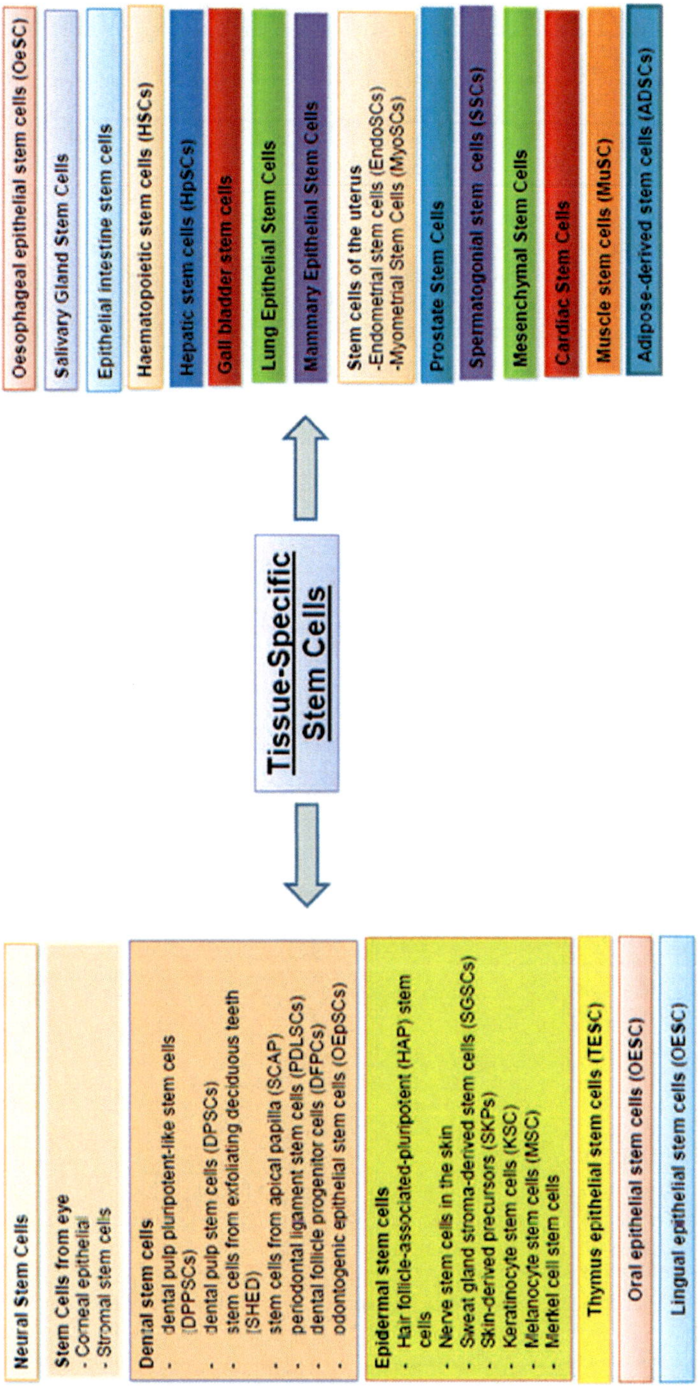

Fig. 1.3 Types of tissue-specific stem cells in adults

Table 1.1 Tissue-specific stem cells (TSSCs) in their respective "niche"s

Tissue-specific stem cells		Location	Origin	Therapeutic usage
Neural stem cells (NSC)		Subventricular zone (SVZ) of brain [50]	Neural [47, 48]	In spinal cord injury and repair, ischemic stroke and neurodegenerative disorder [51–53]
Stem cells in the eye	Corneal epithelial (CESCs)/ Limbal stem cells (LSCs)	Limbus, found in basal layer of limbal epithelium, in the junction between cornea and conjunctiva [54]	Epithelial [55]	Maintenance of the corneal epithelium [54]
	Stromal stem cells	Corneal stroma [56]	Mesenchymal [56]	Remodel pathological stromal tissue [56]
Dental stem cells (DSCs)	Dental pulp pluripotent-like stem cells (DPPSCs)	Dental pulp from third molar [57, 58]	Mesenchymal [59]	Differentiate into the osteogenic, odontogenic, myogenic, adipogenic, neurogenic, and melanocytic cells; transdifferentiate to corneal cells and islet cells of pancreas, holding promise in regenerative dentistry and medicine [57, 59]
	Dental pulp stem cells (DPSCs)			
	Stem cells from exfoliating deciduous teeth (SHED)			
	Stem cells from apical papilla (SCAP)			
	Periodontal ligament stem cells (PDLSCs)			
	Dental follicle progenitor cells (DFPCs)			
	Odontogenic epithelial stem cells (OEpSCs)	Dental lamina [59]	Epithelial [59]	
Stem cells found in the epidermis (skin)	Hair follicle-associated-pluripotent (HAP) stem cells	Bulge area of hair follicle in the skin, under sebaceous gland [60, 61]	Epithelial [62]	Differentiates into keratinocytes, nerve cells, glial cells, smooth and cardiac

(continued)

Table 1.1 (continued)

Tissue-specific stem cells	Location	Origin	Therapeutic usage
			muscle cells, and melanocytes [60]
Nerve stem cells located in the skin	Skin [60]		Shows trans-differentiation in addition to maintaining homeostasis [60]
Stem cells located in the eccrine gland	Dermal papilla [60]		
Skin derived precursors (SKP)			
Keratinocyte stem cells (KSC)	Hair follicle [60]		Differentiate into keratinocytes [63]
Melanocyte stem cells (MeSC)	Skin [60]		Differentiate into melanocytes [64]
Merkel cell stem cells (MerSc)	Epidermis [65]	Epithelial [65]	Differentiate into Merkel cells [65]
Thymus stem cells (TESC)	Thymus [66]	Epithelial [66]	Thymic regeneration [66]
Oral epithelial stem cells (OESC)	Basal layer of mucosa throughout oral cavity [67]	Epithelial [67]	Autologous cell-based therapy for bilateral limbal stem cell deficiency [68]
Lingual epithelial stem cells (LESC)	Epithelial cell layer located at the base of the interpapillary pit of the tongue [69]	Epithelial [69]	Regenerative medicine [69]
Oesophageal epithelial stem cells (OeSC)	Interpapillary zone of oesophagus [70, 71]	Epithelial [70, 71]	Maintenance of the oesophagus [70, 71]
Salivary gland stem cells (SGSC)	Ductal compartment of salivary gland	Epithelial [72]	Renewal of salivary gland tissue [73]
Epithelial intestine stem cells (EISC)	Luminal side of intestine [74, 75]	Epithelial [74]	Renewal and repair [74]
Haematopoietic stem cells (HSCs)	Mobilized peripheral blood [76, 77], the bone marrow [77] and umbilical cord blood [77–79]	Mesodermal [43]	Haematopoietic reconstitution [80]
Gallbladder stem cells (GbSC)	Mucosal crypts of gall bladder [81]	Endodermal [82]	Renewal and repair [81]
Hepatic stem cells (HpSCs)	Proximal biliary structures, meeting point between hepatocyte canaliculi and the bile ducts,	Mesodermal [43]	Renewal and repair [83, 84]

(continued)

Table 1.1 (continued)

Tissue-specific stem cells	Location	Origin	Therapeutic usage
	canals of Hering [83, 84]		
Lung epithelial stem cells (LESC)	Basal and mucous cells of proximal airways and in Clara cells of bronchioles and type II pneumocytes in the alveoli [85]	Epithelial [85]	Renewal and repair [86, 87]
Mammary epithelial stem cells (MaSCs)	Normal mammary gland tissue [88]	Epithelial [89]	Postnatal organogenesis, remodelling; developmental and homeostatic functions [89]
Stem cells of uterus: Endometrial and myometrial stem cells (EndoSCs)	Perivascular regions in human endometrium [90]	Epithelial [91]	Renewal and regeneration [92]
Stem cells of uterus: Myometrial stem cells (MyoSCs)	Myometrium [90]		
Prostate stem cells (PrSC)	Prostate gland [93]	Epithelial [93]	Renewal and regeneration [93]
Spermatogonial stem cells (SSCs)	Seminiferous tubule along with undifferentiated spermatogonia [94, 95]	Epithelial [94, 95]	Developmental and homeostatic functions [94, 95]
Mesenchymal stem cells (MSCs)	Bone marrow [96], peripheral blood [97], adipose tissue, [98]	Mesenchymal [99]	Renewal, regeneration, and plasticity [99]
Cardiac stem cells (CarSCs)	Myocardium [100]	Mesodermal [43]	Renewal and regeneration [101]
Muscle stem cells (MuSC)	Between sarcolemma and basement membrane of the terminally-differentiated muscle fibres [102]	Mesodermal [43]	Renewal and regeneration [102]
Adipose-derived stem cells (ADSCs)	Stromal-vascular portion of tissue [103]	Mesodermal [43]	Autologous transplant in therapy targeted damaged tissue [103]

with the embryonic stem (ES) cells. Such cells are the induced pluripotent cells (iPSC). This induction of pluripotency in stem cells was done by Takahashi et al. in 2006 from mouse embryonic or adult fibroblasts, by introducing four transcription factors, namely, Sox2, Oct3/4, c-Myc, and Klf4 [101]. A similar process of generation of the iPSCs from dermal fibroblasts of adult human was done adding the same four transcription factors by Yamanaka et al. [108]. Thus, pluripotency can be induced in any cells.

The arena of stem cells biology has been strongly driven by this study of iPSCs, which provides novel cells, for exploring the molecular mechanisms of pluripotency, ageing, and cancer biology. Additionally, they can be used as a cellular tool to disentangle the underlying mechanisms of human development and several diseases. The iPSCs also hold potential for different cell-based therapies and regenerative medicine [109].

1.4.2.4 Cancer Stem Cells

The perception of cancer stem cells (CSCs) first evolved in 1990 by John Dick in the late 1990s [110]. According to this proposal, tumours develop from a rare population of cells having stem cell properties, and are termed CSCs. A miniscule portion of cells within tumours are tumorigenic, hinting towards the fact that only the CSCs can produce the malignant cells necessary to repopulate a tumour. CSCs share normal stem cell properties like self-renewal, remaining quiescent and getting activated only upon stimulus. Additionally, they produce cells that are different from them, utilizing common signalling pathways. Mostly they are robust and resistant to treatment, making their management even more difficult [9, 111].

1.4.2.5 Plasticity of Stem Cells or Transdifferentiation

Plasticity of stem cells refers to the flexibility of these cells derived from tissues, to display a phenotypic potential which extends beyond the differentiated cell phenotypes of their residing tissue (Fig. 1.4) [112, 113]. Thus, this property of the stem cells, also termed as transdifferentiation, allows stem cells committed to differentiate to a definite progeny, to produce other specialized cells, when relocated to a new "niche" [113].

Epithelial plasticity is a common phenomenon seen in both embryonic and cancer cells. During embryonic development, numerous cells travel long distances within the embryo, to reach their final destination. These embryonic epithelial cells endure a process of mesenchymal conversion termed epithelial-to-mesenchymal transition (EMT) to increase motility and invasiveness. Similarly, EMT is seen not only in cancer cell but also in CSCs where they escape from the primary tumour and distribute to other parts of the body to inhabit distant organs and form metastases. The reverse of EMT is also noted where both embryonic and cancerous cells revert to their original epithelial phenotype to differentiate into organs or form secondary tumours [114, 115].

One of the most elaborately studied stem cells is the HSC, residents of the BM. The BM also contains MSCs. Very recently it has been treasured that the BM

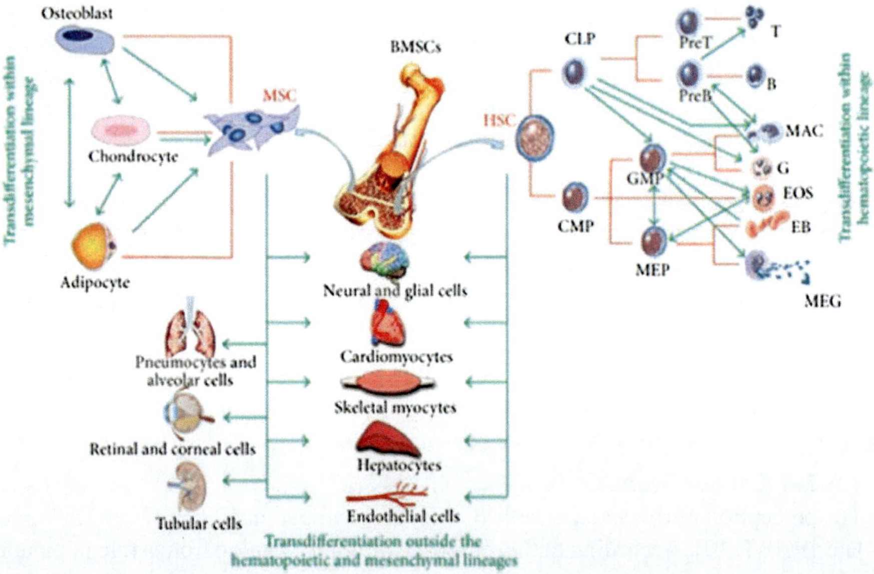

Fig. 1.4 Plasticity of HSC, BM, and MSC. Reproduced from Catacchio I, Berardi S, Reale A, Luisi ADe, Racanelli V, Vacca A, and Ria R, Evidence for Bone Marrow Adult Stem Cell Plasticity: Properties, Molecular Mechanisms, Negative Aspects, and Clinical Applications of Hematopoietic and Mesenchymal Stem Cells Transdifferentiation, Volume 2013, Article ID 589139, 11 pages, under open access article distributed under the Creative Commons Attribution License

also contributes cells in other organs. As an explanation it is emphasized that the plasticity of these cells helps them to act as local stem cells in its new "niche" [116].

MSCs from marrow aspirates differentiate to fibroblastic colonies which further differentiate into bone and cartilage that can be transplanted [117]. BM derived CD34 + cells also contribute in a more delicate way to repair of cardiac function after myocardial infarction [118].

An early demonstration of stem cell plasticity was revealed in the rat liver by Petersen et al. where the oval hepatic cells expressed antigens like CD34, c-kit, Thy-1, and flt-3 traditionally associated with haematopoietic cells. Further investigation revealed that a stem cell associated with the BM has epithelial lineage [119]. Similar observation was noted in humans where hepatocytes were derived from BM cell populations [120]. Thus, there is adequate evidence indicating that organ-specific stem cells need not depend entirely on their own resources for preservation and repair. Such reflection is also manifested in the research arena where there is increased hope in exploiting stem cell plasticity in therapeutic applications [113].

1.5 Discussion

It is evident from the above information that stem cells are found in almost every organ of the body including embryos and the UCB. Their main function is to renew, survive in quiescent condition, repair, and maintain homeostasis. This is why isolation, culture, cryopreservation, and further usage of these cells have been extensively studied and are still continuing. They find usage in clinical research and in therapeutic application.

While the normal stem cells restore balance and become proactive in case of injury or inflammation, its counterpart, the cancer stem cells are fatal because of having similar properties. The self-renewal ability of the CSCs is one of the major reasons for relapse and mortality in cancer-patients. Understanding the CSCs and deciphering their biology can improve management strategies of cancer. The enigma of stem cells, discussed in the following chapters, has come a long way with its prospects and can be miraculous in many therapies.

References

1. Zakrzewski W, Dobrzyński M, Szymonowicz M, Rybak Z (2019) Stem cells: past, present, and future. Stem Cell Res Ther 10:68. https://doi.org/10.1186/s13287-019-1165-5
2. Sobhani A, Khanlarkhani N, Baazm M, Mohammadzadeh F, Najafi A, Mehdinejadiani S, Sargolzaei AF (2017) Multipotent stem cell and current application. Acta Med Iran 55(1):6–23
3. Burns CE, Zon LI (2002) Portrait of a stem cell. Dev Cell 3:612–613
4. Preston SL, Alison MR, Forbes SJ, Direkze NC, Poulsom R, Wright NA (2003) The new stem cell biology: something for everyone. Mol Pathol 56(2):86–96
5. Ramalho-Santos M, Willenbring H (2007) On the origin of the term "stem cell". Cell Stem Cell 1(1):35–38. https://doi.org/10.1016/j.stem.2007.05.013
6. McCulloch EA, Till JE (2005) Perspectives on the properties of stem cells. Nat Med 11 (10):1026–1028
7. Till JE, McCulloch EA (1961) A direct measurement of the radiation sensitivity of normal mouse bone marrow cells. Radiat Res 14:213–222
8. Siminovitch L, McCulloch EA, Till JE (1963) The distribution of colony-forming cells among spleen colonies. J Cell Comp Physiol 62:327–336
9. Sneddon JB, Werb Z (2007) Location, location, location: the cancer stem cell "niche". Cell Stem Cell 1(6):607–611. https://doi.org/10.1016/j.stem.2007.11.009
10. Scadden DT (2014) Nice neighborhood: emerging concepts of the stem cell "niche". Cell 157 (1):41–50. https://doi.org/10.1016/j.cell.2014.02.013
11. Schofield R (1983) The stem cell system. Biomed Pharmacother 37:375–380
12. Quesenberry PJ, Becker PS (1998) Stem cell homing: rolling, crawling, and nesting. Proc Natl Acad Sci U S A 95:15155–15157
13. Cross MA, Enver T (1997) The lineage commitment of haemopoietic progenitor cells. Curr Opin Genet Dev 7:609–613
14. Jensen UB, Lowell S, Watt FM (1999) The spatial relationship between stem cells and their progeny in the basal layer of human epidermis: a new view based on whole-mount labelling and lineage analysis. Development 126:2409–2418
15. Lemischka IR, Moore KA (2003) Stem cells: interactive "niche"s. Nature 425:778–779
16. Fuchs E, Tumbar T, Guasch G (2004) Socializing with the neighbors: stem cells and their "niche". Cell 116:769–778

17. Kelly SS, Sola CB, de Lima M, Shpall E (2009) Ex vivo expansion of cord blood. Bone Marrow Transplant 44(10):673–681. https://doi.org/10.1038/bmt.2009.284
18. Sato T, van Es JH, Snippert HJ, Stange DE, Vries RG, van den Born M, Barker N, Shroyer NF, van de Wetering M, Clevers H (2011) Paneth cells constitute the "niche" for Lgr5 stem cells in intestinal crypts. Nature 469(7330):415–418. https://doi.org/10.1038/nature09637
19. Arwert EN, Hoste E, Watt FM (2012) Epithelial stem cells, wound healing and cancer. Nat Rev Cancer 12(3):170–180. https://doi.org/10.1038/nrc3217
20. Krebsbach PH, Villa-Diaz LG (2017) The role of integrin α6 (CD49f) in stem cells: more than a conserved biomarker. Stem Cells Dev 26(15):1090–1099. https://doi.org/10.1089/scd.2016.0319
21. Xiao Y, Palomero J, Grabowska J, Wang L, de Rink I, van Helvert L, Borst J (2017) Macrophages and osteoclasts stem from a bipotent progenitor downstream of a macrophage/osteoclast/ dendritic cell progenitor. Blood Adv 1(23):1993–2006. https://doi.org/10.1182/bloodadvances.2017008540
22. Bongso A, Lee EH (2005) Stem cells: their definition, classification and sources. In: Stem cells: from bench to bedside. World Scientific Publishing, Singapore
23. Lu F, Zhang Y (2015) Cell totipotency: molecular features, induction, and maintenance. Natl Sci Rev 2(2):217–225
24. Rossant J (2008) Stem cells and early lineage development. Cell 132(4):527–531. https://doi.org/10.1016/j.cell.2008.01.039
25. Takahashi K, Yamanaka S (2006) Induction of pluripotent stem cells from mouse embryonic and adult fibroblast cultures by defined factors. Cell 126(4):663–676
26. Yamanaka S, Blau HM (2010) Nuclear reprogramming to a pluripotent state by three approaches. Nature 465(7299):704–712
27. Seita J, Weissman IL (2010) Hematopoietic stem cell: self-renewal versus differentiation. Wiley Interdiscip Rev Syst Biol Med 2(6):640–653. https://doi.org/10.1002/wsbm.86
28. Majo F, Rochat A, Nicolas M, Jaoudé GA, Barrandon Y (2008) Oligopotent stem cells are distributed throughout the mammalian ocular surface. Nature 456(7219):250–254. https://doi.org/10.1038/nature07406
29. Xavier-Ferrucio J, Krause DS (2018) Concise review: bipotent megakaryocytic-erythroid progenitors: concepts and controversies. Stem Cells 36(8):1138–1145. https://doi.org/10.1002/stem.2834
30. Chan SS, Hagen HR, Swanson SA, Stewart R, Boll KA, Aho J, Thomson JA, Kyba M (2016) Development of bipotent cardiac/skeletal myogenic progenitors from MESP1+ mesoderm. Stem Cell Rep 6(1):26–34. https://doi.org/10.1016/j.stemcr.2015.12.003
31. Visvader JE, Stingl J (2014) Mammary stem cells and the differentiation hierarchy: current status and perspectives. Genes Dev 28(11):1143–1158. https://doi.org/10.1101/gad.242511.114
32. Zhang Y, Bai XF, Huang CX (2003) Hepatic stem cells: existence and origin. World J Gastroenterol 9(2):201–204
33. Miyawaki K, Iwasaki H, Jiromaru T, Kusumoto H, Yurino A, Sugio T, Uehara Y, Odawara J, Daitoku S, Kunisaki Y, Mori Y, Arinobu Y, Tsuzuki H, Kikushige Y, Iino T, Kato K, Takenaka K, Miyamoto T, Maeda T, Akashi K (2017) Identification of unipotent megakaryocyte progenitors in human hematopoiesis. Blood 129(25):3332–3343. https://doi.org/10.1182/blood-2016-09-741611
34. Thomson JA, Itskovitz-Eldor J, Shapiro SS, Waknitz MA, Swiergiel JJ, Marshall VS et al (1998) Embryonic stem cell lines derived from human blastocysts. Science 282 (5391):1145–1147
35. Gearhart J (1998) New potential for human embryonic stem cells. Science 282 (5391):1061–1062
36. Vazin T, Freed WJ (2010) Human embryonic stem cells: derivation, culture, and differentiation: a review. Restor Neurol Neurosci 28(4):589–603. https://doi.org/10.3233/RNN-2010-0543

37. Wolf DP, Morey R, Kang E, Ma H, Hayama T, Laurent LC, Mitalipov S (2017) Concise review: embryonic stem cells derived by somatic cell nuclear transfer: a horse in the race? Stem Cells 35(1):26–34. https://doi.org/10.1002/stem.2496
38. Dulak J, Szade K, Szade A, Nowak W, Józkowicz A (2015) Adult stem cells: hopes and hypes of regenerative medicine. Acta Biochim Pol 62(3):329–337. https://doi.org/10.18388/abp.2015_1023
39. Barker N, Bartfeld S, Clevers H (2010) Tissue-resident adult stem cell populations of rapidly self-renewing organs. Cell Stem Cell 7(6):656–670. https://doi.org/10.1016/j.stem.2010.11.016
40. Nuti N, Corallo C, Chan BM, Ferrari M, Gerami-Naini B (2016) Multipotent differentiation of human dental pulp stem cells: a literature review. Stem Cell Rev Rep 12(5):511–523
41. Chen Z, de Paiva CS, Luo L, Kretzer FL, Pflugfelder SC, Li DQ (2004) Characterization of putative stem cell phenotype in human limbal epithelia. Stem Cells 22(3):355–366
42. Jones KB, Klein OD (2013) Oral epithelial stem cells in tissue maintenance and disease: the first steps in a long journey. Int J Oral Sci 5(3):121–129. https://doi.org/10.1038/ijos.2013.46
43. Can A (2008 Jun 5) A concise review on the classification and nomenclature of stem cells. Turk J Haematol 25(2):57–59
44. Lavker RM, Sun TT (2000 Dec 5) Epidermal stem cells: properties, markers, and location. Proc Natl Acad Sci U S A 97(25):13473–13475
45. Lavker RM, Sun TT, Oshima H, Barrandon Y, Akiyama M, Ferraris C, Chevalier G, Favier B, Jahoda CA, Dhouailly D, Panteleyev AA, Christiano AM (2003 Jun) Hair follicle stem cells. J Investig Dermatol Symp Proc 8(1):28–38
46. Klimczak A, Kozlowska U (2016) Mesenchymal stromal cells and tissue-specific progenitor cells: their role in tissue homeostasis. Stem Cells Int 2016:4285215. https://doi.org/10.1155/2016/4285215
47. Gage FH, Temple S (2013) Neural stem cells: generating and regenerating the brain. Neuron 80(3):588–601. https://doi.org/10.1016/j.neuron.2013.10.037
48. Zhao X, Moore DL (2018) Neural stem cells: developmental mechanisms and disease modeling. Cell Tissue Res 371(1):1–6. https://doi.org/10.1007/s00441-017-2738-1
49. Wang YX, Dumont NA, Rudnicki MA (2014) Muscle stem cells at a glance. J Cell Sci 127 (Pt 21):4543–4548. https://doi.org/10.1242/jcs.151209
50. Chicheportiche A, Ruat M, Boussin FD, Daynac M (2018) Isolation of neural stem and progenitor cells from the adult brain and live imaging of their cell cycle with the FUCCI system. Methods Mol Biol 1686:69–78. https://doi.org/10.1007/978-1-4939-7371-2_5
51. Stenudd M, Sabelström H, Frisén J (2015) Role of endogenous neural stem cells in spinal cord injury and repair. JAMA Neurol 72(2):235–237. https://doi.org/10.1001/jamaneurol.2014.2927
52. Bernstock JD, Peruzzotti-Jametti L, Ye D, Gessler FA, Maric D, Vicario N, Lee YJ, Pluchino S, Hallenbeck JM (2017) Neural stem cell transplantation in ischemic stroke: a role for preconditioning and cellular engineering. J Cereb Blood Flow Metab 37 (7):2314–2319. https://doi.org/10.1177/0271678X17700432
53. Marsh SE, Blurton-Jones M (2017) Neural stem cell therapy for neurodegenerative disorders: the role of neurotrophic support. Neurochem Int 106:94–100. https://doi.org/10.1016/j.neuint.2017.02.006
54. Nowell CS, Radtke F (2017) Corneal epithelial stem cells and their "niche" at a glance. J Cell Sci 130(6):1021–1025. https://doi.org/10.1242/jcs.198119
55. Hayashi R, Yamato M, Saito T, Oshima T, Okano T, Tano Y, Nishida K (2008) Enrichment of corneal epithelial stem/progenitor cells using cell surface markers, integrin alpha6 and CD71. Biochem Biophys Res Commun 367(2):256–263
56. Funderburgh JL, Funderburgh ML, Du Y (2016) Stem cells in the limbal stroma. Ocul Surf 14 (2):113–120. https://doi.org/10.1016/j.jtos.2015.12.006
57. Ranganathan K, Lakshminarayanan V (2012) Stem cells of the dental pulp. Indian J Dent Res 23(4):558. https://doi.org/10.4103/0970-9290.104977

58. Atari M, Barajas M, Hernández-Alfaro F, Gil C, Fabregat M, FerrésPadró E, Giner L, Casals N (2011) Isolation of pluripotent stem cells from human third molar dental pulp. Histol Histopathol 26(8):1057–1070. https://doi.org/10.14670/HH-26.1057
59. Padma Priya S, Higuchi A, Abu Fanas S, Pooi Ling M, Kumari Neela V, Sunil PM, Saraswathi TR, Murugan K, Alarfaj AA, Munusamy MA, Kumar S (2015) Odontogenic epithelial stem cells: hidden sources. Lab Investig 95(12):1344–1352. https://doi.org/10.1038/labinvest.2015.108
60. Amoh Y, Hoffman RM (2017) Hair follicle-associated-pluripotent (HAP) stem cells. Cell Cycle 16(22):2169–2175. https://doi.org/10.1080/15384101.2017.1356513
61. Kajiura S, Mii S, Aki R, Hamada Y, Arakawa N, Kawahara K, Li L, Katsuoka K, Hoffman RM, Amoh Y (2016) Protocols for cryopreservation of intact hair follicle that maintain pluripotency of nestin-expressing hair-follicle-associated pluripotent (HAP) stem cells. Methods Mol Biol 1453:173–178. https://doi.org/10.1007/978-1-4939-3786-8_18
62. Yu H, Fang D, Kumar SM, Li L, Nguyen TK, Acs G, Herlyn M, Xu X (2006) Isolation of a novel population of multipotent adult stem cells from human hair follicles. Am J Pathol 168 (6):1879–1888
63. Cotsarelis G, Sun TT, Lavker RM (1990) Label-retaining cells reside in the bulge area of pilosebaceous unit: implications for follicular stem cells, hair cycle, and skin carcinogenesis. Cell 61(7):1329–1337
64. Nishimura EK, Jordan SA, Oshima H, Yoshida H, Osawa M, Moriyama M, Jackson IJ, Barrandon Y, Miyachi Y, Nishikawa S (2002) Dominant role of the "niche" in melanocyte stem-cell fate determination. Nature 416(6883):854–860
65. Wright MC, Reed-Geaghan EG, Bolock AM, Fujiyama T, Hoshino M, Maricich SM (2015) Unipotent, Atoh1+ progenitors maintain the Merkel cell population in embryonic and adult mice. J Cell Biol 208(3):367–379. https://doi.org/10.1083/jcb.201407101
66. Blackburn CC, Manley NR, Palmer DB, Boyd RL, Anderson G, Ritter MA (2002) One for all and all for one: thymic epithelial stem cells and regeneration. Trends Immunol 23(8):391–395
67. Jones KB, Furukawa S, Marangoni P, Ma H, Pinkard H, D'Urso R, Zilionis R, Klein AM, Klein OD (2019) Quantitative Clonal Analysis and Single-Cell Transcriptomics Reveal Division Kinetics, Hierarchy, and Fate of Oral Epithelial Progenitor Cells. Cell Stem Cell 24 (1):183–192.e8. https://doi.org/10.1016/j.stem.2018.10.015
68. Utheim TP (2015) Concise review: transplantation of cultured oral mucosal epithelial cells for treating limbal stem cell deficiency-current status and future perspectives. Stem Cells 33 (6):1685–1695. https://doi.org/10.1002/stem.1999
69. Hisha H, Tanaka T, Ueno H (2016) Lingual epithelial stem cells and organoid culture of them. Int J Mol Sci 17(2):E168. https://doi.org/10.3390/ijms17020168
70. Seery JP, Watt FM (2000) Asymmetric stem-cell divisions define the architecture of human oesophageal epithelium. Curr Biol 10(22):1447–1450
71. Zhang Y, Jiang M, Kim E, Lin S, Liu K, Lan X, Que J (2017) Development and stem cells of the esophagus. Semin Cell Dev Biol 66:25–35. https://doi.org/10.1016/j.semcdb.2016.12.008
72. Nam H, Kim JH, Hwang JY, Kim GH, Kim JW, Jang M, Lee JH, Park K, Lee G (2018) Characterization of primary epithelial cells derived from human salivary gland contributing to in vivo formation of acini-like structures. Mol Cells 41(6):515–522. https://doi.org/10.14348/molcells.2018.0060
73. Maimets M, Rocchi C, Bron R, Pringle S, Kuipers J, Giepmans BN, Vries RG, Clevers H, de Haan G, van Os R, Coppes RP (2016) Long-term in vitro expansion of salivary gland stem cells driven by Wnt signals. Stem Cell Rep 6(1):150–162. https://doi.org/10.1016/j.stemcr.2015.11.009
74. Barker N (2014) Adult intestinal stem cells: critical drivers of epithelial homeostasis and regeneration. Nat Rev Mol Cell Biol 15(1):19–33. https://doi.org/10.1038/nrm3721
75. Heath JP (1996) Epithelial cell migration in the intestine. Cell Biol Int 20(2):139–146
76. Boitano AE, Wang J, Romeo R, Bouchez LC, Parker AE, Sutton SE, Walker JR, Flaveny CA, Perdew GH, Denison MS, Schultz PG, Cooke MP (2010) Aryl hydrocarbon receptor

antagonists promote the expansion of human hematopoietic stem cells. Science 329 (5997):1345–1348. https://doi.org/10.1126/science.1191536
77. Vo LT, Daley GQ (2015) De novo generation of HSCs from somatic and pluripotent stem cell sources. Blood 125(17):2641–2648. https://doi.org/10.1182/blood-2014-10-570234
78. Flores-Guzmán P, Fernández-Sánchez V, Mayani H (2013) Concise review: ex vivo expansion of cord blood-derived hematopoietic stem and progenitor cells: basic principles, experimental approaches, and impact in regenerative medicine. Stem Cells Transl Med 2 (11):830–838. https://doi.org/10.5966/sctm.2013-0071
79. Pineault N, Abu-Khader A (2015) Advances in umbilical cord blood stem cell expansion and clinical translation. Exp Hematol 43(7):498–513. https://doi.org/10.1016/j.exphem.2015.04.011
80. Kohli L, Passegué E (2014) Surviving change: the metabolic journey of hematopoietic stem cells. Trends Cell Biol 24(8):479–487. https://doi.org/10.1016/j.tcb.2014.04.001
81. Carpino G, Cardinale V, Gentile R, Onori P, Semeraro R, Franchitto A, Wang Y, Bosco D, Iossa A, Napoletano C, Cantafora A, D'Argenio G, Nuti M, Caporaso N, Berloco P, Venere R, Oikawa T, Reid L, Alvaro D, Gaudio E (2014) Evidence for multipotent endodermal stem/progenitor cell populations in human gallbladder. J Hepatol 60(6):1194–1202. https://doi.org/10.1016/j.jhep.2014.01.026
82. Manohar R, Li Y, Fohrer H, Guzik L, Stolz DB, Chandran UR, LaFramboise WA, Lagasse E (2015) Identification of a candidate stem cell in human gallbladder. Stem Cell Res 14 (3):258–269. https://doi.org/10.1016/j.scr.2014.12.003
83. Theise ND, Saxena R, Portmann BC, Thung SN, Yee H, Chiriboga L, Kumar A, Crawford JM (1999) The canals of Hering and hepatic stem cells in humans. Hepatology 30(6):1425–1433
84. Kuwahara R, Kofman AV, Landis CS, Swenson ES, Barendswaard E, Theise ND (2008) The hepatic stem cell "niche": identification by label-retaining cell assay. Hepatology 47 (6):1994–2002. https://doi.org/10.1002/hep.22218
85. Berika M, Elgayyar ME, El-Hashash AH (2014) Asymmetric cell division of stem cells in the lung and other systems. Front Cell Dev Biol 2:33. https://doi.org/10.3389/fcell.2014.00033
86. Li F, He J, Wei J, Cho WC, Liu X (2015) Diversity of epithelial stem cell types in adult lung. Stem Cells Int 2015:728307. https://doi.org/10.1155/2015/728307
87. Bishop AE (2004) Pulmonary epithelial stem cells. Cell Prolif 37(1):89–96
88. Shi AP, Fan ZM, Ma KW, Jiang YF, Wang L, Zhang KW, Fu SB, Xu N, Zhang ZR (2017) Isolation and characterization of adult mammary stem cells from breast cancer-adjacent tissues. Oncol Lett 14(3):2894–2902. https://doi.org/10.3892/ol.2017.6485
89. Lloyd-Lewis B, Harris OB, Watson CJ, Davis FM (2017) Mammary stem cells: premise, properties, and perspectives. Trends Cell Biol 27(8):556–567. https://doi.org/10.1016/j.tcb.2017.04.001
90. Teixeira J, Rueda BR, Pru JK (2008) Uterine stem cells. In: StemBook. Harvard Stem Cell Institute, Cambridge, MA
91. Jin S (2019) Bipotent stem cells support the cyclical regeneration of endometrial epithelium of the murine uterus. Proc Natl Acad Sci U S A 116(14):6848–6857. https://doi.org/10.1073/pnas.1814597116
92. Santamaria X, Mas A, Cervelló I, Taylor H, Simon C (2018) Uterine stem cells: from basic research to advanced cell therapies. Hum Reprod Update 24(6):673–693. https://doi.org/10.1093/humupd/dmy028
93. Leong KG, Wang BE, Johnson L, Gao WQ (2008) Generation of a prostate from a single adult stem cell. Nature 456(7223):804–808. https://doi.org/10.1038/nature07427
94. Komeya M, Ogawa T (2015) Spermatogonial stem cells: Progress and prospects. Asian J Androl 17(5):771–775. https://doi.org/10.4103/1008-682X.154995
95. de Rooij DG (2017) The nature and dynamics of spermatogonial stem cells. Development 144 (17):3022–3030. https://doi.org/10.1242/dev.146571
96. Li H, Ghazanfari R, Zacharaki D, Lim HC, Scheding S (2016) Isolation and characterization of primary bone marrow mesenchymal stromal cells. Ann N Y Acad Sci 1370:109–118

97. Ab Kadir R, Zainal Ariffin SH, Megat Abdul Wahab R, Kermani S, Senafi S (2012) Characterization of mononucleated human peripheral blood cells. Sci World J 2012:843843. https://doi.org/10.1100/2012/843843
98. Wagner W, Wein F, Seckinger A, Frankhauser M, Wirkner U, Krause U, Blake J, Schwager C, Eckstein V, Ansorge W (2005) Comparative characteristics of mesenchymal stem cells from human bone marrow, adipose tissue, and umbilical cord blood. Exp Hematol 33:1402–1416
99. Mushahary D, Spittler A, Kasper C, Weber V, Charwat V (2018 Jan) Isolation, cultivation, and characterization of human mesenchymal stem cells. Cytometry A 93(1):19–31. https://doi.org/10.1002/cyto.a.23242
100. Bearzi C, Rota M, Hosoda T, Tillmanns J, Nascimbene A, De Angelis A, Yasuzawa-Amano S, Trofimova I, Siggins RW, Lecapitaine N, Cascapera S, Beltrami AP, D'Alessandro DA, Zias E, Quaini F, Urbanek K, Michler RE, Bolli R, Kajstura J, Leri A, Anversa P (2007) Human cardiac stem cells. Proc Natl Acad Sci U S A 104(35):14068–14073
101. Zhang Y, Mignone J, MacLellan WR (2015) Cardiac regeneration and stem cells. Physiol Rev 95(4):1189–1204. https://doi.org/10.1152/physrev.00021.2014
102. Morgan JE, Partridge TA (2003) Muscle satellite cells. Int J Biochem Cell Biol 35 (8):1151–1156
103. Gentile P, Piccinno MS, Calabrese C (2019) Characteristics and potentiality of human adipose-derived stem cells (hASCs) obtained from enzymatic digestion of fat graft. Cell 8 (3):E282. https://doi.org/10.3390/cells8030282
104. Broxmeyer HE, Douglas GW, Hangoc G et al (1989) Human umbilical cord blood as a potential source of transplantable hematopoietic stem/progenitor cells. Proc Natl Acad Sci U S A 86(10):3828–3832
105. Nakahata T, Ogawa M (1982) Hemopoietic colony-forming cells in umbilical cord blood with extensive capability to generate mono- and multipotential hemopoietic progenitors. J Clin Invest 70(6):1324–1328
106. Ballen KK, Gluckman E, Broxmeyer HE (2013) Umbilical cord blood transplantation: the first 25 years and beyond. Blood 122(4):491–498. https://doi.org/10.1182/blood-2013-02-453175
107. Roura S, Pujal JM, Gálvez-Montón C, Bayes-Genis A (2015) The role and potential of umbilical cord blood in an era of new therapies: a review. Stem Cell Res Ther 6:123. https://doi.org/10.1186/s13287-015-0113-2
108. Takahashi K, Tanabe K, Ohnuki M, Narita M, Ichisaka T, Tomoda K, Yamanaka S (2007) Induction of pluripotent stem cells from adult human fibroblasts by defined factors. Cell 131 (5):861–872
109. Bragança J, Lopes JA, Mendes-Silva L, Almeida Santos JM (2019) Induced pluripotent stem cells, a giant leap for mankind therapeutic applications. World J Stem Cells 11(7):421–430. https://doi.org/10.4252/wjsc.v11.i7.421
110. Dick JE (2008) Stem cell concepts renew cancer research. Blood 112(13):4793–4807. https://doi.org/10.1182/blood-2008-08-077941
111. Dawood S, Austin L, Cristofanilli M (2014) Cancer stem cells: implications for cancer therapy. Oncology (Williston Park) 28(12):1101–1107, 1110
112. Eisenberg LM, Eisenberg CA (2003) Stem cell plasticity, cell fusion, and transdifferentiation. Birth Defects Res C Embryo Today 69(3):209–218
113. Poulsom R, Alison MR, Forbes SJ, Wright NA (2002) Adult stem cell plasticity. J Pathol 197 (4):441–456
114. Nieto MA (2013) Epithelial plasticity: a common theme in embryonic and cancer cells. Science 342(6159):1234850. https://doi.org/10.1126/science.1234850
115. Fang D, Kitamura H (2018) Cancer stem cells and epithelial-mesenchymal transition in urothelial carcinoma: possible pathways and potential therapeutic approaches. Int J Urol 25 (1):7–17. https://doi.org/10.1111/iju.13404
116. Anderson DJ, Gage FH, Weissman IL (2001) Can stem cells cross lineage boundaries? Nat Med 7(4):393–395

117. Friedenstein AJ, Chailakhyan RK, Gerasimov UV (1987) Bone marrow osteogenic stem cells: in vitro cultivation and transplantation in diffusion chambers. Cell Tissue Kinet 20(3):263–272
118. Kocher AA, Schuster MD, Szabolcs MJ, Takuma S, Burkhoff D, Wang J, Homma S, Edwards NM, Itescu S (2001) Neovascularization of ischemic myocardium by human bone-marrow-derived angioblasts prevents cardiomyocyte apoptosis, reduces remodeling and improves cardiac function. Nat Med 7(4):430–436
119. Petersen BE, Bowen WC, Patrene KD, Mars WM, Sullivan AK, Murase N, Boggs SS, Greenberger JS, Goff JP (1999) Bone marrow as a potential source of hepatic oval cells. Science 284(5417):1168–1170
120. Alison MR, Poulsom R, Jeffery R, Dhillon AP, Quaglia A, Jacob J, Novelli M, Prentice G, Williamson J, Wright NA (2000) Hepatocytes from non-hepatic adult stem cells. Nature 406 (6793):257

Sources, Isolation and culture of stem cells?

2

Contents

Abstract

The process of isolation and culture of the stem cells finds importance in both biomedical aspect and drug discovery. In this chapter, we discuss about the different types of stem cells, their sources, methods of isolation and culture. We also discuss the three-dimensional (3D) culture of biomaterials, which aid in the 3D cell culture and their subsequent changes with respect to the biophysical or biochemical properties over time (i.e., fourth dimension) which is the basis of 4D culture. The 3D cultures mimic the real-life system of cells in their microenvironment in vivo and their communication with other cell types and the extracellular matrix (ECM) which finds importance in the giving near perfect to actualization results in the tests for drug discovery and is finding importance in today's biology.

Keywords

Stem cells · Embryonic stem cells · Adult stem cells · 3D culture · Induced pluripotent cells

S. Chowdhury, S. Ghosh, *Stem Cells*, https://doi.org/10.1007/978-981-16-1638-9_2

Abbreviations

3D	Three-dimensional
4D	Four-dimensional
AFP	α-fetoprotein
ALDH	Aldehyde dehydrogenase
ASCs	Adult or postnatal stem cells
BM	Bone marrow
BSA	Bovine serum albumin
CALCR	Calcitonin receptor
CaSC	Cancer stem cell niche
CAV1	Caveolin 1
CB	Cord blood
CESCs	Corneal epithelial stem cells
CFU	Colony forming unit
CK	Cytokeratin
CSF	Colony-stimulating factor
CT	Cholera toxin
CXCR4	Chemokine receptor type 4
DFPCs	Dental follicle progenitor cells
DMEM	Dulbecco's modified Eagle's medium
DMSO	Dimethyl sulfoxide
DPSCs	Dental pulp stem cells
ECM	Extracellular matrix
EDTA	Ethylene diamine tetra acetic acid
EG	Embryonic germ
EGF	Epidermal growth factor
EMT	Epithelial-to-mesenchymal transition
EpCAM	Epithelial cell adhesion molecule
ESCs	Embryonic stem cells
FBS	Foetal bovine serum
FCS	Foetal calf serum
FGF	Fibroblast growth factor
GIFM	Genetic inducible fate mapping
H	Hepes
HAP	Hair follicle-associated-pluripotent stem cells
HBEC	Human breast epithelial cells
HBSS	Hanks balanced salt solution
hESC	Human ESC
HPCs	Haematopoietic progenitor cells
HSCs	Haematopoietic stem cells
HSPCs	Haematopoietic stem and progenitor cells
IBL	Interpapillary zone
ICM	Inner cell mass

IMDM	Iscove's modified Dulbecco's medium
INS	Insulin
iPSC	Induced pluripotent cells
ISC	Epithelial intestine stem cells
IVF	In vitro fertilization
LEC	Limbal epithelial crypts
LSCs	Limbal stem cells
MaSCs	Adult mammary stem cells
MEF	Mouse embryonic fibroblast
MNCs	Mononuclear cells
MSCs	Mesenchymal stem cells
MuSc	Muscle stem cells
NCAM	Neural cell adhesion molecule
NCAM1	Neural cell adhesion molecule 1
Neu5Ac	*N*-acetylneuraminic acid
Neu5Gc	Glycolylneuraminic acid
NSC	Neural stem cells
OEE	Oral epithelial equivalents
OeSC	Oesophageal stem cells
PB	peripheral blood
PBL	Papillae of the supporting connective tissue
PBS	Phosphate-buffered saline
PC	Podocalyxin
PDL	Periodontium including cementum
PDLSCs	Periodontal ligament stem cells
PEG	Polyethylene glycol
SCAP	Stem cells from apical papilla
SDC3 and SDC4	Syndecans 3 and 4
SGSCs	Salivary gland stem cells
SGSCs	Sweat gland stroma-derived stem cells
SHED	Stem cells from exfoliating deciduous teeth
Sias	Sialic acids
SKPs	Skin-derived precursors
SVZ	subventricular zone
TA	Transit-amplifying cells
TEPA	Tetraethylenepentamine
TSSCs	Tissue specific stem cells
UCB	Umbilical cord blood
VCAM1	Vascular cell adhesion molecule 1
α-MEM	Minimal essential medium alpha

2.1 Introduction

The sources of the stem cells are various and therefore, based on the source, each type has a different type of isolation and culture protocol. Isolation of a pure population of cells help to improve the procedure for in vitro culture. In two-dimensional (2-D) cultures, stem cells are propagated as a monolayer in plastic culture plates using undefined or xenogenic materials including attachment substrates, cytokines growth factors and serum [1].

Although 2D cell culture has a wide usage, it lacks the sophistication necessary to mimic the stem cell niche. The three-dimensional (3D) stem cell culture system offers dynamic, specialized microenvironments, which controls the fate of the stem cells in vivo. 3D culture system promotes complex spatial interactions between cells, extracellular components, and gradients of nutrients, oxygen, and waste [1, 2].

The biomaterials which promote 3D cell culture and subsequent changes in their biophysical or biochemical properties over a period of time (i.e., the fourth dimension) form the basis of four-dimensional culture (4D). They provide suitable microenvironments for the culture of stem cells and control the signals for modulating their behaviour [3]. In this chapter, the diverse sources of the stem cells have been discussed along with their isolation and culture protocol.

2.2 Embryonic Stem Cells (ESCs)

The embryonic stem cells (ESCs) exhibit pluripotency with the ability of indefinite self-renewal and capacity to differentiate into every somatic cell types of the embryo. The crucial features of ESCs include limitless in vitro proliferation without any differentiation, diploid karyotype just like any normal cell, in vivo differentiation into teratomas comprising cells of all three primary germ layers and the ability to differentiate into a wide variety of cell types. ESCs express markers of pluripotency and contribute to both somatic tissues and the germline. They help to understand the mechanisms of the development of specific cells and formation of organ structures. The ability of self-renewal and plasticity of ESCs enable us to generate unlimited number of specific cell types in vitro, thereby opening new avenues for regenerative medicine [4, 5].

History

The concept of pluripotent cells in early embryos has been *in vogue* for years. The primary experimental isolation of mouse ESCs was reported in 1981 [6, 7]. The ESCs from the primates were first isolated from the rhesus monkey [5]. Human ESCs (hESCs) were first isolated from the ICM of blastocysts from the left over of in vitro fertilization (IVF) [8] and the germ cells referred to as embryonic germ (EG) cells, derived from spontaneously aborted foetuses [9].

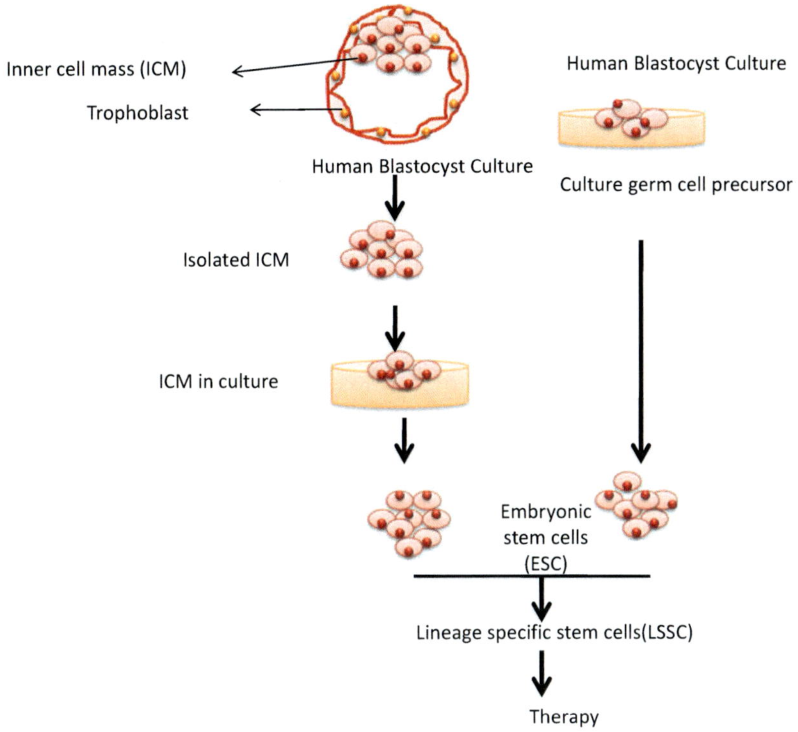

Fig. 2.1 Embryonic stem cell culture

Source

The process of fertilization of the sperm and ovum is followed by the formation of a diploid zygote. Numerous mitotic divisions of this zygote during initial embryogenesis form a structure termed blastocyst. The blastocyst consists of the embryoblast and the trophoblast referring to the inner layer and outer layer of cells, respectively. The trophectoderm (or the outer cell mass) forms the extra-embryonic tissue, which form the future placenta, chorion, and umbilical cord. The embryoblast (or the ICM) develops to form the embryo. The ICM of the blastocysts contains the ESCs [10].

At present the two sources of hESCs (Fig. 2.1) are the ICM of blastocysts resulting from in vitro fertilization (IVF) [11] and the germ cells referred to as EG cells, derived from spontaneously aborted foetuses [9]. Experimental evidences show that ESCs can also be derived from late morulae [12].

Isolation

The pluripotent hESCs express the transcription factors Nanog, Oct 4, and Sox 2 along with tumour rejection antigens namely, Tra–1–81 and Tra–1–60. These cells are also positive for the stage-specific embryonic antigens, SSEA 3 and SSEA 4 but do not express SSEA 1 [13, 14].

For clinical purposes, the hESCs can be obtained from human embryos at cleavage stage, produced by IVF. The embryos are cultured to the blastocyst stage and ICMs are isolated from the trophectoderm by immune-surgery or mechanical dissection [8, 11]. hESC lines were obtained by immune-surgical method using animal-derived products like anti-human serum antibodies and complement from guinea pig [15]. Due to prior contact to animal-derived products and probable infections, these cells were rendered unsuitable for transplantation therapies. Therefore, isolation of ICM from the trophectoderm either by mechanical or enzymatic manner which avoids contact between ICM and animal products can be advantageous for impending clinical applications [16]. Laser beams can also be used to obtain hESC lines. A small opening is created at the zona pellucida which encapsulates the blastocyst and is followed by laser-assisted isolation of the ICM [17].

Culture of the hESC: In Vitro Proliferation

When cultured on mouse embryonic fibroblast (MEF) feeders, the pluripotent ESCs exhibit normal karyotype with the ability of forming large teratomas (germ cell tumours), when grafted to severe combined immunodeficient (SCID) mice [4].

Primary hESC lines have been supported by MEF layers for propagation in the primitive undifferentiated state [8, 11, 18]. While attempting to develop xeno-free hESC culture system, numerous methods using cell types derived from human sources including fibroblast feeder cells from fallopian tube epithelium, foetal foreskin, muscle, bone marrow (BM), or amniotic epithelium have been used [4]. HESC can also be cultured in feeder-free environments, containing extracellular matrices like fibronectin and matrigel. In order to maintain hESCs in an undifferentiated state in feeder-free conditions, culture medium conditioned with feeder fibroblast cells and basic fibroblast growth factor (FGF) were used [19, 20].

In the feeder-free culture system, hESC typically differentiate to stromal-like cells or fibroblast which serve as supportive cells for maintaining the growth of hESC in undifferentiated condition [21, 22]. However, these feeder layers are mortal and have to be replaced, the process of which is exhaustive. Several efforts have been made to eliminate the feeder cells in hESC culture and utilize the factors secreted from MEF feeder layers, for maintaining the self-renewal of hESC. Suppression of bone morphogenetic protein (BMP) signalling along with higher concentrations of bFGF in serum-free media sustain proliferation of hESC in undifferentiated condition. Treatments of hESC with signalling molecules like activin A and transforming growth factor-beta 1 (TGF-β1) along with mixtures of human recombinant proteins have also been used. In serum-free culture of hESC, basal medium like Dulbecco's modified Eagle's medium (DMEM)/F12 containing glucose, amino acids, insulin (INS), vitamins, transferrin, selenium in a balanced salt solution is used. GABA, lipids, cholesterol, β mercaptoethanol, pipecolic, and ascorbic acid are also included [4, 13].

Three-Dimensional (3D) Culture of the hESC

The perfect hESC culture must reiterate the niche and corporal milieu of its residence. Mostly, hESC grown in colonies restrict themselves to narrow size during the passaging procedures to avert differentiation. This attests to the fact that cell–cell interaction and other signals within the colonies play pivotal role in maintaining the hESCs in undifferentiated state. Cell signalling, gene expression, and association of cells in 2D cultures are distinctly different from 3D culture system, which resemble the in vivo hESC environment more meticulously [23, 24]. 3D culture greatly augments self-renewal and cell proliferation in comparison to 2D tissue culture surfaces. Synthetic polyamide matrix scaffolds have been used to immitate the fibrillar network of the basement membrane, which enhanced ability of the cells to proliferate and it was interrelated with the cellular mechano-transduction by the stimulation of small GTPase Rac, phosphoinositide 3-kinase (PI3K) pathway and improved expression of Nanog [25].

Polymeric fibrous scaffold comprising of cellulose acetate forms a meshwork that allows surface immobilization of numerous extra cellular matrix (ECM) molecules and growth factors. Such 3D matrix helps in the study of uncommitted human EG cell derivatives, which share some properties with hESC. The improved interaction of cells in 3D cultures lead to improved maintenance of multipotency, propagation, and cell survival [26].

These hESCs can also be cultured in the biocompatible poly-glycerol-co-sebacate acrylate (PGSA) elastomers that are photo-polymerized to form permeable scaffolds but the undifferentiated state cannot be maintained [27]. Till date the most pertinent matrix developed for the culture of hESC in undifferentiated state consists of hyaluronic acid (HA) hydrogel. It imitates the chief components of ECM found plentiful in embryos and stem cell niches, and are capable of maintaining hESCs in undifferentiated state while conserving their complete differentiation capacity [4, 27].

Culture of hESC: In Vitro Differentiation

Hypothetically, hESCs can differentiate into all somatic cell types of an organism namely, ectoderm, mesoderm, and endoderm. However, a deeper understanding of the intricate mechanisms governing the growth of stem cells, their fate, generation of tissue and organogenesis is still required. When the hESCs are accumulated into 3-D embroid body (EB) structures, an inductive influence guides in vitro differentiation of numerous cell lineages. Cell-specific endodermal differentiation using hESC has been achieved to get hepatocytes [28], insulin manufacturing β cells [29], and epithelium of lungs [21, 30] for the progress of probable cell replacement therapy in acute hepatitis/liver failure, diabetes type I, and pulmonary diseases [4].

Cell-specific mesodermal differentiation using hESCs include mesodermal differentiation of mesenchymal precursors, cardiomycetes, haematopoietic and endothelial cells. The mesenchymal precursors differentiated to osteocytes, chondrocytes, and skeletal myoblasts under well-defined conditions [31, 32] having potential for treatment of osteoarthritis and osteoporosis. The hESC derived cardiomyocytes hold therapeutic potential for heart transplantation and ischaemic heart disease

[33, 34]. The blood-forming haematopoietic cells [35–37] and endothelial cells [38] could be used for blood transfusion, analysing defects of haematopoietic and vascular system or for cell transplantation in immunodeficiency disorders [4].

Cell-specific ectodermal differentiation forms the ectoderm, neural tube, and crest. These structures form cells of external sense organs, epidermis, and the central and peripheral nervous system [10]. The cultured hESCs can be stimulated to differentiate into the non-neuronal derivatives of the ectoderm like the keratinocytes [39, 40]. Differentiation of hESC to retinal pigment epithelial cells has been attained with an aim to combat age-linked macular degeneration, a leading reason of human blindness, due to loss and dysfunction of retinal pigment epithelium [41].

The hESCs can be differentiated to form the neuroectoderm with the purpose of treating neurodegenerative diseases. Isolation of expandable neural progenitor cells from hESCs can be done in vitro [18]. A novel population of neural stem cells termed as neural rosette cells (RNSC), having an unique gene profile, derived from hESCs, are also isolated [42].

2.3 Adult Stem Cells

As the name suggests the postnatal or adult stem cells (ASCs) are found in sites which are not embryonic. The types of these cells are site or organ specific including the stem cells from the umbilical cord blood (UCB) and various others tissues as introduced in Chap. 1. A brief description of each of the types along with their isolation and culture protocol is discussed below.

2.3.1 Umbilical Cord Blood Stem Cells

The blood which is found in the placenta and the attached umbilical cord after childbirth is the umbilical cord blood (UCB). Human UCB, a potent source of stem cells, has the ability of therapeutic reconstitution. These cells are multipotent progenitors, capable of self-renewal as can be seen by their in vitro replating efficiency in colony assay where they also differentiate to form the multipotent, erythroid (BFU-E), granulocyte-macrophage (CFU-GM), and other progenitors. The occurrence of progenitor cells in cord blood (CB) is more than that of marrow and peripheral blood (PB). Progenitor stem cells isolated from human UCB can be sustained in long-term liquid culture systems for several weeks, signifying their origin from more primitive cells [43, 44].

UCB is a substitute source of haematopoietic stem cell (HSC) in those having haematologic diseases and can be cured by allogeneic HSC transplantation [45]. The stem cells from UCB hold potential for treating many genetic diseases, haematologic malignancies, and immune deficiencies. UCB can also be of clinical use for a sick sibling or relative [46].

History

The site of haematopoiesis shifts from the yolk sac of the foetus to foetal liver and finally to the adult BM during foetal development. Initially, foetal liver cells were used as a source of HSCs but were later discarded due to frequent failure. CB was proved to be a better source of progenitor cells [44]. In 1988, a patient suffering from Fanconi's anaemia received the first CB transplant from his human leukocyte antigen (HLA)-identical sibling [47, 48].

Source

Umbilical cord is the source of the UCB stem cells. Immediately after parturition, the umbilical cord is clamped doubly 5–7 cm from the umbilicus and split between the clamps. Blood is collected from the maternal (placental) end of the umbilical cord. Blood can also be obtained from the removed placenta [44].

Isolation

The UCB is collected in sterile container containing anticoagulants like heparin, ethylene diamine tetra acetic acid (EDTA) or Alsever's solution. Mononuclear cells (MNCs) are isolated using Ficoll/Hypaque by density gradient centrifugation [44, 49]. Among many other markers, two major markers of UCB stem cells are CD34 and aldehyde dehydrogenase (ALDH), where the former is a cell surface marker while the latter is an enzyme whose presence attests to the functional normal nature of the cells [49, 50].

Culture

To improve UCB transplantation, different methods have been employed, including liquid suspension culture with numerous cytokines and expansion factors, co-culture with stromal elements and unremitting perfusion systems. Ex vivo expansion strategies of UCB are very similar to that used from BM and mobilized PB derived HSC culture. There are two goals of ex-vivo culture and expansion of CB. The primary focus of expansion includes generation of adequate number of HSCs to enhance the graft for transplantation. Another important objective is to produce more lineage-committed progenitors that would permit rapid recovery from pancytopenia, thus reducing early morbidity and mortality [51].

The ex vivo culture of UCB stem cells began with culturing MNCs in medium supplemented with human urinary bladder carcinoma cell line 5637 (CM 5637) and recombinant preparations of interleukin 3, granulocyte (G)-macrophage (M) colony-stimulating factor (CSF), G-CSF, M-CSF, and erythropoietin (Epo) to support the clonal proliferation of haematopoietic progenitors [44].

For studying the differentiation potential, the colony forming unit (CFU)-GM assay was performed in agar (0.3%) culture medium containing McCoy's SA medium, supplemented with essential and nonessential amino acids, serine, glutamine, vitamins, asparagine, sodium pyruvate, and heat-inactivated foetal bovine serum (FBS), instead of CM 5637, GM-CSF, interleukin 3, M-CSF or G-CSF. Colonies (>40 cells per aggregate) and clusters (3–40 cells per aggregate) were calculated after 7 and 14 days of incubation. The assay for CFU-GEMM, BFU-E, and CFU-GM was performed in Iscove's modified Dulbecco's medium (IMDM) supplemented with methyl cellulose, FBS, 2-mercaptoethanol, Epo, with or without hemin or CM 5637. Colonies were also counted after 14 days incubation [44]. At present, many different strategies are used for ex vivo expansion of UCB stem cells like suspension culture, stromal co-culture and continuous perfusion culture system [51].

In liquid suspension culture, cells from UCB are often cultured with combinations of growth factors, cytokines, and other growth-inducing compounds in flasks, to simulate the natural milieu of BM where haematopoiesis occurs. The haematopoietic progenitors (primarily $CD34^+$ or $CD133^+$) from UCB are isolated and cultured in medium containing a combination of growth factors like IL-3, SCF, IL-6, and G-CSF [52], TPO, SCF, and G-CSF [53, 54], and Flt-3 ligand (FL), IL-3, SCF, IL-6, and G-CSF [55, 56].

The haematopoietic microenvironment or the stem cell "niche," contains the haematopoietic and non-haematopoietic (cellular as well as extracellular) components, provides molecular signals which direct haematopoiesis, regulating differentiation and the maturation of HSCs [57–59]. When these cells are cultured ex vivo, they lose this provision provided by their niche, depending on exogenous supply of specified growth factors and cytokines present in the culture media [51]. In stromal co-culture, the UCB cells are co-cultured with the mesenchymal stem cells (MSCs) even from allogeneic source to reinstate some interaction occurring between the microenvironment of the BM stroma and the HSCs [60–65].

For the stromal co-culture, isolated MNCs from UCB are co-cultured with previously established MSC monolayers in medium supplemented with FBS and growth factor cocktail like TPO, SCF, and G-CSF [66]. After 7 days, the non-adherent fraction of cells are removed from the co-culture and subjected to a secondary expansion on an additional MSC monolayer. The primary adherent layer, which is composed of HSC and MSC, is replenished with growth factor containing fresh medium. This culture is continued for another 7 days (14 days) [51, 53].

"Bioreactors" or automatic, continuous perfusion culture systems, are also used for HSC expansion ex vivo, rather than the use of "static" culture (bags or culture flasks) [54]. These systems allow larger volumes and provide better gas exchange and nutrient delivery. The products secreted from mature granulocytes and macrophages are toxic to progenitors [67] and mature macrophages can directly harm haematopoietic progenitors and cultured stroma [68]. Thus, continuous perfusion of the medium which removes mature cells could reduce the harmful effects of the toxic by-products [51].

Recent expansion techniques include usage of the copper chelator tetraethylenepentamine (TEPA) and notch ligand-based cultures [48] (Fig. 2.2) which decreases the differentiation of haematopoietic cells from UCB cultured ex vivo, resulting in favoured expansion of early progenitors [69]. A clinically pertinent Notch mediated ex vivo expansion system for human UCB progenitors has resulted in increment in the number of progenitor cells having the ability to repopulate [70].

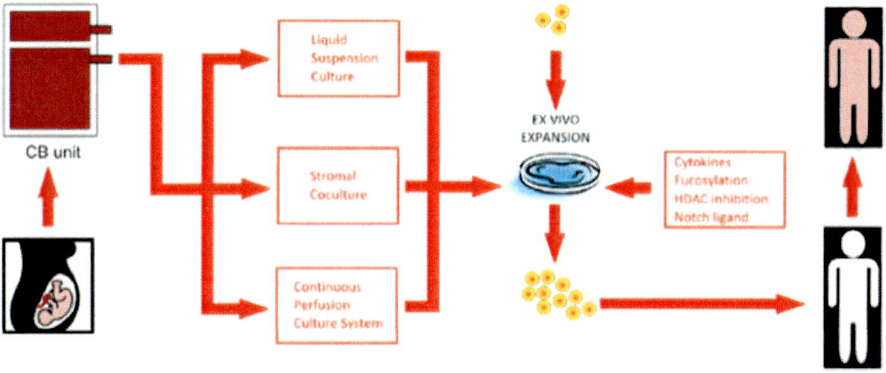

Fig. 2.2 Ex vivo expansion technique for cord blood transplantation. Abbreviations used: *CB* cord blood, *HDAC* histone deacetylase. (Adapted with permission from Munoz, J., Shah, N., Rezvani, K., Hosing, C., Bollard, C.M., Oran, B., Olson, A., Popat, U., Molldrem, J., McNiece, I.K. and Shpall, E.J. (2014), Concise Review: Umbilical Cord Blood Transplantation: Past, Present, and Future. STEM CELLS Translational Medicine, 3: 1435–1443. doi:10.5966/sctm.2014-0151)

2.3.2 Neural Stem Cells (NSCs)

The stem cells which can afford the flexibility to generate, repair, and change the nervous system function are the neural stem cells (NSCs). This contradicts our earlier belief that the nervous system could not regenerate. NSCs are multipotent and produce more restricted neural progenitor cells which, in turn, divide a few times to produce differentiated progeny [71]. The study of these cells has therapeutic implication in neuronal debilitating diseases like Parkinson's disease, Amyotrophic lateral sclerosis, Alzheimer's disease, and cell-based therapies against stroke [72].

History

In the mouse cerebral cortex, it was seen that several progenitor cells at mid-gestation produced small clones, having some capacity for self-renewal, a fundamental property of stem cells. These discoveries guided the concept that the development of both peripheral and central nervous system was similar to that of the blood system and relied on multipotent NSCs [71, 73, 74].

Isolation and Culture

One of the sources of NSCs is the subventricular zone (SVZ) of the brain, which can be isolated, minced into minor pieces with a scalpel, and transferred to tube containing SVZ dissociation solution. It can be centrifuged using Percoll for getting a clean population of cells [75, 76] and identified using specific markers [77].

Cells can be cultured in medium comprising bFGF and/or epidermal growth factor (EGF) as mitogens, to generate multipotent clonal aggregates termed neurospheres which are powerful tools for the study of neural stem cells (NSCs) [78]. Neural tissues can also be cultured using different types of scaffold like hydrogels, fibrous materials, solid porous polymers, and decellularized tissues as well as microfluidic devices and lab-on-a-chip systems [79].

2.3.3 Stem Cells of the Eye: Corneal Epithelial and Stromal Stem Cells

The epithelium of the cornea is a protective barrier on the anterior ocular surface, necessary for sustaining the transparency of the cornea as well as visual perception. The corneal epithelium is generated from self-renewing stem cells during homeostasis and repair, a process which continue throughout the lifetime of the organism. These cells are termed corneal epithelial stem cells (CESCs) or limbal stem cells (LSCs) (Fig. 2.3), the functional activity of which is firmly regulated by the neighbouring microenvironment (or niche) which offers a range of signals to maintain this progenitor population [80]. The corneal stroma also contains mesenchymal cells near the limbal basement membrane having characteristics of ASCs. These corneal stromal stem cells or "niche cells" support the viability of the limbal epithelial stem cells [81].

History

In 1971, it was proposed that the LSC reside in a series of radially oriented fibrovascular ridges of the human limbus (palisades of Vogt, [82]). In 2005, histological examination of human limbus detected the presence of limbal epithelial crypts (LEC) which harbour the LSCs. LEC are more prevalent in the superior or inferior limbus [83–85].

Source

The CESCs reside exclusively at the limbus, at the basal layer of limbal epithelium, present at the junction between the conjunctiva and cornea [80, 86]. Cells isolated from this area can produce long-term proliferative clones in vitro (holoclones) that can reconstitute the cornea upon grafting [87, 88].

Isolation

Human corneoscleral rims are normally collected from cadaveric donor eyes and stored in optisol for about 8 days before use. The limbal tissues are isolated from these corneoscleral rims. Excised tissues are incubated in DMEM containing dispase II at 37 °C for 1 h. To create single cell suspensions, the epithelial layers are separated under dissecting microscope and treated with trypsin–EDTA solution (Invitrogen). Enzymatic activity is stopped by adding an equal volume of 10% FBS containing DMEM. The CESCs are $CD49f^{bri}CD71^{dim}$ [86].

Culture

Corneal cells cultivated from biopsy sample (1 mm^2) collected from the limbus of the healthy eye of patients in suspension culture showed that the corneal progenitors are confined in the limbus [89]. For colony-forming assays, the feeder layer is

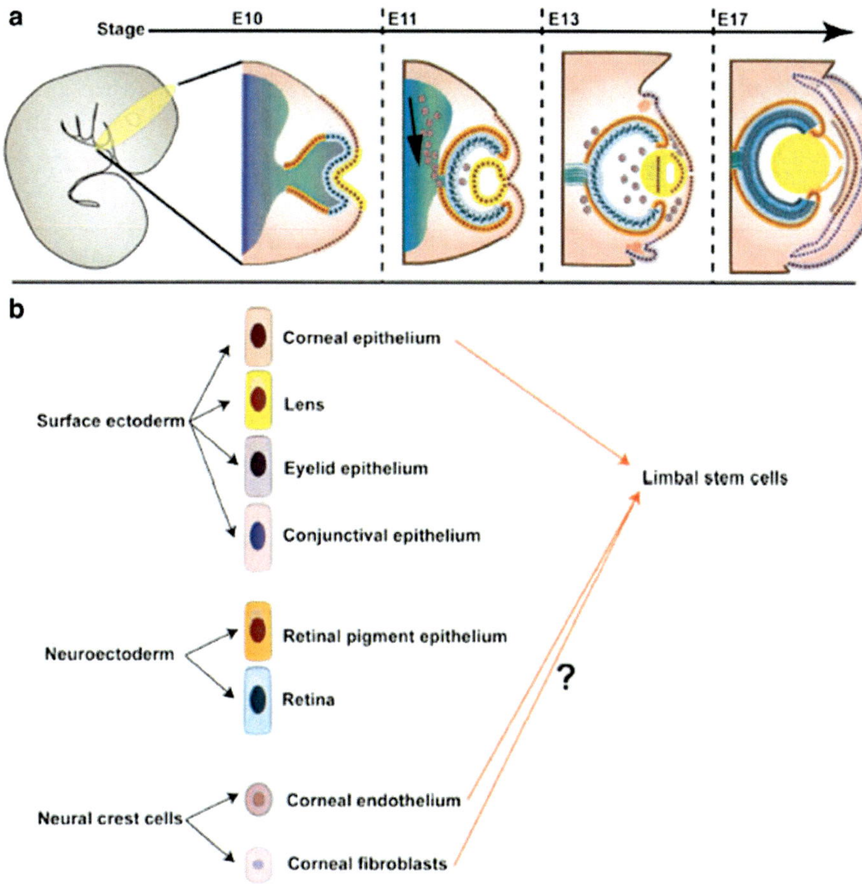

Fig. 2.3 Origin and development of LSC. (**a**) Schematic illustration of corneal development. From left to right: Extending neuroepithelium induces the surface ectoderm to form the lens, cornea, conjunctiva, and eyelid. Migrating neural crest cells differentiate into corneal endothelium and stromal fibroblasts. (**b**) Contribution of diverse embryonic layers to the formation of the LSC niche. Current models suggest that LSC could be developmental descendants of the surface ectoderm as well as the periocular mesenchyme (Reproduced from Gonzalez G, Sasamoto Y, Ksander BR, Frank MH, Frank NY. Limbal stem cells: identity, developmental origin, and therapeutic potential. *Wiley Interdiscip Rev Dev Biol.* 2018;7(2):10.1002/wdev.303. doi:10.1002/wdev.303 with permission)

arranged on tissue culture dishes, followed by seeding of flow-cytometrically sorted cell populations ($CD49f^{dim}/CD71^{dim}$, $CD49f^{bri}/CD71^{dim}$, $CD49f^{dim}/CD71^{bri}$, and $CD49f^{bri}/CD71^{bri}$) at a density of 500–1000 cells per dish. After 10–14 days, the dishes are fixed, stained with rhodamine B and colony formation assessed under dissecting microscope [86].

A three-dimensional (3D) bio-scaffold used in cell culture is HyStem-C hydrogel which is a commercially available substrate. It supports numerous kinds of stem cell expansion ex vivo like hESCs, human MSCs, hepatic progenitors, and cardiac progenitor cells to name a few. HA hydrogel scaffolds based xeno-free culture system has also been devised for ex vivo cultivation of CESCs using a definite supplement hormonal epithelial medium (SHEM) to negate the drawbacks of the xenobiotic components in the media. The conventional FBS is substituted with the KnockOut Serum Replacement [90].

2.3.4 Dental Stem Cells

The dental stem cells (DSCs) comprise of stem cell populations of mesenchymal and epithelial origin. Seven different types of DSCs have been studied so far [91].

They can be classified in two major groups, dental pulp-related stem cells and periodontium-related stem cells. Dental pulp-related stem cells are dental pulp stem cells (DPSCs), stem cells from human exfoliated deciduous teeth (SHED) and stem cells from apical papilla (SCAP), and dental pulp pluripotent-like stem cells (DPPSCs). Periodontium-related are periodontal ligament stem cells (PDLSCs), dental follicle progenitor cells (DFPCs), and Odontogenic epithelial stem cells (OEpSCs) [92] (Fig. 2.4).

The DPSCs demonstrate high self-renewal, proliferative, and multi-lineage differentiation ability. They are employed in the arenas of tissue engineering and regenerative treatment [93]. DPSCs are ectoderm-derived stem cells, developing from migrating neural crest cells and exhibit MSC like properties [94]. The DPPSCs

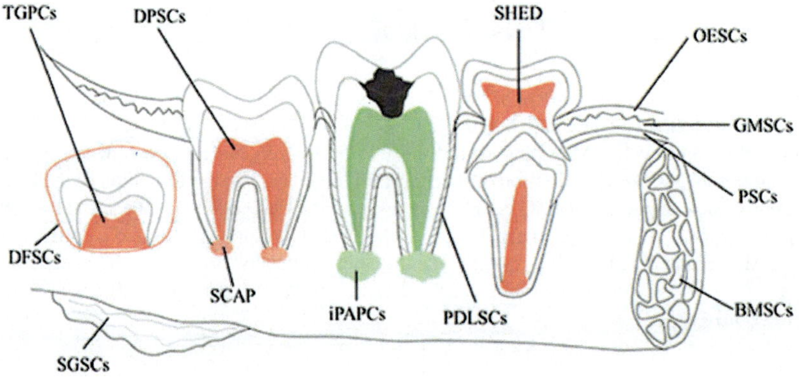

Fig. 2.4 Source of the stem cells from dental tissues. Cell types include tooth germ progenitor cells (TGPCs); dental follicle stem cells (DFSCs); stem cells of the apical papilla (SCAP); salivary gland stem cells (SGSCs); dental pulp stem cells (DPSCs); inflamed periapical progenitor cells (iPAPCs); stem cells from human exfoliated deciduous teeth (SHED); periodontal ligament stem cells (PDLSCs); bone marrow stem cells (BMSCs); oral epithelial stem cells (OESCs); gingival-derived mesenchymal stem cells (GMSCs); and periosteal stem cells (PSCs, adapted from open access publication Yang B, Qiu Y, Zhou N, Ouyang H, Ding J, Cheng B and Sun J (2017) Application of Stem Cells in Oral Disease Therapy: Progresses and Perspectives. *Front. Physiol.* 8:197. doi: 10.3389/fphys.2017.00197)

are stem cells found in human dental pulp of the third molar with embryonic phenotypes [95].

These progenitor cells are MSCs and have the ability to differentiate into osteogenic, odontogenic, myogenic, adipogenic, neurogenic, and melanocytic cells and can transdifferentiate to corneal cells and islet cells of pancreas, thus holding promise in regenerative dentistry and medicine [93].

The odontogenic epithelial stem cells (OEpSCs) are demarcated as stem cells involved in tooth development (odontogenesis). They are of epithelial/outer ectodermal in origin and can interact reciprocally with odontogenic MSCs of ectomesenchymal origin [91].

History

DPSCs were first isolated by Gronthos*et al.* in 2000 from permanent teeth [96]. Subsequently SHED was isolated [97]. In 2004, PDLSCs were isolated as the MSC population from periodontium [98, 99] followed by discovery of SCAP [100, 101]. DPPSCs were isolated from the third molars [95].

Source

The process of tooth development begins at sixth week of gestation. As the teeth are formed, a portion of the primitive ectomesenchyme is enclosed within the prospective teeth, to form the dental pulp, which denotes a source of all dental stem cells. The usual source of the DSCs is the dental pulp from the third molar [93, 95].

The sources of OEpSCs in humans are located in the active dental lamina in postnatal life, in fragments of dental lamina (the gubernaculum cord), in the epithelial cell rests of Malassez, and in the reduced enamel epithelium [91].

Isolation

The DPPSCs are Oct4+, SSEA-4+, Nanog+, HNF3beta+, FLK-1+, Nestin+, Sox2+, Lin28+, CD13+, c-Myc+, CD105+, CD34−, CD45−, CD90low, CD29+, CD73low, STRO-1low, and CD146− [102]. For isolation of DPSCs, the third molars from normal humans are collected. Tooth surfaces are dressed and cut around the cementum-enamel junction to expose the pulp chamber. The pulp tissue is moderately separated from the crown and root and digested in collagenase type I and dispase at 37 °C for 1 h. Single-cell suspensions can be obtained by passing the cells through a 70-mm strainer [96].

A similar procedure is followed for isolation of SHED where the source is normal, exfoliated, deciduous incisors collected from 7- to 8-year-old children under permitted guidelines [97]. For isolation of PDLSCs, the periodontium including cementum (PDL) is collected from premolars or third molars extracted from healthy donors aged 12–29 years. PDL tissues are minced in minimal essential medium alpha (α-MEM) comprising collagenase type I and dispase and processed as stated above [96, 99]. SCAP is isolated from root apical papilla of normal human impacted third molars, by gently separating from the surface of root, followed by mincing and digestion in a solution of collagenase type I and dispase [100] to get single-cell suspensions [96].

Culture

Single-cell suspensions obtained from the dental pulp are seeded into 6-well plates in alpha modification of Eagle's medium containing 20% FCS, glutamine, L-ascorbic acid 2-phosphate and streptomycin, and then incubated in 5% CO_2 at 37 °C. Colonies are assessed after 14 days of cultures [95, 96, 100]. For inducing osteogenic differentiation, recombinant human BMP-4 is used while for the neural differentiation, EGF, FGF, Neurobasal A, penicillin, and B27 supplement are used to culture cells attached to 0.1% gelatin-coated dishes. For the sphere-like cell-cluster formation, 3% rat serum and B27 are added [97].

DPSCs can be expanded and cultured in serum-free medium supplemented with FGF and EGF [103]. In 3D culture, DPSCs immobilized in alginate hydrogel microspheres showed enhanced potential of osteogenic differentiation, maintaining greater cell viability compared to 2D cultures [94, 104]. 3D cultures of DPPSCs can also be done using the Cell Carrier 3D glass scaffold can be used [95].

2.3.5 Stem Cells from the Epidermis (Skin)

The mammalian epidermis is a rich source of several self-renewing compartments containing epithelial stem cell populations [105, 106] like keratinocyte progenitor cells from hair follicle [107], melanocyte progenitor cells [108], nerve stem cells of the skin [109], stem cells of the eccrine gland [110], skin derived precursors (SKPs) found in dermal papilla and nestin-expressing hair follicle associated pluripotent (HAP) stem cells present in the bulge area of hair follicle [111, 112]. Thus, as a system of self-regeneration, the hair follicle and its associated connective tissue sheath are potent source of epithelial, mesenchymal, melanocyte, and neural stem cells [107, 108, 113–115].

The nerve stem cells in the skin, stem cells of the eccrine gland, and SKPs located in the dermal papilla differentiate to innumerable types of cells while the keratinocyte progenitors in the hair follicle and melanocyte-progenitor cells differentiate only to keratinocytes and melanocytes, respectively [116].

2.3.5.1 Hair Follicle Associated Pluripotent (HAP) Stem Cells

Hair follicle associated pluripotent (HAP) stem cells are a population of multipotent follicular epithelial stem cells with the ability to regenerate parts of hair follicle and the epidermis. They are the most primitive stem cells present in the skin, differentiating into glial cells, nerve cells, cardiac muscle cells, smooth muscle cells, keratinocytes, and melanocytes. Transplanted HAP stem cells help in the recovery of spinal-cord and peripheral-nerve injuries, having the potential of regenerating the heart [116].

History

Existence of the hair follicle has been known since 1990. In 2003, it was reported that a novel type of stem cell residing in the hair follicle expresses the brain stem-cell marker nestin and was referred to as the HAP stem cells [116].

Source

The hair follicle is an outstanding system to explore the fate of the epithelial stem cells [117]. The HAP stem cells are situated in the skin, in the bulge area of the hair follicle, below the sebaceous gland [116, 118].

Isolation

Human hair follicles are collected from the tissues of human scalp. Tissues are rinsed, cleared from adipose tissues, trimmed to small pieces, and exposed to enzymatic dissociation in dispase in DMEM at 4 °C for 24 h. After treatment, the epidermis is skinned off and hair follicles are pulled from the dermis. Hair follicles are rinsed in phosphate-buffered saline (PBS) and observed under an inverted microscope. To prepare single cell suspension from the follicular epithelium, the hair follicles are treated with 0.25% trypsin/EDTA at 37 °C 30 min. This cell suspension is sieved and cells are counted for further assays [106, 119, 120].

Culture

Single cell suspension is cultured in tissue culture flasks containing media supplemented with knockout DMEM/F-12 medium, β-mercaptoethanol, knockout serum replacer, nonessential amino acids, L-glutamine, and basic FGF [121, 122]. The medium is pre-conditioned by using it as growth medium for MEF feeder layer for 48 h [123]. hESC medium conditioned by MEF is added to fresh hESC medium (3:1), and supplemented with bFGF prior to use. For analysing the property of self-renewal, detached cells from the hair spheres are successively diluted in hESC medium in uncoated 96-well plates and each well is assessed microscopically. Cells are cultured on MEF feeder layers which are mitotically inactivated [106, 121, 122].

For examining the differentiation potential, the hair spheres and attached cells are dissociated into single cells enzymatically and plated onto Matrigel or fibronectin coated tissue culture-grade plastic (melanogenic differentiation) in differentiation medium [106].

2.3.5.2 Nerve Stem Cells of the Skin

HAP stem cells cultured in the 3D histoculture supported by Gelfoam® yield P-111 tubulin positive fibres, containing HAP stem cells expressing nestin-driven green fluorescent protein (ND-GFP), extending up to 500 mm from nerve stump. The growing fibres had growth cones on their tips, expressing F-actin signifying that they were budding axons. The results proposed that one of the functions of HAP stem cells is the growth of the hair follicle sensory nerve [109, 124].

2.3.5.3 Stem Cells of the Eccrine Gland

Nestin positive stem cells found in human sweat gland stroma are termed sweat gland stroma derived stem cells (SGSCs). They are a source of unique stem cells with the ability of notable self-renewal, multilineage differentiation potential and high proliferation capacity, holding promise in autologous application in clinical therapies. The SGSCs also have some immunological potential as suggested by their ability to secrete few prominent cytokines [110].

History

Initial experiments to address the existence of sweat gland progenitors were done by Lobitz, Holyoke, & Montagna, 1954. Further studies analysed the regenerative potential of sweat glands in vitro by showing proliferation of dissociated gland cells in cell culture and in vivo by demonstrating the contribution of sweat gland for epidermal reconstitution after superficial skin injuries [117, 125].

Source

The precise localization of the SGSCs are the apocrine and eccrine sweat glands of human axillary skin [110].

Isolation and Culture

Sweat glands are isolated mechanically from a pre-digested and dissociated axillary skin biopsy under a light microscope. These sweat glands and associated outgrowing cells are cultured in DMEM containing 20% FBS and antibiotics. The culture medium is replaced in 3 days, and cells are split once they are confluent. These cells can be trypsinized and counted [126, 127].

2.3.5.4 Skin Derived Precursors (SKP)

A population of multipotent precursors has been isolated from dermis of adult skin that produces neuron-specific proteins upon differentiation. These cells are the skin-derived precursors (SKPs) and produce both neural and mesodermal progeny. In culture, they can proliferate and differentiate to produce glia, neurons, smooth muscle cells, and adipocytes [111].

History

SKPs were reported from the rodents and dermis of adult skin recently [111, 112, 128]. The incidence of pluripotent neural crest stem cells associated with the dermal papillae of adult mammalian hair follicles were reported later and were claimed to be SKPs recently [112].

Source

Human SKPs are found to reside within the dermis throughout adulthood and can be isolated from the foreskin [129, 130].

Isolation and Culture

In order to culture the human SKPs, pieces of scalp tissue (~6 mm^2) are isolated by neurosurgery, washed with Hanks balanced salt solution (HBSS), and cut into minor pieces. These pieces are digested enzymatically using 0.1% trypsin, followed by washing. Dissociated cells are suspended in culture medium containing leukaemia inhibitory factor (LIF) and cultured. Cells can be sub-cultured by fractional dissociation of clusters that is formed every 7–10 days [111, 129, 131].

2.3.5.5 Keratinocyte Stem Cells (KSC)

The keratinocyte stem cells (KSCs) are the progenitor population of cells responsible for epidermal renewal during the life of any individual. Two populations have been isolated, one within the hair follicle [116] while another is located at the basal layer and is surrounded by their direct progeny, transient amplifying cells or keratinocyte progenitors [132].

Source

The KSCs are found within the hair follicle of human skin (Fig. 2.5) in the upper part [116].

Isolation

Initial culture of keratinocytes usually begins from healthy donor human skin with informed consent. The epidermis is detached from the dermis using dispase. The cells are then dissociated using trypsin EDTA followed by counting and testing of viability by trypan blue assay. Freshly isolated keratinocytes are incubated with a fluorescent tagged murine anti-human CD71 and rat anti-human CD49f antibody at 4 °C for 30 min separately. The CD49f^{high}CD71^{low} KSCs can be isolated by flow-cytometry and used for culture [132, 133].

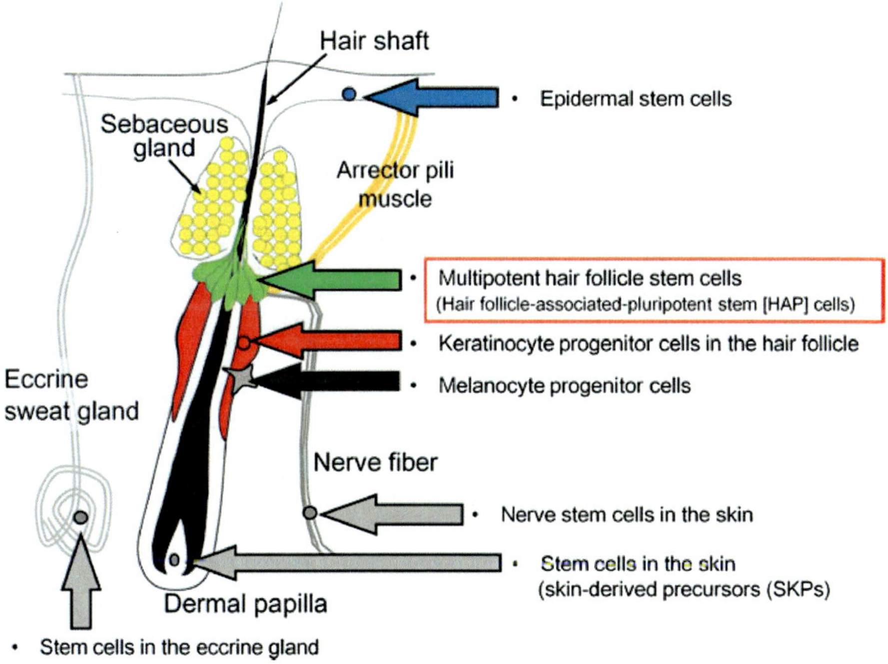

Fig. 2.5 The location of the numerous types of stem cells in the skin. Reproduced with permission from Amoh Y, Hoffman RM. Hair follicle-associated-pluripotent (HAP) stem cells. *Cell Cycle.* 2017;16(22):2169-2175. doi:10.1080/15384101.2017.1356513

Culture

For clonogenic assay, the sorted KSCs can be seeded on a feeder layer of inactivated human fibroblasts, pre-conditioned in keratinocyte medium DMEM and Ham's F12 (3:1), containing 10% FCS, EGF, adenine, hydrocortisone, INS, triiodo-L-thyronine, antibiotics, and cholera toxin (CT). The following day, the cells are irradiated and put in 14-day culture, changing the medium thrice every week. For clone staining, cells were fixed and stained with rhodamine.

For reconstruction of human epidermis, the cells are cultured in the aforestated keratinocyte medium for 7 days followed by elevation at the air/liquid interface for another 7 days in a differentiation medium (DMEM) with 8 mg/mL bovine serum albumin (BSA), INS, hydrocortisone and antibiotics [132].

2.3.5.6 Melanocyte Stem Cells

The pigment cells or melanocytes present in the hair follicles, proliferate and differentiate, in close association to hair regeneration cycle [108, 134]. The melanocyte stem cells, originally derived from the neural crest, are the unipotent progenitor population responsible for generation of melanocytes [108, 135].

Source

The melanocyte stem cells can be identified from the lower permanent portion of the hair follicles, throughout the hair cycle [108].

Isolation and Culture

The scalp tissues are extracted during punch biopsy and extra adipose tissue is eliminated. The strips are gently collected and separated under sterile conditions in physiologic solution (0.9% NaCl). They are centrifuged and either put into suspension culture or mechanically penetrated into the scalp of patients [136].

2.3.5.7 Merkel Cell Stem Cells

Merkel cells are a population of exclusive cells found in the epidermis, associated with the nerve endings. They play pivotal role in the perception of touch. These cells often transform to cancerous cells causing Merkel cell carcinoma, a violent type of skin cancer without any treatment. They need the transcription factor *Atoh1* for their specification. A subpopulation of cells expressing *Atoh1* was identified from the whisker and hair follicles of mouse skin which solely formed the Merkel cells during development. Eradicating *Atoh1*-positive cells from adult mouse skin cause perpetual reduction in the number of Merkel cells, indicating that no other stem cell populations can produce them [137].

2.3.6 Thymus Epithelial Stem Cells (TESC)

Thymus is a primary lymphoid organ composed of two discrete anatomical regions, the outer cortex and the inner medulla, playing a major role in establishing the central T-cell tolerance for self-antigens. The thymic epithelial stem or progenitor cells (TESC) are either bi- or unipotent in the postnatal thymus [138].

History

The concept of a common thymic TESC from which both thymic and medullary cells arise was proposed in 1988 [139].

Isolation and Culture

The thymus cells are isolated by the mechanical disruption followed by enzymatic digestion and flow cytometry [140, 141]. These cells grow in 2D system as well as in 3D system. Different 3D cultures have been used for the clonogenic expansion of the TESCs; semi-solid Matrigel culture supplemented with serum and growth factor like keratinocyte growth factor along with MEF or culturing total $CD45^-$ thymic cell population including the endothelial and the mesenchymal cells in liquid medium supplemented with EGF and basic FGF under low adherent conditions is prevalent [138].

2.3.7 Oral Epithelial Stem Cells (OESC)

The oral mucosa serves as a barricade to the regular physical and chemical insults caused by mastication, food, and micro-organisms and is one of the fastest dividing tissues of the body. Any damage to this barrier can lead to significant morbidity. The squamous epithelium covering the oral mucosa depends on the epithelial stem cells for tissue renewal. These stem cells are termed as oral epithelial stem cells (OESCs). The study of OESCs is supreme to understanding their role in homeostasis and disease [142, 143]. OESCs are an excellent source of autologous cell therapy for severe bilateral deficiency of limbal stem cells [144].

Source

OESCs are present broadly within the basal layer of the mucosa, distributed throughout the oral cavity [145].

Isolation

The OESCs can be isolated from $\alpha6\beta4^+CD71^-$ gingival keratinocytes which have high colony forming efficiency and can form oral epithelial equivalents (OEE) which is a fully stratified epithelium derived from in vitro cultured isolated oral epithelial cells [143, 146, 147].

Culture

K5 and β1 integrin expressing tongue epithelial cells produce more holoclones in culture than the epithelial cells which express lower levels of these markers [148]. A collagen IV matrix used for enrichment of OESCs showed that the adherent cells had the highest colony forming efficiencies in vitro [146]. Cells in the gingival and buccal epithelium which express high levels of neurotrophin receptor p75 have superior in vitro proliferative capacity but are slowly cycling in vivo, and can recapitulate OEEs ex vivo. These cells are present in the tips of the dermal papillae and the rete ridges, signifying that p75 precisely labels OESCs [143, 149].

2.3.8 Lingual Epithelial Stem Cells (LESC)

The epithelial stem cell population accountable for maintenance of the lingual epithelial cells of the tongue are the lingual epithelial stem cells (LESC). These primitive cells are Bmi1-positive, located at the second or third layer of the epithelial cell layer at the base of the interpapillary pit (IPP). They can be cultured as organ culture, on extracellular matrix (ECM), organotypic raft culture and organoid culture for enhanced knowledge for regenerative medicine and to learn about the regulatory mechanisms of lingual epithelium [150].

2.3.9 Oesophageal Epithelial Stem Cells (OeSC)

The oesophagus serves as the channel to transport food from the mouth to the stomach. The heterogeneous epithelial basal layer of the oesophagus comprises of two discrete zones, one overlying the papillae of the supporting connective tissue (PBL) and the second one covering the interpapillary zone (IBL). The IBL zone is rich in oesophageal stem cells (OeSC), accountable for the maintenance of the oesophagus [151, 152].

History

The presence of OeSCs in the basal cells in the human oesophagus were detected in 2000 [151].

Source

Human OeSCs, are post-mitotic amplifying, critically differentiating cells and are situated in distinct anatomical regions. IBL has more OeSCs, divide occasionally and asymmetrically, and found as large, actively growing clones in culture. The PBL cells proliferate symmetrically and more frequently than the IBL cells, having lower proportion of clonogenic cells. The latter are the transit amplifying cells [151].

Isolation and Culture

Primary epithelial cells can be isolated from the mucosal layer of the oesophagus by enzymatic dissociation using dispase and Trypsin/EDTA. Undamaged mucosal sheets endured two trypsin incubations to yield a single-cell suspension. The resulting cell suspensions are filtered through cell strainer and centrifuged. The pellets are resuspended in medium and seeded onto 6-well collagen-coated plates and maintained at 37 °C, 5% CO2. The medium is replaced every 2–3 days and cells are passaged at 60% confluence [153, 154].

2.3.10 Salivary Gland Stem Cells (SGSCs)

Salivary glands maintain the vigour of the oral cavity and are regularly damaged by therapeutic radiation for head and neck cancer as well as by autoimmune diseases like Sjögren's syndrome. Regenerative approaches based on the activation of endogenous stem cells or transplant of exogenous stem cells hold promise in restoring the structure and function of these organs [155]. The salivary gland stem cells (SGSCs) are the eventual source for renewal of salivary gland tissue [156].

History

The presence of SGSCs were reported from the intercalated ducts of the adult salivary glands in 1993 [155, 157].

Source

The ductal compartment in the salivary gland is the location which harbours the stem cells [156].

Isolation and Culture

The SGScs are isolated by the mechanical and enzymatic tissue digestion of the salivary glands followed by identification [158–160] and suspending the cells in medium and cultured [161]. Human single clonal SGSCs can also be cultured in the nanofibrous microwells through priming and differentiation for 3D spheroid cultures [162].

2.3.11 Epithelial Intestine Stem Cells (ISC)

In adult mammals, the intestinal epithelium is the most hastily self-renewing tissue. It is presently believed that four to six crypt stem cells exist immediately above the Paneth cells in the small intestine termed as the epithelial intestine stem cells (ISCs, [163]). Small populations of these ASCs are accountable for the extraordinary ability of the epithelial lining of the intestine to be competently renewed and repaired throughout life [164].

History

ISCs produce a population of dynamically proliferating progenitors called the transit amplifying cells, which rapidly expand through multiple rounds of cell division as they move up as an articulate column toward the crypt or the villus border [165].

Source

The luminal side of the intestine is the source of the ISCs [164, 165] and are identified by the presence of pleckstrin homology-like domain family A member 1 (PHLDA1+), CD49f [166, 167] along with CD44, CD24, CD166 and lack of GRP78 and c-kit [168].

Isolation

The luminal side of the intestine is scraped, washed, cut into pieces and further washed and incubated in EDTA and PBS for 20 min on ice for isolation of ISCs. This is followed by removal of EDTA, addition of cold PBS and releasing of the crypts by manual shaking of the suspension. Supernatant is collected and sieved. The crypt-containing fractions are collected and centrifuged at 110 *g* for 5 min [169, 170].

Culture

The crypt-containing fractions are rooted in polyethylene glycol (PEG) gels or Matrigel, which are cast into 20 μL droplets at the bottom of wells in 24-well plate. Subsequent to polymerization at 37 °C for 15 min, the gels are overlaid with 500 μL of expansion medium (Advanced DMEM/F12 containing HEPES, Glutamax, penicillin-streptomycin and supplemented with growth factors). To induce the differentiation of stem cells and organoid formation, these cells are cultured in fresh medium containing Noggin, EGF, and R-spondin [169, 170].

2.3.12 Haematopoietic Stem Cells (HSCs)

Blood is a dynamic tissue in the mammals, having a diverse array of the cells. While mature blood cells have short lifespans, the accountability of maintaining blood homeostasis lies entirely on a rare residents of somatic stem cells, termed the haematopoietic stem cells (HSCs) [171]. HSCs are multipotent, maintain themselves, undergo self-renewal, retaining their ability to differentiate throughout life [172, 173]. The phenomenon of formation of the cellular components of blood often termed as haematopoiesis, occurs in all vertebrates and is accountable for the production and renewal of all types of the blood cells throughout the life of an animal [174].

During embryonic development, haematopoiesis initiates in the yolk sac, aorta-gonad-mesonephros region, and placenta. Later the HSCs move through the foetal liver and lastly migrate to the BM. In adults, HSCs are primarily found in specialized niches in the BM cavity which continues to be the site of haematopoiesis throughout the life of the individual [171, 174]. The BM niche plays a pivotal role in maintaining and differentiation of the HSCs [173, 175].

History

The role of the BM was unknown in the distant past and in 1868 it was first described as the blood-forming tissue [176]. In the twentieth century the perception of "hematopoietic inductive microenvironment" was hypothesized, where stem cells may migrate to the marrow and begin to proliferate and produce the blood cells [177, 178]. This work was later supported by Paul Ehrlich [176].

Source

Human $CD34^+$ cells contain both stem (HSCs) and haematopoietic progenitor cells (HPCs) [179, 180]. The source of HSCs is mobilized PB [181, 182], BM [182], and the umbilical cord blood [182–184]. One milliliter of UCB contains about 25,000 HPCs [183]. HSCs can also be cultured from ESCs, human pluripotent cells [185], and induced pluripotent cells (iPSCs) [186].

Isolation

The common cell surface markers of HScs are CD34 [187, 188], CDw90 [189, 190], CD117 [191], CD133 [192, 193], and ALDH [194, 195] to name a few. The HSCs can be isolated using Ficoll, by density gradient centrifugation followed by washing with medium. The mononuclear cells (MNCs) obtained contain a population of HSCs which can be flow-cytometrically sorted or by using magnetic beads [49, 196]. Since the evolving stage of the haematopoietic cell commands its biological behaviour, primitive subpopulations of the $CD34^+$ cells, e.g., $CD34^+Rh^{low}$ [197], $CD34^+$ $CD38^-$ [198, 199], CD34 + CD45RA-CD71- and CD34 + CD45RA-CD71-CD90+ cells have greater expansion potentials than their more mature counterparts [200].

Culture

HSC transplantation can reinstate a new functional haematopoietic system in patients having a non-functional system of his own, especially in case of haematological malignancies. To improve transplantation, a necessity for betterment of in vitro culture and proliferation of HSCs still persists [180].

The ability of long-term reconstitution of the haematopoietic system by the HSCs demonstrate their multipotent nature and ability to self-renew. HPCs generate the haematopoietic colonies when cultured in semisolid media, in the presence of recombinant stimulatory cytokines [201–203].

For ex vivo culture of the HSCs or HPCs several culture conditions have been put forward over the past few decades. Two significant considerations in such protocols are the cell population and the particular culture conditions to be used [204]. MNCs especially $CD34^+$ cells are selected for the purpose. The culture supplements normally contain the molecules favouring the growth of the HSCs and HPCs, namely the early and late acting cytokines. The use of different type of stromal cell also enhances the expansion. The cytokines involved in in vitro haematopoiesis include FLT3 ligand (FL), stem cell factor (SCF), and thrombopoietin (TPO) among many [181]. 3D culture of human HSCs can also be done in degradable zwitterionic hydrogel to attain considerable expansion of phenotypically primitive $CD34^+$ cord blood and BM derived HSCs [205].

2.3.13 Gallbladder Stem Cells

The biliary tree is divided into two parts—the intrahepatic and the extrahepatic biliary systems. The latter consists of the cystic duct, gallbladder, and the common bile duct. The biliary system is a channel for transportation of the bile to the intestine from the liver. The gallbladder stores bile, regulates the content and concentration, thereby playing an important role in digestion. The residing stem cells of the biliary system hold importance in biliary diseases [206]. Gallbladders are a profound source of stem or progenitor cells that can be propagated in culture as organoids [207].

History
The gallbladder stem cells were identified and isolated initially from mouse and then human [206, 208]

Source
The gallbladder stem cells can be isolated from foetal liver within 19–23 weeks of gestation [208].

Isolation
Gallbladder mucosa was exposed and placed in HBSS. Bile was washed off by gently rubbing the mucosal surface with blunt-ended forceps. Gallbladder was treated with Collagenase II, Hyaluronidase, DNaseI Trypsin /EDTA treatment to obtain a cell suspension [208].

Culture
Cell suspension was plated on irradiated rat feeder layer which select for epithelial growth [208]. Cell pellets can also be mixed with cold Matrigeland plated in 24-well plates. Tissue culture media (AdDMEM/F12) with supplements like human recombinant growth factors (HGF, EGF and FGF10) were used. Medium was replaced every 3 days, and organoids were split every 4–5 days by mechanical dissociation [207].

2.3.14 Hepatic Stem Cells (HpSCs)

The human liver comprises of two pluripotent progenitors, namely the hepatic stem cells (HpSCs) and hepatoblasts (HBs) [209]. In adult liver, the credible stem cell niche is the proximal biliary structure, the canals of Hering, the meeting point between hepatocyte canaliculi, and the bile duct [210, 211].

History
The existence of HpSCs were first reported in mouse. Experimental evidences from their work suggested that a fraction of adult mouse hepatocytes have growth potential similar to that of HSCs [212].

Source
The HpSCs can be isolated from intact livers of cadaveric neonatal, paediatric as well as adult donors [209].

Isolation

Liver cell suspension can be prepared using collagenase digestion of perfused liver tissue at 37 °C [213]. Once the liver tissue is digested and the cells are released, the hepatocytes can be separated by centrifugation, and the pellets obtained are washed with ice-cold buffer solution to purify the cells. The cell viability and yield are then evaluated for culture [214].

Culture

Primary human hepatocytes can be cultured by seeding the hepatocytes in hepatocyte culture medium containing FBS onto type I collagen-coated plates. The medium is replaced 6 h after seeding. The hepatic progenitor cells are seeded at 400,000 cells per well on coating with 4.0 $\mu g/cm^2$ bovine fibronectin in 24-well plates and cultured with chemically defined medium supplemented with activin A and FGF2 and consisting of 50% IMDM, 50% F12 Nutrient Mixture supplemented with INS, transferrin, monothioglycerol, and BSA. To confirm hepatic differentiation, the cells are differentiated toward hepatocyte-like cells by dissociating them into single cells and seeding them with Geltrex in 24-well plates. The cells are cultured with mTeSR1 medium in 4% O_2 and 5% CO_2 approximately for 24 h, when the cells are sub-confluent. The medium is replaced with RPMI1640 medium containing B27, activin A, BMP4, and FGF2 in ambient O_2 and 5% CO_2 [215].

2.3.15 Lung Epithelial Stem Cells

Lung is a multifaceted organ, lined throughout with the epithelial cells whose integrity maintains the normal lung function. Analogous to other adult tissues, epithelial stem/progenitor cells in the adult lung are crucial for the development, tissue homeostasis, and injury repair of the lungs. These cells are a subset of undifferentiated cells that undergo asymmetric cell division during normal lung morphogenesis [216].

Source

The stem or progenitor cells of the pulmonary epithelium are present in the basal and mucous cells of the proximal airways, Clara cells in the bronchioles, and type II pneumocytes of the alveoli [217, 218].

Isolation and Culture

The lung peripheral tissue from the donor is mildly homogenized and then digested in DNAse and Pronase for 24 h at 4 °C. The processed tissue is plated on culture flasks for 30 min at 37 °C. Following this, the non-adherent cells are transferred to culture on human Collagen-IV-coated flasks and sustained in Small Airway Growth Media. Primary endothelial cells are isolated from the larger vessels of the donor lungs using the same digestion protocol. The endothelial population is sorted for CD31+ cells by flow-cytometry and cultured on Gelatin-coated flasks in EGM2 until utilized for lung scaffold recellularization. For in vitro co-culture, endothelial cells are plated on 0.4 μm transwell insert (Corning, 07200161), placed above the epithelial cell culture, with or without the addition of VEGF, for 7 days [219, 220].

For Air-Liquid Interface culture, the primary epithelial cells at third passage are plated on 0.4 μm Transwell inserts coated with collagen IV and maintained in submerged culture for 5 days. Media is substituted with PneumaCult™-ALI medium (Stemcell Technologies, 05001) in the basal chamber only, and maintained for 21 days at Air-Liquid interface, with every alternate day media change.

For the 3-dimentional sphere assay, the primary basal epithelial cells at the third passage are filtered and transferred to a 0.4 μm Transwell insert (5000 cells/90uL of 50:50 matrigel-to-SAGM substrate). After confirming the single cell suspension by light microscopy, cultures can be maintained with SAGM in the basal chamber for about 7 days [221].

2.3.16 Mammary Epithelial Stem Cells (MaSCs)

Adult mammary stem cells (MaSCs) control the postnatal organogenesis and remodelling of the mammary gland. The longevity and potential of these progenitor cells have important implications for breast cancer. Unlike other organs, the mammary gland primarily develops postnatally and thus stem cells in the adult mammary gland serve both developmental and homeostatic functions [222].

History

The genetic scrutiny of adjoining regions of individual human mammary ducts in the same breast signified their clonal origin and hinted toward the presence of multipotent mammary epithelial cells in human breast [223].

Source

MaSCs can be isolated from the normal mammary gland tissue surrounding the primary breast tumours by sorting immunomagnetically [224].

Isolation

Normal tissue isolated from the mastectomy samples and normal tissue of reduction mammoplasties can be identified by specific biomarkers [225], crushed and dissociated in Ham's F12/DMEM, supplemented with Hepes (H), BSA, CT, INS, hydrocortisone, collagenase and hyaluronidase for 18 h at 37 °C. The epithelial-cell rich pellet is then collected by centrifuging, followed by washing with F12/DME/H. The supernatant from the first centrifugation is saved for future to serve as a source of mammary stromal cells (fibroblasts), which is then concentrated by centrifugation and cultured in F12/DME/H supplemented with 5% FBS and INS [226, 227].

Culture

For primary culture, isolated cells are suspended in F12/DME/H containing FBS, INS, BSA, Hydrocortisone, and CT, and then plated on dehydrated rat-tail collagen coated tissue culture flasks. After 36 h, this medium is changed with the same fresh medium lacking FBS [228].

For seeding the cells in 3D collagen gel cultures, sorted human breast epithelial cells (HBEC) or HBEC obtained from primary cultures are mixed at a density of 5×10^2–10^4 cells/mL of gel, with a solution of rat-tail collagen dissolved in 0.1% acetic acid, 10X F12/DME/H, NaOH and serum free medium, hormones, and growth factors. Aliquots of this solution (0.5 mL) are then pipetted into individual wells of 24-well plates with irradiated human mammary fibroblast feeder layer and overlaid with 0.3 mL of cell-free collagen gel. The human mammary fibroblasts are irradiated and seeded in F12/DME/H containing 5% FBS and INS. Before adding primary layer of the cell-free collagen gel, feeder layers are rinsed twice with F12/DME/H. Once the collagen layer containing the suspended HBEC gels, 0.5 mL of the serum free medium is added on top. All 3D cultures are incubated in humidified atmosphere consisting of 5% CO2 and 95% air for 14 days at 37 °C. Colonies are usually scored after removing the gel from the well, mildly detaching and disposing the cell-free collagen gel underlay and placing the HBEC-containing gel on a plastic grid. The number of colonies (>3 cells) present is enumerated directly under dissecting microscope [226, 227].

2.3.17 Uterine Stem Cells: Endometrial and Myometrial Stem Cells (EndoSCs and MyoSCs)

The human uterus is a reproductive organ and has two sets of stem cells, namely the endometrial and the myometrial stem cells (EndoSCs and MyoSCs). The EndoSCs include endothelial, stromal, and epithelial cells, and contribute to the rapid regeneration of the endometrium [229].

History

The migration of stem cells from BM to the endometrium was first reported by Taylor's group [230]. The first indication of the presence of MyoSC population relates to the embryological development of the female reproductive tract [231].

Source

The EndoSCs have two sources, the tissue-resident progenitor cells located in defined niches around the perivascular regions of human endometrium, and exogenous cells derived from the BM with a provisional capacity to engraft. The MyoScs are residents of the MyoSC niche and are fewer in number [229].

Isolation and Culture

Due to the lack of precise phenotypic markers for human EndoSCs, they are defined by functional properties like clonogenicity [232]. These stem cells can be cultured using standard procedure of isolation of cells by mechanical force and enzymatic digestion followed by culture [231, 232].

2.3.18 Prostate Stem Cells

The adult prostate gland encompasses a simple columnar epithelium with a scarce neuroendocrine cell component, which originates from a rare, relatively dormant stem cell population and maintains the glandular homeostasis throughout life [233].

Isolation and Culture

The methods for isolating prostate stem cells have primarily exploited flow cytometry and 3D spheroid culture techniques [233]. Although the residing stem cells are quiescent in vivo, when positioned in a 3D matrix culture without any niche restraints, they undergo asymmetric division, producing progenitor cells that rapidly proliferate and differentiate [234].

2.3.19 Spermatogonial Stem Cells (SSCs)

The spermatogonial stem cells (SSCs) comprise of a heterogenous population of self-renewing stem cells, playing a pivotal role in spermatogenesis. Primarily, SSCs are the mother cells from where spermatogenesis begins. Following this commitment to differentiate, the succeeding cellular processes proceed in a stringently methodical way, leading to the formation of sperm in the seminiferous tubules. Secondly, the SSCs sustain spermatogenesis all through the life, owing to their role as stem cells. Thirdly, disruption of spermatogenesis, whether due to physiological or other reasons can be regulated or restored by SSCs. Finally, SSCs maintain the integrity of the DNA, so that it can be effectively transmitted to the subsequent generation. Thus, the SSCs ensure the continuation of life [235]. The study of the SSCs has important implications in deciphering male infertility [236].

Source

The population of SSCs that can both self-renew and differentiate to spermatogonia are found along with the undifferentiated spermatogonia. Largely, these tissue stem cells are localized in a niche where they favourably self-renew, whereas the daughter cells that migrate out of the niche differentiate. SSCs are found in seminiferous tubules, located in areas of the basal lamina opposite to the interstitial tissue, especially where the interstitial arterioles and venules are present [235, 236].

Isolation

The expression of the inhibitor of differentiation 4 (ID4) [235], paired box 7(PAX7) [237], and B cell specific Moloney murine leukaemia virus integration site1 (BMI1) are markers for SSCs [238].

A testis cell fraction enriched for SSCs can be isolated from the testes after birth and in adult stages. The isolation protocol includes dissociating the seminiferous tubules into a single cell suspension, using mechanical disruption and enzymes. From this suspension, Thy1+ cell fraction is sorted which contains almost the entire SSC population and can be used for further studies [239].

Culture

At present it is possible to culture and propagate SSCs of mice and some other species. Two significant attributes which contributed to the progress of these culture methods are the establishment of the spermatogonial transplantation technique and identification of glial cell line-derived neurotrophic factor (GDNF) as the key regulatory factor for the self-renewal of SSCs. Spermatogonial transplantation technique involved dissociation of testis cells from donor mouse into single cells and injecting it in the seminiferous tubules of the recipient mice, which were pre-treated with busulfan to ablate endogenous spermatogenesis. The spermatogenic colonies generated are proportional to the number of cells transplanted. Thus, this method, called spermatogonial transplantation, is a novel and unique functional assay for SSCs, with each colony in the transplanted testis corresponding to a single SSC or a unit of SSCs [235]. Although the cultures permit the propagatation of SSCs from many animals, complete spermatogenesis could not be achieved in in vitro condition [236].

In in vitro culture, MEF feeder layer is prepared and Thy1+ cells isolated from the testis of mouse are layered. The cells can be cultured in a serum-free medium containing recombinant human GDNF and recombinant human bFGF. The SSCs are observed within cell clumps which are loosely attached to the underlying STO feeders [239, 240].

2.3.20 Mesenchymal Stem Cells (MSCs)

The MSCs are adult, non-haematopoietic, multipotent progenitor cells having the ability of self-renewal, differentiating into many specialized cell types and immunomodulatory properties. The differentiation ability is usually restricted to osteogenic, chondrogenic, and adipogenic potentials as well as endodermal (hepatocytes) and ectodermal (neurocytes) lineages [241, 242].

History

A set of discrete non-haematopoietic stem cell population having multi-lineage differentiation potential was first reported to be residing in the BM by Friedenstein et al. in 1976. These stromal cells were termed as mesenchymal stem cells [243].

Source

The usual source of MSC in man is BM which contains a rare population of MSCs, identified as non-haematopoietic skeletal progenitor cells with chief importance for the haematopoietic microenvironment [244, 245]. Other sources include multiple tissues [242] like PB [246], adipose tissue [247, 248], dental tissues [249], foreskin and skin [250, 251], salivary gland [252], limb buds [253], synovial fluid [254], menstrual blood [255], amniotic fluid [256], amniotic membrane [257], and other perinatal tissues [258–261] like umbilical cord tissue (Wharton's jelly, WJ) [261] and UCB [262–264], endometrium [265] and sub-amniotic umbilical cord lining membrane [266]. The cell surface markers of MSCs include positive expression of

CD73, CD90, and CD105, and negative expression of CD34, CD14, CD45, and human leukocyte antigen-DR (HLA-DR) [267].

Isolation

The major constraints in isolation of MSCs are the absence of standard protocols and absence of a universal marker. MSCs can be isolated as a plastic-adherent cell population with simple procedures like tissue mincing, optional enzymatic digestion, and cell outgrowth on a plastic surface [244]. The procedure is divided into enzymatic and explant protocols. The explant culture (Fig. 2.6a) is a primitive technique of cell isolation and cultivation in vitro. The tissue is rinsed to eliminate blood cells, split mechanically into small pieces by cutting or chopping and then placed into plastic culture vessels with growth medium. The MSC grow out from tissue pieces onto the culture dish surface. After several days the tissue pieces can be removed. Enzymatic protocol (Fig. 2.6b) incudes steps where the chopped tissue is incubated with enzyme solution to degrade the ECM. Single cells or small cellular aggregates are released from the tissue which are transferred to medium containing culture dishes. The explant method yields a more homogeneous cell population showing higher cell viability and proliferation rates when compared to the enzymatic method [241]. Another method of isolation of MSCs from BM, PB and synovial fluid involve Ficoll density gradient method with slight modification and seeding into culture plates [258, 261]. The MSCs from BM can also be isolated using a novel marrow filter device [268].

Culture

MSCs isolated from diverse sources can be cultured using media like DMEM [252, 268, 269], DMEM-F12 [251, 257], αMEM [246, 254], DMEM-LG [253, 258], DMED-HG [250, 266], and RPMI (Roswell Park Memorial Institute) medium [270].

The initial culture media is supplemented with 10% FBS [252, 268, 269], new-born calf serum [246] or FCS [252]. The oxygen concentration also determines the proliferation of the MSCs [270]. The expansion of MSCs is also observed when cultured in DMEM containing low glucose and growth factors like EGF, FGF, and B27 [250]. The most common culture medium for culturing and expanding MSCs in vitro is DMEM containing 10% FBS [242]. The confluent adherent cell population represent the MSCs [268].

2.3.21 Cardiac Stem Cells

In contrast to the existing belief that the heart loses the capability of reviving after birth, present research shows that adult heart is able to regenerate, although the degree of the new cardiac myocyte formation may vary. This renewal and regeneration is attributed to the endogenous cardiac progenitors or stem cells [271].

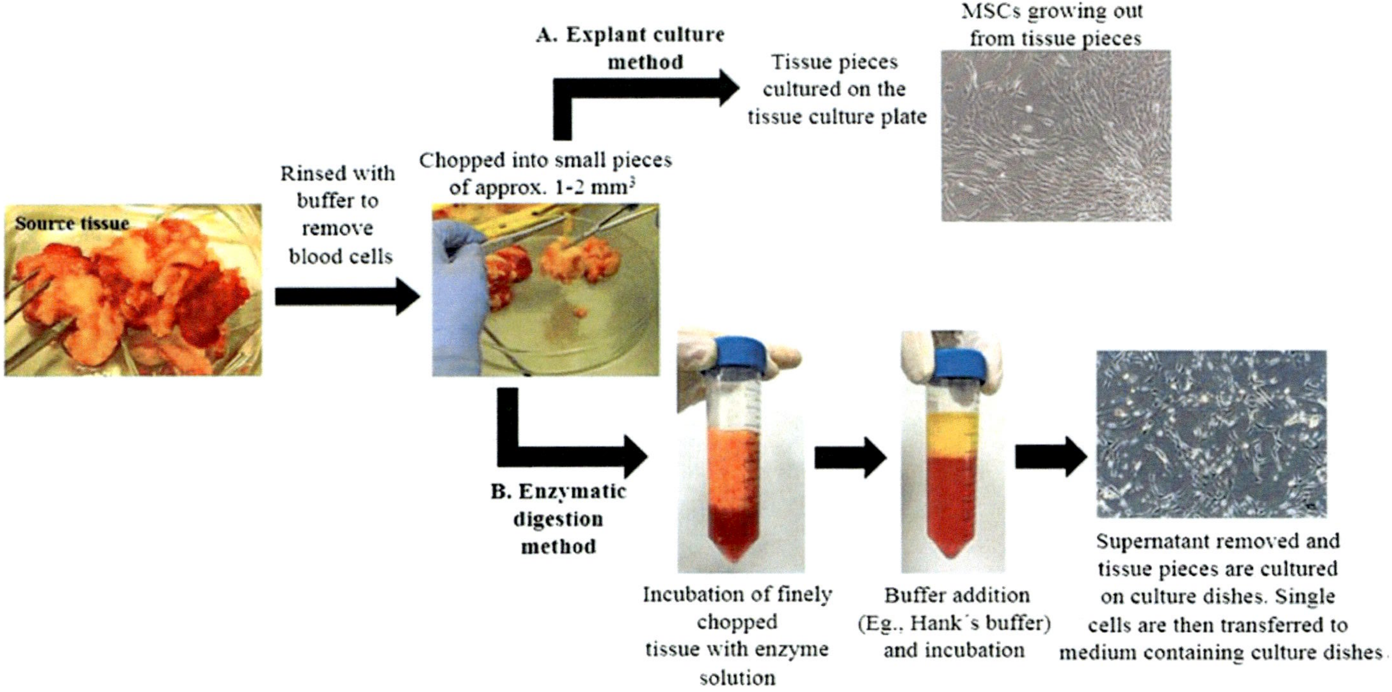

Fig. 2.6 Representation of isolation of the MSCs. (**a**) explant culture method and (**b**) enzymatic method (Mushahary D, Spittler A, Kasper C, Weber V, Charwat V. Isolation, cultivation, and characterization of human mesenchymal stem cells. *Cytometry A*. 2018;93(1):19-31. doi:10.1002/cyto.a.23242) with permission

History

Cardiac stem cells were first isolated by Beltrami et al. as cells expressing c-kit, in rats in 2003. These multipotent tissue resident stem cells were found to be self-renewing, proliferative and gave rise to cardiomyocytes [272]. Similar progenitor population was isolated from humans by Bearzi et al. in 2007 [273].

Source

The cardiac stem cells are isolated from the myocardium present in the human heart [273]. The cardiac progenitors or stem cells can be identified by the existence of a set of cell surface molecular markers like c-kit, SCA-1 (Ly6A), and ABC transporter cells [274, 275]

Isolation and Culture

The cardiac stem cells were isolated by the enzymatic dissociation of myocardial samples, followed by sorting of c-*kit*$^{+}$ cells using immunobeads and flow cytometry. These cells are plated at a low density to obtain multicellular clones from the single founder cells [273].

2.3.22 Muscle Stem Cells (MuSC)

Muscle stem cells (MuSC) or satellite cells are adult stem cells, found in skeletal muscle tissue. They are responsible for self-renewal and differentiating into skeletal muscle cells. Normally, they are in quiescent state but are triggered in response to muscle injury and regenerate damaged muscle tissue [276].

History

The existence of MuScs were first reported by Partridge and Sloper in 1977 when their first engraftment assays, were done by transplanting minced muscles using H^3-thymidine into wild-type muscles [277].

Source

MuSCss are residents of a dedicated niche beneath the basal lamina of myofibres [276]. They are mononucleated, dormant, myogenic cells, located in between the sarcolemma and the basement membrane of terminally-differentiated muscle fibres [278].

Isolation and Culture

Myofibres are isolated mechanically using enzymatic digestion. The isolated fibres are put into culture in DMEM containing FBS, 1% chicken embryo extract, glucose, L-glutamine, and sodium pyruvate [279–282].

2.3.23 Adipose Stem Cells

Human adipose-derived stem cells (ADSCs) are the adult stem cells isolated from fat tissue during procedures like liposuction, abdominoplasty, or breast reduction. They are found in the stromal-vascular region, and is isolated using an amalgamation of washing steps and enzymatic digestion. These cells also express genes that constitute the core circuitry of self-renewal like OCT4, NANOG, SOX2, and neurogenic lineage genes such as PAX6, NEUROD1, and SOX3. They find importance in several therapeutic approaches intended to treat tissue damage, the autologous transplantation of adipose tissue being an added advantage [283].

2.4 Induced Pluripotent Cells

As introduced in Chap. 1, induced pluripotency is a phenomenon where differentiated cells are reprogrammed, to an embryonic-like state, either by the transfer of nuclear contents into oocytes or by fusing with ESCs. The resultant de-differentiated cells are termed the induced pluripotent cells (iPSCs) and the pluripotency is induced using four transcription factors, namely, Sox2, Oct3/4, Klf4, and c-Myc [284].

History

Pluripotency was first induced in mouse embryonic or adult fibroblasts by Takahashi et al. in 2006. In addition to the above-mentioned transcription factors, Nanog was also used. However, it was dispensable [284]. A similar process of generation of the iPSCs from dermal fibroblasts of adult humans and somatic cells were done separately by adding the four above-mentioned transcription factors by Yamanaka et al. and Thomson et al., respectively [284, 285].

Source

The iPSCs can be created practically from any embryonic or adult somatic cell type, without destruction of the human blastocysts. In addition, the iPSCs can also be generated from the somatic cells harvested from patients or normal individuals [284–286].

Generation of iPSCs and Culture

Induction of the iPSCs needs retroviruses with high transduction. For generation of human iPSCs, the retroviruses containing human Klf4, Sox2, Oct3/4, and c-Myc has been introduced into human dermal fibroblasts obtained from human facial skin. These cells were harvested by trypsinization 6 days after transduction, and plated onto mitomycin C-treated SNL feeder cells and DMEM containing 10% FBS. The following day, the medium is replaced by a medium used for primate ES cell culture containing 4 ng/mL basic FGF. After about 3 weeks, the colonies resembling human ESC colonies formed [284].

Application of iPSCs

The study of iPSCs has boosted the field of stem cell biology, offering new cells, to study the molecular mechanisms of human development, cancer biology, pluripotency, and ageing. Thus, the iPSCs or cells derived from them hold enormous potential for cell-based therapies, and for discovering personalized pharmacological treatments for disorders [286].

2.5 3D Stem Cell Culture

Cells in vivo are in a 3D microenvironment with specific cell–cell/molecular interactions, conducting bio-physical and bio-mechanical signals, that regulates cellular physiology and biological functions of adhesion, gene expression, migration, growth, and differentiation. Therefore, 3D cell culture scaffolds and models are simulating normal cell morphology, proliferation, differentiation, and migrations and therefore replacing 2D cell culture systems. Some cellular processes of differentiation and morphogenesis for tissue engineering reveal preferences for 3D instead of 2D. 3D cell culture systems are finding immense importance in drug discovery and tissue engineering as they provide far more accurate physiological processes as in vivo systems.

3D cell culture systems show contrast when compared to 2D monolayer cultures, based on the profiles of the growth of cells, proliferation, population, gene and protein expression. In real-life scenario, the cells are in the microenvironment of other cells and in contact with an ECM as is mimicked in a 3D manner. But the 2D cell culture does not address concerns and cellular interaction with cells and factors in the 3D microenvironment of cells. Therefore, 3D cell culture are more accurate in predicting effects of drugs and agents in cells and finds importance in drug discovery models.

A traditional 2D culture involves growth of cells in monolayer on the tissue culture polystyrene plastic flasks or glass while 3D cell cultures grow the cells into 3D spheroids or aggregates using a matrix or scaffold or in a scaffold-free condition. Matrix/scaffold materials include biological scaffold systems like hyaluronic acid, Cultrex® basement membrane extract (BME; Trevigen), BD Matrigel™ basement membrane matrix (BD Sciences), and synthetic scaffolds including PEG, polylactide-co-glycolide, polyvinyl alcohol, and polycaprolactone materials. Scaffold-free 3D cell spheroids are obtained as suspensions by different methods including hanging drop method, forced floating method or agitation-based approaches wherein the cells grow certainly in a 3D environment, with normal interaction with ECM, and their microenvironment.

In contrast to the 2D monolayer culture, in 3D culture systems, cells form spheroids or aggregates in a matrix or in a suspension culture. 3D spheroids have been classified into grape-like, round, and stellate structures first reported from 3D spheroids formed by breast cancer cell lines. The round-type, spheroid colony displays strong cell to cell adhesion with nuclei at about the centre of the colony.

The mass-type spheroids reveal round colony shapes outlines, disorganized nuclei, aggregation of cells in the centre, and strong cellular communication, reported to overexpress keratin 18 (KRT18) and luminal keratin 8 (KRT8). The grape-like spheroids colony looks like a bunch of grapes with meagre cell to cell interactions. The stellate-type of spheroids colony reveals invasive structure with stellate projections that can bridge multiple colonies and invade matrixes.

Although 3D culture systems mimic better cell–cell and cell–ECM interactions in comparison to 2D monolayer, 3D systems are devoid of complex vascular systems necessary for nutrients, oxygenation, and waste removal. Cells in the 3D culture execute these functions only by diffusion and prove to be problem for larger spheroids (Fig. 2.7).

A 3D cell spheroid reveals proliferation zones of the cell and the distribution of diffused oxygen, CO_2 and nutrition at different depths, from the surface of the aggregate/spheroid in dissimilar nutritional states and at different stages of the cell cycle mimics the real-life in vivo scenario while a 2D monolayer culture system reveals uniform rich oxygenation and nutrition to every cell [272] (Fig. 2.8).

2.6 4D Culture

Stem cells in 3D environment of the body, that changes with time, allow cellular functions and processes like differentiation and migration. Hydrogel-based biomaterials mimic this aspect of in vivo scenario due to their content of water, soft tissue-like elasticity, and tunable biochemical content. Hydrogels are able to direct stem cell function and fate.

Four-dimensional (4D) biomaterials or 3D matrices of hydrogel-based biomaterials found application in study of fates of the ESCs, MSCs, iPSCs, and progenitor cells. 4D material enables techniques of degradation and non-covalent interactions and finds application in far more precise mimic of actual cellular microenvironment with changing time [287, 288] (Fig. 2.9).

2.7 Discussion

Stem cells have the unique potential of maintaining stemness and their differentiation into different lineages. The isolation and culture of stem cells holds great promise in biomedical and clinical research. We have highlighted the different stem cells and their isolations, markers and their cultures. We have also discussed in brief the latest developments in stem cell culture encompassing the 3D and 4D stem cell cultures which are far more suited for drug testing since they mimic the real-life in vivo situation of cells in their microenvironment and interactions with cells and molecules and ECMs and their changes with time. This is a step towards better understanding of drug reactions on stem cells and search of targets in drug research.

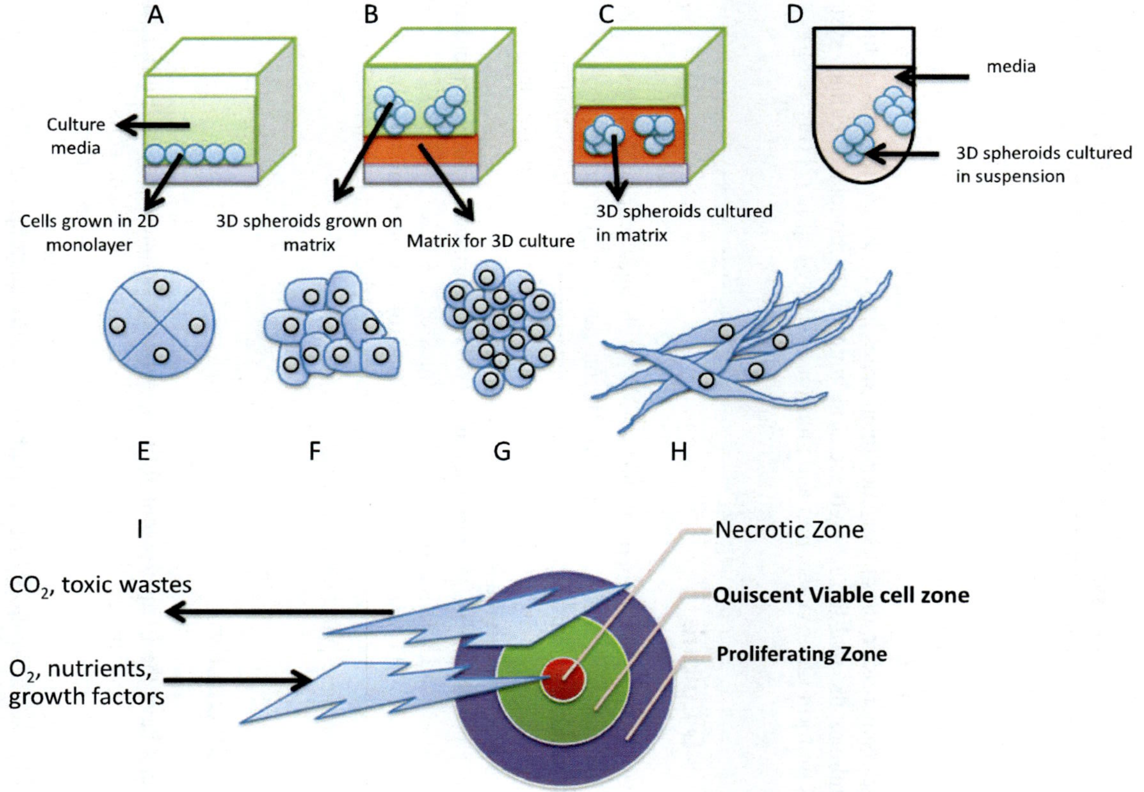

Fig. 2.7 (**a**) Cells grown in 2D monolayer; (**b**) 3D spheroids grown on matrix; (**c**) 3D spheroids cultured in matrix; (**d**) 3D spheroids cultured in suspension; (**e**–**h**) Shapes of cells in 3D: Round, Mass, grapes, Stellate; (**i**) Distribution of CO_2, toxic wastes, O_2, nutrients, growth factors in a 3D cultured cell population

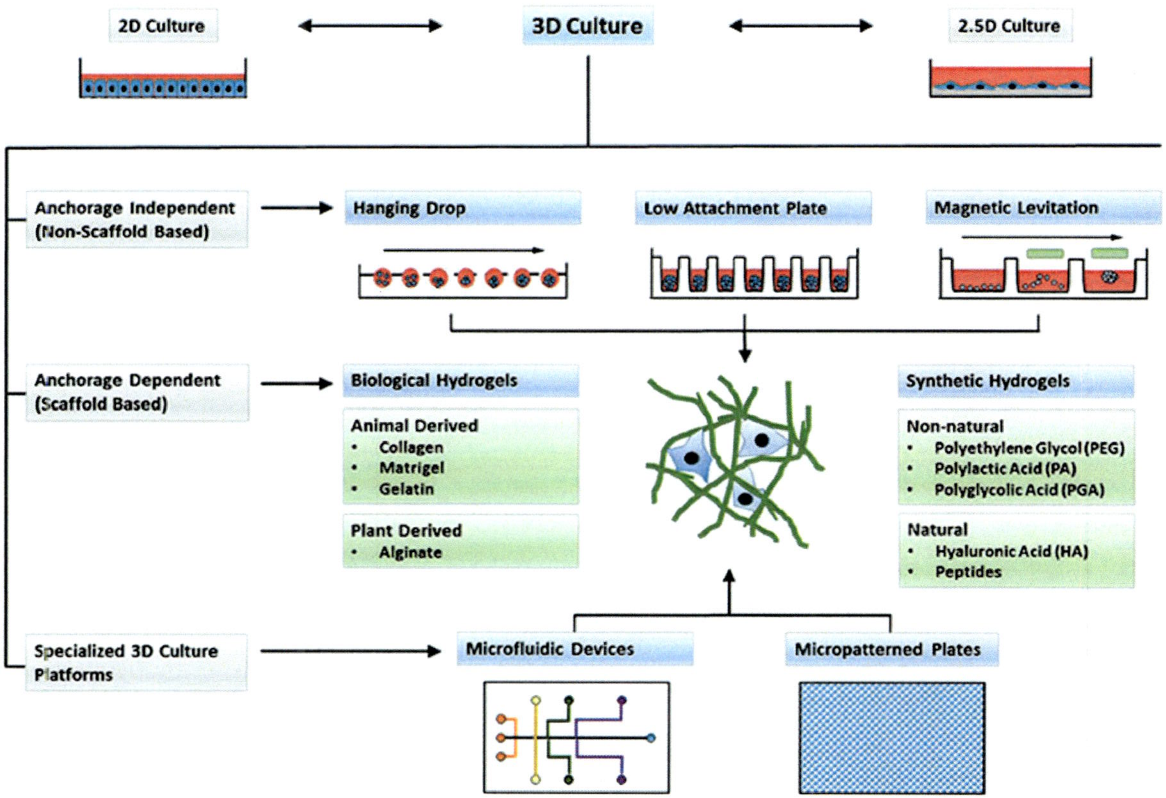

Fig. 2.8 Types of 3D culture. As compared to the 2D monolayers or 2.5D cultures in which cells are plated on top of a thick layer of ECM, 3D cultures form a complex structure. Cells form spheroids (anchorage independent systems) or can be encapsulated in cell culture scaffolds (anchorage dependent). Microfluidic

Fig. 2.8 (continued) devices and micropatterned plates with ECM components, and cultures in which formed spheroids are embedded in ECM scaffolds, can form hybrid culture systems that combine the advantages of both systems to form a complex microenvironment for 3D cell culture. Adapted from Open access publication Langhans SA, 2018, Three-Dimensional in Vitro Cell Culture Models in Drug Discovery and Drug Repositioning. Front. Pharmacol. 9:6. doi: 10.3389/fphar.2018.00006 [273]

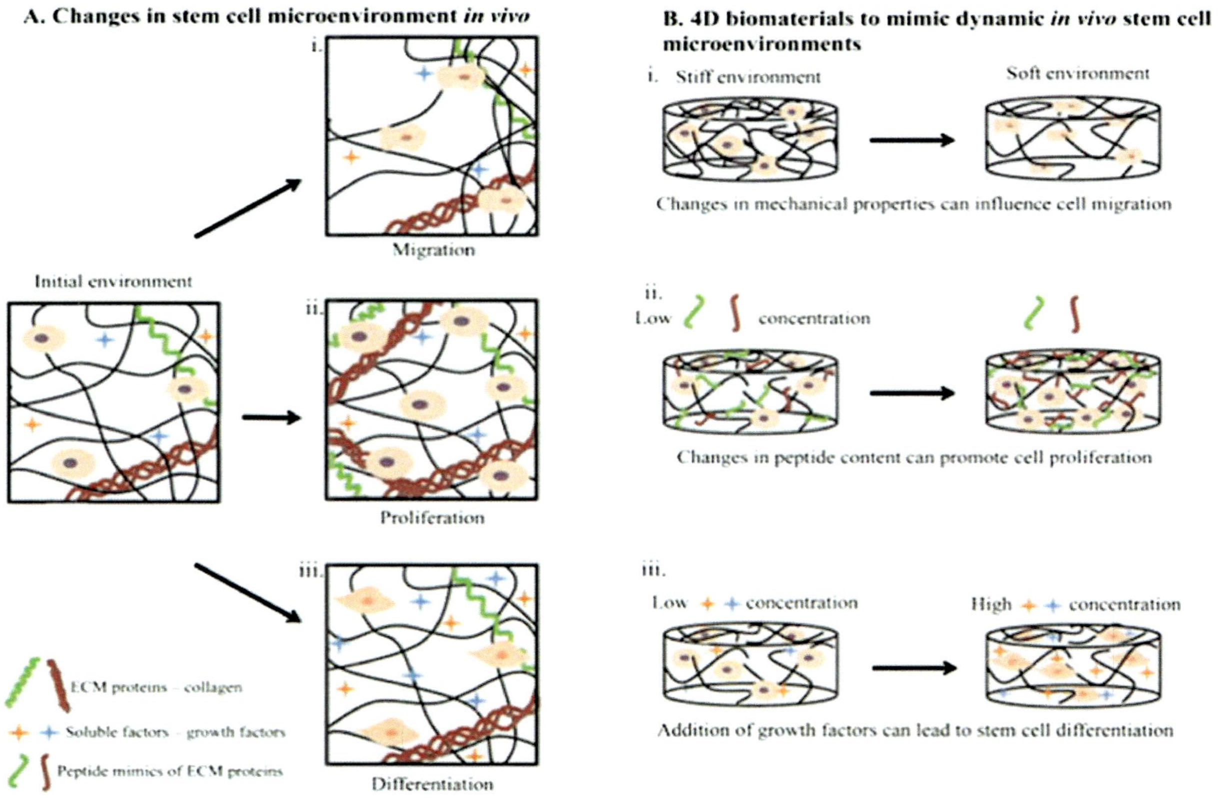

Fig. 2.9 Changes in the stem cell microenvironments in vivo captured by 4D biomaterials in vitro. (**a**) The microenvironment of stem cells in the body changes with time (i) migration in response to matrix density, (ii) proliferation in response to matrix remodelling, and (iii) differentiation in response to soluble growth

Fig. 2.9 (continued) factors. (**b**) Creating materials that capture such changes aids in the study of how cells respond to microenvironment remodelling events, such as wound healing or disease progression, toward ultimately directing these processes. 4D hydrogel-based biomaterials are designed to enable (i) changes in the mechanical properties of synthetic matrices by addition or removal of crosslinks, influencing cell migration throughout the materials. At higher crosslink densities and moduli, cells are entrapped within hydrogel-based matrices, whereas at lower crosslink densities and moduli cell migration seen. (ii) Variation in biochemical content within the hydrogels through addition or removal of biochemical moieties like integrin-binding peptides or protein fragments promote cell proliferation. (iii) Addition or sequestration of growth factors swollen within or tethered to the synthetic matrix can regulate stem cell differentiation. Reproduced with permission from Hilderbrand AM, Ovadia EM, Rehmann MS, Kharkar PM, Guo C, Kloxin AM. Biomaterials for 4D stem cell culture. *CurrOpin Solid State Mater Sci*. 2016;20(4):212–224. doi:10.1016/j.cossms.2016.03.002

References

1. McKee C, Chaudhry GR (2017) Advances and challenges in stem cell culture. Colloids Surf B Biointerfaces 159:62–77. https://doi.org/10.1016/j.colsurfb.2017.07.051
2. Lee J, Cuddihy MJ, Kotov NA (2008) Three-dimensional cell culture matrices: state of the art. Tissue Eng Part B Rev 14(1):61–86. https://doi.org/10.1089/teb.2007.0150
3. Hilderbrand AM, Ovadia EM, Rehmann MS, Kharkar PM, Guo C, Kloxin AM (2016) Biomaterials for 4D stem cell culture. Curr Opin Solid State Mater Sci 20(4):212–224. https://doi.org/10.1016/j.cossms.2016.03.002
4. Vazin T, Freed WJ (2010) Human embryonic stem cells: derivation, culture, and differentiation: a review. Restor Neurol Neurosci 28(4):589–603. https://doi.org/10.3233/RNN-2010-0543
5. Wolf DP, Morey R, Kang E, Ma H, Hayama T, Laurent LC, Mitalipov S (2017) Concise review: embryonic stem cells derived by somatic cell nuclear transfer: a horse in the race? Stem Cells 35(1):26–34. https://doi.org/10.1002/stem.2496
6. Evans MJ, Kaufman MH (1981) Establishment in culture of pluripotential cells from mouse embryos. Nature 292:154–156
7. Martin GR (1981) Isolation of a pluripotent cell line from early mouse embryos cultured in medium conditioned by teratocarcinoma stem cells. Proc Natl Acad Sci U S A 78:7634–7638
8. Thomson JA, Itskovitz-Eldor J, Shapiro SS, Waknitz MA, Swiergiel JJ, Marshall VS et al (1998) Embryonic stem cell lines derived from human blastocysts. Science 282 (5391):1145–1147
9. Gearhart J (1998) New potential for human embryonic stem cells. Science 282 (5391):1061–1062
10. Gilbert SF (2010) Developmental biology, 9th edn. Sinauer Associates, Sunderland, MA
11. Thomson JA, Kalishman J, Golos TG, Durning M, Harris CP, Becker RA, Hearn JP (1995) Isolation of a primate embryonic stem cell line. Proc Natl Acad Sci U S A 92(17):7844–7848
12. Strelchenko N, Verlinsky O, Kukharenko V, Verlinsky Y (2004) Morula-derived human embryonic stem cells. Reprod Biomed Online 9(6):623–629
13. Desai N, Rambhia P, Gishto A (2015) Human embryonic stem cell cultivation: historical perspective and evolution of xeno-free culture systems. Reprod Biol Endocrinol 13:9. https://doi.org/10.1186/s12958-015-0005-4
14. Martí M, Mulero L, Pardo C, Morera C, Carrió M, Laricchia-Robbio L, Esteban CR, Izpisua Belmonte JC (2013) Characterization of pluripotent stem cells. Nat Protoc 8(2):223–253. https://doi.org/10.1038/nprot.2012.154
15. Ellerström C, Strehl R, Moya K, Andersson K, Bergh C, Lundin K et al (2006) Derivation of a xeno-free human embryonic stem cell line. Stem Cells 24(1):2170–2176
16. Strom S, Inzunza J, Grinnemo KH, Holmberg K, Matilainen E, Stromberg AM et al (2007) Mechanical isolation of the inner cell mass is effective in derivation of new human embryonic stem cell lines. Hum Reprod 22(12):3051–3058
17. Turetsky T, Aizenman E, Gil Y, Weinberg N, Shufaro Y, Revel A et al (2008) Laser-assisted derivation of human embryonic stem cell lines from IVF embryos after preimplantation genetic diagnosis. Hum Reprod 23(1):46–53
18. Reubinoff BE, Itsykson P, Turetsky T, Pera MF, Reinhartz E, Itzik A et al (2001) Neural progenitors from human embryonic stem cells. Nat Biotechnol 19(12):1134–1140
19. Van Hoof D, Braam SR, Dormeyers W, WardvanOostwaard D, Heck AJ, Krijgsveld J et al (2008) Feeder-free monolayer cultures of human embryonic stem cells express an epithelial plasma membrane protein profile. Stem Cells 26(11):2777–2781
20. Xu C, Inokuma MS, Denham J, Golds K, Kundu P, Gold JD et al (2001) Feeder-free growth of undifferentiated human embryonic stem cells. Nat Biotechnol 19(10):971–974
21. Wang D, Haviland DL, Burns AR, Zsigmond E, Wetsel RA (2007) A pure population of lung alveolar epithelial type II cells derived from human embryonic stem cells. Proc Natl Acad Sci U S A 104(11):4449–4454

22. Stojkovic P, Lako M, Stewart R, Przyborski S, Armstrong L, Evans J et al (2005) An autogeneic feeder cell system that efficiently supports growth of undifferentiated human embryonic stem cells. Stem Cells 23(3):306–314
23. Lee GY, Kenny PA, Lee EH, Bissell MJ (2007) Three-dimensional culture models of normal and malignant breast epithelial cells. Nat Methods 4(4):359–365
24. Postovit LM, Seftor EA, Seftor RE, Hendrix MJ (2006) A three-dimensional model to study the epigenetic effects induced by the microenvironment of human embryonic stem cells. Stem Cells 24(3):501–505
25. Nur-E-Kamal A, Ahmed I, Kamal J, Schindler M, Meiners S (2006) Three-dimensional nanofibrillar surfaces promote self-renewal in mouse embryonic stem cells. Stem Cells 24 (2):426–433
26. Yim EK, Leong KW (2005) Proliferation and differentiation of human embryonic germ cell derivatives in bioactive polymeric fibrous scaffold. J Biomater Sci Polym Ed 16 (10):1193–1217
27. Gerecht S, Burdick JA, Ferreira LS, Townsend SA, Langer R, Vunjak-Novakovic G (2007) Hyaluronic acid hydrogel for controlled self-renewal and differentiation of human embryonic stem cells. Proc Natl Acad Sci U S A 104(27):11298–11303
28. Agarwal S, Holton KL, Lanza R (2008) Efficient differentiation of functional hepatocytes from human embryonic stem cells. Stem Cells 26(5):1117–1127
29. Assady S, Maor G, Amit M, Itskovitz-Eldor J, Skorecki KL, Tzukerman M (2001) Insulin production by human embryonic stem cells. Diabetes 50(8):1691–1697
30. Samadikuchaksaraei A, Cohen S, Isaac K, Rippon HJ, Polak JM, Bielby RC et al (2006) Derivation of distal airway epithelium from human embryonic stem cells. Tissue Eng 12 (4):867–875
31. Barberi T, Willis LM, Socci ND, Studer L (2005) Derivation of multipotent mesenchymal precursors from human embryonic stem cells. PLoS Med 2(6):e161
32. Sottile V, Thomson A, McWhir J (2003) *In vitro* osteogenic differentiation of human ES cells. Cloning Stem Cells 5(2):149–155
33. Kehat I, Kenyagin-Karsenti D, Snir M, Segev H, Amit M, Gepstein A et al (2001) Human embryonic stem cells can differentiate into myocytes with structural and functional properties of cardiomyocytes. J Clin Invest 108(3):407–414
34. Laflamme MA, Chen KY, Naumova AV, Muskheli V, Fugate JA, Dupras SK et al (2007) Cardiomyocytes derived from human embryonic stem cells in pro-survival factors enhance function of infarcted rat hearts. Nat Biotechnol 25(9):1015–1024
35. Chadwick K, Wang L, Li L, Menendez P, Murdoch B, Rouleau A et al (2003) Cytokines and BMP-4 promote hematopoietic differentiation of human embryonic stem cells. Blood 102 (3):906–915
36. Kaufman DS, Hanson ET, Lewis RL, Auerbach R, Thomson JA (2001) Hematopoietic colony-forming cells derived from human embryonic stem cells. Proc Natl Acad Sci U S A 98(19):10716–10721
37. Vodyanik MA, Bork JA, Thomson JA, Slukvin II (2005) Human embryonic stem cell-derived CD34+ cells: efficient production in the coculture with OP9 stromal cells and analysis of lymphohematopoietic potential. Blood 105(2):617–626
38. Levenberg S, Golub JS, Amit M, Itskovitz-Eldor J, Langer R (2002) Endothelial cells derived from human embryonic stem cells. Proc Natl Acad Sci U S A 99(7):4391–4396
39. Aberdam E, Barak E, Rouleau M, de LaForest S, Berrih-Aknin S, Suter DM et al (2008) A pure population of ectodermal cells derived from human embryonic stem cells. Stem Cells 26 (2):440–444
40. Metallo CM, Ji L, de Pablo JJ, Palecek SP (2007) Retinoic acid and bone morphogenetic protein signalling synergize to efficiently direct epithelial differentiation of human embryonic stem cells. Stem Cells 26(2):372–380

41. Gong J, Sagiv O, Cai H, Tsang SH, Del Priore LV (2008) Effects of extracellular matrix and neighboring cells on induction of human embryonic stem cells into retinal or retinal pigment epithelial progenitors. Exp Eye Res 86(6):957–965
42. Elkabetz Y, Panagiotakos G, Al Shamy G, Socci ND, Tabar V, Studer L (2008) Human ES cell-derived neural rosettes reveal a functionally distinct early neural stem cell stage. Genes Dev 22(2):152–165
43. Nakahata T, Ogawa M (1982) Hemopoietic colony-forming cells in umbilical cord blood with extensive capability to generate mono- and multipotential hemopoietic progenitors. J Clin Invest 70(6):1324–1328
44. Broxmeyer HE, Douglas GW, Hangoc G et al (1989) Human umbilical cord blood as a potential source of transplantable hematopoietic stem/progenitor cells. Proc Natl Acad Sci U S A 86(10):3828–3832
45. Ballen KK, Gluckman E, Broxmeyer HE (2013) Umbilical cord blood transplantation: the first 25 years and beyond. Blood 122(4):491–498. https://doi.org/10.1182/blood-2013-02-453175
46. Roura S, Pujal JM, Gálvez-Montón C, Bayes-Genis A (2015) The role and potential of umbilical cord blood in an era of new therapies: a review. Stem Cell Res Ther 6:123. https://doi.org/10.1186/s13287-015-0113-2
47. Gluckman E, Broxmeyer HA, Auerbach AD, Friedman HS, Douglas GW, Devergie A, Esperou H, Thierry D, Socie G, Lehn P et al (1989) Hematopoietic reconstitution in a patient with Fanconi's anemia by means of umbilical-cord blood from an HLA-identical sibling. N Engl J Med 321(17):1174–1178
48. Munoz J, Shah N, Rezvani K, Hosing C, Bollard CM, Oran B, Olson A, Popat U, Molldrem J, McNiece IK, Shpall EJ (2014) Concise review: umbilical cord blood transplantation: past, present, and future. Stem Cells Transl Med 3(12):1435–1443. https://doi.org/10.5966/sctm.2014-0151
49. Chowdhury S, Bhattacharya K, Mandal C, Mondal S, Sarkar S, Chandra S, Banerjee S, Mandal C (2018) A glycomic approach towards identification of signature molecules in $CD34^+$ haematopoietic stem cells from umbilical cord blood. Adv Exp Med Biol 1112:309–318. https://doi.org/10.1007/978-981-13-3065-0_21
50. Storms RW, Green PD, Safford KM, Niedzwiecki D, Cogle CR, Colvin OM, Chao NJ, Rice HE, Smith CA (2005) Distinct hematopoietic progenitor compartments are delineated by the expression of aldehyde dehydrogenase and CD34. Blood 106(1):95–102
51. Kelly SS, Sola CB, de Lima M, Shpall E (2009) Ex vivo expansion of cord blood. Bone Marrow Transplant 44(10):673–681. https://doi.org/10.1038/bmt.2009.284
52. Purdy MH, Hogan CJ, Hami L, McNiece I, Franklin W, Jones RB et al (1995) Large volume *ex vivo* expansion of CD34-positive hematopoietic progenitor cells for transplantation. J Hematother 4:515–525
53. McNiece I, Jones R, Cagnoni P, Bearman S, Nieto Y, Shpall EJ (1999) Ex-vivo expansion of hematopoietic progenitor cells: preliminary results in breast cancer. Hematol Cell Ther 41:82–86
54. Pecora AL, Stiff P, Jennis A, Goldberg S, Rosenbluth R, Price P et al (2000) Prompt and durableengraftment in two older adult patients with high risk chronic myelogenous leukemia (CML) using *ex vivo* expanded and unmanipulated unrelated umbilical cord blood. Bone Marrow Transplant 25:797–799
55. Glimm H, Eaves CJ (1999) Direct evidence for multiple self-renewal divisions of human *in vivo* repopulating hematopoietic cells in short-term culture. Blood 94:2161–2168
56. Glimm H, Oh IH, Eaves CJ (2000) Human hematopoietic stem cells stimulated to proliferate *in vitro* lose engraftment potential during their S/G(2)/M transit and do not reenter G(0). Blood 96:4185–4193
57. Fuchs E, Tumbar T, Guasch G (2004) Socializing with the neighbors: stem cells and their niche. Cell 116:769–778
58. Lemischka IR, Moore KA (2003) Stem cells: interactive niches. Nature 425:778–779
59. Schofield R (1983) The stem cell system. Biomed Pharmacother 37:375–380

60. Hackney JA, Charbord P, Brunk BP, Stoeckert CJ, Lemischka IR, Moore KA (2002) A molecular profile of a hematopoietic stem cell niche. Proc Natl Acad Sci U S A 99:13061–13066
61. Etheridge SL, Spencer GJ, Heath DJ, Genever PG (2004) Expression profiling and functional analysis of wntsignaling mechanisms in mesenchymal stem cells. Stem Cells 22:849–860
62. Kadereit S, Deeds LS, Haynesworth SE, Koc ON, Kozik MM, Szekely E et al (2002) Expansion of LTCICs and maintenance of p21 and BCL-2 expression in cord blood CD34 (+)/CD38(−) early progenitors cultured over human MSCs as a feeder layer. Stem Cells 20:573–582
63. Rattis FM, Voermans C, Reya T (2004) Wntsignaling in the stem cell niche. Curr Opin Hematol 11:88–94
64. Robinson SN, Ng J, Niu T, Yang H, McMannis JD, Karandish S, Kaur I, Fu P, Del Angel M, Messinger R, Flagge F, de Lima M, Decker W, Xing D, Champlin R, Shpall EJ (2006) Superior ex vivo cord blood expansion following co-culture with bone marrow-derived mesenchymal stem cells. Bone Marrow Transplant 37(4):359–366
65. Zhang J, Niu C, Ye L, Huang H, He X, Tong WG et al (2003) Identification of the haematopoietic stem cell niche and control of the niche size. Nature 425:836–841
66. Jaroscak J, Goltry K, Smith A, Waters-Pick B, Martin PL, Driscoll TA et al (2003) Augmentation of umbilical cord blood (UCB) transplantation with *ex vivo*-expanded UCB cells: results of a phase 1 trial using the AastromReplicell System. Blood 101:5061–5067
67. Tsai S, Emerson SG, Sieff CA, Nathan DG (1986) Isolation of a human stromal cell strain secreting hemopoietic growth factors. J Cell Physiol 127:137–145
68. Meagher RC, Salvado AJ, Wright DG (1988) An analysis of the multilineage production of human hematopoietic progenitors in long-term bone marrow culture: evidence that reactive oxygen intermediates derived from mature phagocytic cells have a role in limiting progenitor cell self renewal. Blood 72:273–281
69. de Lima M, McMannis J, Gee A, Komanduri K, Couriel D, Andersson BS, Hosing C, Khouri I, Jones R, Champlin R, Karandish S, Sadeghi T, Peled T, Grynspan F, Daniely Y, Nagler A, Shpall EJ (2008) Transplantation of ex vivo expanded cord blood cells using the copper chelator tetraethylenepentamine: a phase I/II clinical trial. Bone Marrow Transplant 41 (9):771–778. https://doi.org/10.1038/sj.bmt.1705979
70. Delaney C, Heimfeld S, Brashem-Stein C, Voorhies H, Manger RL, Bernstein ID (2010) Notch-mediated expansion of human cord blood progenitor cells capable of rapid myeloid reconstitution. Nat Med 16(2):232–236. https://doi.org/10.1038/nm.2080
71. Gage FH, Temple S (2013) Neural stem cells: generating and regenerating the brain. Neuron 80(3):588–601. https://doi.org/10.1016/j.neuron.2013.10.037
72. Takagi Y (2016) History of neural stem cell research and its clinical application. Neurol Med Chir (Tokyo) 56(3):110–124. https://doi.org/10.2176/nmc.ra.2015-0340
73. Anderson SA, Eisenstat DD, Shi L, Rubenstein JL (1997) Interneuron migration from basal forebrain to neocortex: dependence on Dlx genes. Science 278(5337):474–476
74. Cattaneo E, McKay R (1990) Proliferation and differentiation of neuronal stem cells regulated by nerve growth factor. Nature 347(6295):762–765
75. Yuan SH, Martin J, Elia J, Flippin J, Paramban RI, Hefferan MP, Vidal JG, Mu Y, Killian RL, Israel MA, Emre N, Marsala S, Marsala M, Gage FH, Goldstein LS, Carson CT (2011) Cell-surface marker signatures for the isolation of neural stem cells, glia and neurons derived from human pluripotent stem cells. PLoS One 6(3):e17540. https://doi.org/10.1371/journal.pone.0017540
76. Chicheportiche A, Ruat M, Boussin FD, Daynac M (2018) Isolation of neural stem and progenitor cells from the adult brain and live imaging of their cell cycle with the FUCCI system. Methods Mol Biol 1686:69–78. https://doi.org/10.1007/978-1-4939-7371-2_5
77. Zhang J, Jiao J (2015) Molecular biomarkers for embryonic and adult neural stem cell and neurogenesis. Biomed Res Int 2015:727542. https://doi.org/10.1155/2015/727542

78. Belenguer G, Domingo-Muelas A, Ferrón SR, Morante-Redolat JM, Fariñas I (2016) Isolation, culture and analysis of adult subependymal neural stem cells. Differentiation 91 (4-5):28–41. https://doi.org/10.1016/j.diff.2016.01.005
79. Murphy AR, Laslett A, O'Brien CM, Cameron NR (2017) Scaffolds for 3D in vitro culture of neural lineage cells. Acta Biomater 54:1–20. https://doi.org/10.1016/j.actbio.2017.02.046
80. Nowell CS, Radtke F (2017) Corneal epithelial stem cells and their niche at a glance. J Cell Sci 130(6):1021–1025. https://doi.org/10.1242/jcs.198119
81. Funderburgh JL, Funderburgh ML, Du Y (2016) Stem cells in the limbal stroma. Ocul Surf 14 (2):113–120. https://doi.org/10.1016/j.jtos.2015.12.006
82. Davanger M, Evensen A (1971) Role of the pericorneal papillary structure in renewal of corneal epithelium. Nature 229(5286):560–561
83. Grieve K, Ghoubay D, Georgeon C, Thouvenin O, Bouheraoua N, Paques M et al (2015) Three-dimensional structure of the mammalian limbal stem cell niche. Exp Eye Res 140:75–84
84. Pajoohesh-Ganji A, Pal-Ghosh S, Simmens SJ, Stepp MA (2006) Integrins in slow-cycling corneal epithelial cells at the limbus in the mouse. Stem Cells 24(4):1075–1086
85. Gonzalez G, Sasamoto Y, Ksander BR, Frank MH, Frank NY (2018) Limbal stem cells: identity, developmental origin, and therapeutic potential. Wiley Interdiscip Rev Dev Biol 7 (2):303. https://doi.org/10.1002/wdev.303
86. Hayashi R, Yamato M, Saito T, Oshima T, Okano T, Tano Y, Nishida K (2008) Enrichment of corneal epithelial stem/progenitor cells using cell surface markers, integrin alpha6 and CD71. Biochem Biophys Res Commun 367(2):256–263
87. Daniels JT, Dart JK, Tuft SJ, Khaw PT (2001) Corneal stem cells in review. Wound Repair Regen 9(6):483–494
88. Daniels JT, Notara M, Shortt AJ, Secker G, Harris A, Tuft SJ (2007) Limbal epithelial stem cell therapy. Expert Opin Biol Ther 7(1):1–3
89. Pellegrini G, Traverso CE, Franzi AT, Zingirian M, Cancedda R, De Luca M (1997) Long-term restoration of damaged corneal surfaces with autologous cultivated corneal epithelium. Lancet 349(9057):990–993
90. Chen D, Qu Y, Hua X, Zhang L, Liu Z, Pflugfelder SC, Li DQ (2017) A hyaluronan hydrogel scaffold-based xeno-free culture system for ex vivo expansion of human corneal epithelial stem cells. Eye (Lond) 31(6):962–971. https://doi.org/10.1038/eye.2017.8
91. Padma Priya S, Higuchi A, Abu Fanas S, Pooi Ling M, Kumari Neela V, Sunil PM, Saraswathi TR, Murugan K, Alarfaj AA, Munusamy MA, Kumar S (2015) Odontogenic epithelial stem cells: hidden sources. Lab Investig 95(12):1344–1352. https://doi.org/10.1038/labinvest.2015.108
92. Pilbauerová N, Suchánek J (2018) Cryopreservation of dental stem cells. Acta Med (Hradec Kralove) 61(1):1–7. https://doi.org/10.14712/18059694.2018.16
93. Ranganathan K, Lakshminarayanan V (2012) Stem cells of the dental pulp. Indian J Dent Res 23(4):558. https://doi.org/10.4103/0970-9290.104977
94. Nuti N, Corallo C, Chan BM, Ferrari M, Gerami-Naini B (2016) Multipotent differentiation of human dental pulp stem cells: a literature review. Stem Cell Rev Rep 12(5):511–523
95. Atari M, Barajas M, Hernández-Alfaro F, Gil C, Fabregat M, FerrésPadró E, Giner L, Casals N (2011) Isolation of pluripotent stem cells from human third molar dental pulp. Histol Histopathol 26(8):1057–1070. https://doi.org/10.14670/HH-26.1057
96. Gronthos S, Mankani M, Brahim J, Robey PG, Shi S (2000) Postnatal human dental pulp stem cells (DPSCs) *in vitro* and in vivo. Proc Natl Acad Sci U S A 97(25):13625–13630
97. Miura M, Gronthos S, Zhao M, Lu B, Fisher LW, Robey PG, Shi S (2003) SHED: stem cells from human exfoliated deciduous teeth. Proc Natl Acad Sci U S A 100(10):5807–5812
98. Seo BM, Miura M, Gronthos S, Bartold PM, Batouli S, Brahim J, Young M, Robey PG, Wang CY, Shi S (2004) Investigation of multipotent postnatal stem cells from human periodontal ligament. Lancet 364(9429):149–155

99. Nagata M, Iwasaki K, Akazawa K, Komaki M, Yokoyama N, Izumi Y, Morita I (2017) Conditioned medium from periodontal ligament stem cells enhances periodontal regeneration. Tissue Eng Part A 23(9-10):367–377. https://doi.org/10.1089/ten.TEA.2016.0274
100. Sonoyama W, Liu Y, Fang D, Yamaza T, Seo BM, Zhang C, Liu H, Gronthos S, Wang CY, Wang S, Shi S (2006) Mesenchymal stem cell-mediated functional tooth regeneration in swine. PLoS One 1:e79
101. Huang GT, Sonoyama W, Liu Y, Liu H, Wang S, Shi S (2008) The hidden treasure in apical papilla: the potential role in pulp/dentin regeneration and bioroot engineering. J Endod 34 (6):645–651. https://doi.org/10.1016/j.joen.2008.03.001
102. Liu J, Yu F, Sun Y, Jiang B, Zhang W, Yang J, Xu GT, Liang A, Liu S (2015) Concise reviews: characteristics and potential applications of human dental tissue-derived mesenchymal stem cells. Stem Cells 33(3):627–638. https://doi.org/10.1002/stem.1909
103. Bonnamain V, Thinard R, Sergent-Tanguy S, Huet P, Bienvenu G, Naveilhan P, Farges JC, Alliot-Licht B (2013) Human dental pulp stem cells cultured in serum-free supplemented medium. Front Physiol 4:357. https://doi.org/10.3389/fphys.2013.00357
104. Kanafi MM, Ramesh A, Gupta PK, Bhonde RR (2014) Dental pulp stem cells immobilized in alginate microspheres for applications in bone tissue engineering. Int Endod J 47(7):687–697. https://doi.org/10.1111/iej.12205
105. Blanpain C, Lowry WE, Geoghegan A, Polak L, Fuchs E (2004) Self-renewal, multipotency, and the existence of two cell populations within an epithelial stem cell niche. Cell 118 (5):635–648
106. Yu H, Fang D, Kumar SM, Li L, Nguyen TK, Acs G, Herlyn M, Xu X (2006) Isolation of a novel population of multipotent adult stem cells from human hair follicles. Am J Pathol 168 (6):1879–1888
107. Cotsarelis G, Sun TT, Lavker RM (1990) Label-retaining cells reside in the bulge area of pilosebaceous unit: implications for follicular stem cells, hair cycle, and skin carcinogenesis. Cell 61(7):1329–1337
108. Nishimura EK, Jordan SA, Oshima H, Yoshida H, Osawa M, Moriyama M, Jackson IJ, Barrandon Y, Miyachi Y, Nishikawa S (2002) Dominant role of the niche in melanocyte stem-cell fate determination. Nature 416(6883):854–860
109. Mii S, Duong J, Tome Y, Uchugonova A, Liu F, Amoh Y, Saito N, Katsuoka K, Hoffman RM (2013) The role of hair follicle nestin-expressing stem cells during whisker sensory-nerve growth in long-term 3D culture. J Cell Biochem 114(7):1674–1684. https://doi.org/10.1002/jcb.24509
110. Nagel S, Rohr F, Weber C, Kier J, Siemers F, Kruse C, Danner S, Brandenburger M, Matthiessen AE (2013) Multipotent nestin-positive stem cells reside in the stroma of human eccrine and apocrine sweat glands and can be propagated robustly *in vitro*. PLoS One 8(10): e78365. https://doi.org/10.1371/journal.pone.0078365
111. Toma JG, Akhavan M, Fernandes KJ, Barnabé-Heider F, Sadikot A, Kaplan DR, Miller FD (2001) Isolation of multipotent adult stem cells from the dermis of mammalian skin. Nat Cell Biol 3(9):778–784
112. Fernandes KJ, McKenzie IA, Mill P, Smith KM, Akhavan M, Barnabé-Heider F, Biernaskie J, Junek A, Kobayashi NR, Toma JG, Kaplan DR, Labosky PA, Rafuse V, Hui CC, Miller FD (2004) A dermal niche for multipotent adult skin-derived precursor cells. Nat Cell Biol 6 (11):1082–1093. Erratum in: Nat Cell Biol. 2005 May;7(5):531
113. Cotsarelis G (2006) Epithelial stem cells: a folliculocentric view. J Invest Dermatol 126 (7):1459–1468
114. Fuchs E, Horsley V (2008) More than one way to skin. Genes Dev 22(8):976–985. https://doi.org/10.1101/gad.1645908
115. Tiede S, Kloepper JE, Bodò E, Tiwari S, Kruse C, Paus R (2007) Hair follicle stem cells: walking the maze. Eur J Cell Biol 86(7):355–376
116. Amoh Y, Hoffman RM (2017) Hair follicle-associated-pluripotent (HAP) stem cells. Cell Cycle 16(22):2169–2175. https://doi.org/10.1080/15384101.2017.1356513

117. Lavker RM, Sun TT, Oshima H, Barrandon Y, Akiyama M, Ferraris C, Chevalier G, Favier B, Jahoda CA, Dhouailly D, Panteleyev AA, Christiano AM (2003) Hair follicle stem cells. J Investig Dermatol Symp Proc 8(1):28–38
118. Kajiura S, Mii S, Aki R, Hamada Y, Arakawa N, Kawahara K, Li L, Katsuoka K, Hoffman RM, Amoh Y (2016) Protocols for cryopreservation of intact hair follicle that maintain pluripotency of nestin-expressing hair-follicle-associated pluripotent (HAP) stem cells. Methods Mol Biol 1453:173–178. https://doi.org/10.1007/978-1-4939-3786-8_18
119. Xu X, Lyle S, Liu Y, Solky B, Cotsarelis G (2003) Differential expression of cyclin D1 in the human hair follicle. Am J Pathol 163(3):969–978
120. Roh C, Tao Q, Photopoulos C, Lyle S (2005) In vitro differences between keratinocyte stem cells and transit-amplifying cells of the human hair follicle. J Invest Dermatol 125 (6):1099–1105
121. Zhang SC, Wernig M, Duncan ID, Brüstle O, Thomson JA (2001) In vitro differentiation of transplantable neural precursors from human embryonic stem cells. Nat Biotechnol 19 (12):1129–1133
122. Thomson JA, Itskovitz-Eldor J, Shapiro SS, Waknitz MA, Swiergiel JJ, Marshall VS, Jones JM (1998) Embryonic stem cell lines derived from human blastocysts. Science 282 (5391):1145–1147. Erratum in: Science 1998 Dec 4;282(5395):1827
123. Xu C, Inokuma MS, Denham J, Golds K, Kundu P, Gold JD, Carpenter MK (2001) Feeder-free growth of undifferentiated human embryonic stem cells. Nat Biotechnol 19(10):971–974
124. Mii S, Duong J, Tome Y, Uchugonova A, Liu F, Amoh Y, Saito N, Katsuoka K, Hoffman RM (2016) Nestin-expressing hair-follicle-associated pluripotent (HAP) stem cells promote whisker sensory-nerve growth in long-term 3D-Gelfoam® histoculture. Methods Mol Biol 1453:39–47. https://doi.org/10.1007/978-1-4939-3786-8_6
125. Tadeu AM, Horsley V (2014) Epithelial stem cells in adult skin. Curr Top Dev Biol 107:109–131. https://doi.org/10.1016/B978-0-12-416022-4.00004-4
126. Danner S, Kremer M, Petschnik AE, Nagel S, Zhang Z, Hopfner U, Reckhenrich AK, Weber C, Schenck TL, Becker T, Kruse C, Machens HG, Egaña JT (2012) The use of human sweat gland-derived stem cells for enhancing vascularization during dermal regeneration. J Invest Dermatol 132(6):1707–1716. https://doi.org/10.1038/jid.2012.31
127. Petschnik AE, Klatte JE, Evers LH, Kruse C, Paus R, Danner S (2010) Phenotypic indications that human sweat glands are a rich source of nestin-positive stem cell populations. Br J Dermatol 162(2):380–383. https://doi.org/10.1111/j.1365-2133.2009.09512.x
128. Hunt DP, Jahoda C, Chandran S (2009) Multipotent skin-derived precursors: from biology to clinical translation. Curr Opin Biotechnol 20(5):522–530. https://doi.org/10.1016/j.copbio.2009.10.004
129. Wang X, Dong S, Wu Y (2019) Isolation and cultivation of skin-derived precursors. Methods Mol Biol 1879:149–152. https://doi.org/10.1007/7651_2018_141
130. Rodrigues RM, De Kock J, Branson S, Vinken M, Meganathan K, Chaudhari U, Sachinidis A, Govaere O, Roskams T, De Boe V, Vanhaecke T, Rogiers V (2014) Human skin-derived stem cells as a novel cell source for in vitro hepatotoxicity screening of pharmaceuticals. Stem Cells Dev 23(1):44–55. https://doi.org/10.1089/scd.2013.0157
131. Biernaskie JA, McKenzie IA, Toma JG, Miller FD (2006) Isolation of skin-derived precursors (SKPs) and differentiation and enrichment of their Schwann cell progeny. Nat Protoc 1 (6):2803–2812
132. Metral E, Bechetoille N, Demarne F, Damour O, Rachidi W (2018) Keratinocyte stem cells are more resistant to UVA radiation than their direct progeny. PLoS One 13(9):e0203863. https://doi.org/10.1371/journal.pone.0203863
133. Li A, Simmons PJ, Kaur P (1998) Identification and isolation of candidate human keratinocyte stem cells based on cell surface phenotype. Proc Natl Acad Sci U S A 95(7):3902–3907
134. Slominski A, Paus R (1993) Melanogenesis is coupled to murine anagen: toward new concepts for the role of melanocytes and the regulation of melanogenesis in hair growth. J Invest Dermatol 101(1 Suppl):90S–97S

135. Tanimura S, Tadokoro Y, Inomata K, Binh NT, Nishie W, Yamazaki S, Nakauchi H, Tanaka Y, McMillan JR, Sawamura D, Yancey K, Shimizu H, Nishimura EK (2011) Hair follicle stem cells provide a functional niche for melanocyte stem cells. Cell Stem Cell 8 (2):177–187. https://doi.org/10.1016/j.stem.2010.11.029
136. Gentile P, Scioli MG, Bielli A, Orlandi A, Cervelli V (2017) Stem cells from human hair follicles: first mechanical isolation for immediate autologous clinical use in androgenetic alopecia and hair loss. Stem Cell Investig 4:58. https://doi.org/10.21037/sci.2017.06.04
137. Wright MC, Reed-Geaghan EG, Bolock AM, Fujiyama T, Hoshino M, Maricich SM (2015) Unipotent, Atoh1+ progenitors maintain the Merkel cell population in embryonic and adult mice. J Cell Biol 208(3):367–379. https://doi.org/10.1083/jcb.201407101
138. Blackburn CC, Manley NR, Palmer DB, Boyd RL, Anderson G, Ritter MA (2002) One for all and all for one: thymic epithelial stem cells and regeneration. Trends Immunol 23(8):391–395
139. Lampert IA, Ritter MA (1998) The origin of the diverse epithelial cells of the thymus: is there a common stem cell? Thymus Update 1:2–25
140. McLelland BT, Gravano D, Castillo J, Montoy S, Manilay JO (2011) Enhanced isolation of adult thymic epithelial cell subsets for multiparameter flow cytometry and gene expression analysis. J Immunol Methods 367(1-2):85–94. https://doi.org/10.1016/j.jim.2011.02.008
141. Gray DH, Chidgey AP, Boyd RL (2002) Analysis of thymic stromal cell populations using flow cytometry. J Immunol Methods 260(1-2):15–28
142. Papagerakis S, Pannone G, Zheng L, About I, Taqi N, Nguyen NP, Matossian M, McAlpin B, Santoro A, McHugh J, Prince ME, Papagerakis P (2014) Oral epithelial stem cells - implications in normal development and cancer metastasis. Exp Cell Res 325(2):111–129. https://doi.org/10.1016/j.yexcr.2014.04.021
143. Jones KB, Klein OD (2013) Oral epithelial stem cells in tissue maintenance and disease: the first steps in a long journey. Int J Oral Sci 5(3):121–129. https://doi.org/10.1038/ijos.2013.46
144. Utheim TP (2015) Concise review: transplantation of cultured oral mucosal epithelial cells for treating limbal stem cell deficiency-current status and future perspectives. Stem Cells 33 (6):1685–1695. https://doi.org/10.1002/stem.1999
145. Jones KB, Furukawa S, Marangoni P, Ma H, Pinkard H, D'Urso R, Zilionis R, Klein AM, Klein OD (2019) Quantitative clonal analysis and single-cell transcriptomics reveal division kinetics, hierarchy, and fate of oral epithelial progenitor cells. Cell Stem Cell 24(1):183–192. e8. https://doi.org/10.1016/j.stem.2018.10.015
146. Igarashi T, Shimmura S, Yoshida S, Tonogi M, Shinozaki N, Yamane GY (2008) Isolation of oral epithelial progenitors using collagen IV. Oral Dis 14(5):413–418. https://doi.org/10.1111/j.1601-0825.2007.01390.x
147. Calenic B, Ishkitiev N, Yaegaki K, Imai T, Costache M, Tovaru M, Tovaru S, Parlatescu I (2010) Characterization of oral keratinocyte stem cells and prospects of its differentiation to oral epithelial equivalents. Rom J Morphol Embryol 51(4):641–645
148. Luo X, Okubo T, Randell S, Hogan BL (2009) Culture of endodermal stem/progenitor cells of the mouse tongue. In Vitro Cell Dev Biol Anim 45(1-2):44–54. https://doi.org/10.1007/s11626-008-9149-2
149. Nakamura T, Endo K, Kinoshita S (2007) Identification of human oral keratinocyte stem/progenitor cells by neurotrophin receptor p75 and the role of neurotrophin/p75 signaling. Stem Cells 25(3):628–638
150. Hisha H, Tanaka T, Ueno H (2016) Lingual epithelial stem cells and organoid culture of them. Int J Mol Sci 17(2):E168. https://doi.org/10.3390/ijms17020168
151. Seery JP, Watt FM (2000) Asymmetric stem-cell divisions define the architecture of human oesophageal epithelium. Curr Biol 10(22):1447–1450
152. Zhang Y, Jiang M, Kim E, Lin S, Liu K, Lan X, Que J (2017) Development and stem cells of the esophagus. Semin Cell Dev Biol 66:25–35. https://doi.org/10.1016/j.semcdb.2016.12.008
153. Zhao R, Quaroni L, Casson AG (2012) Identification and characterization of stemlike cells in human esophageal adenocarcinoma and normal epithelial cell lines. J Thorac Cardiovasc Surg 144(5):1192–1199

154. Maghsoudlou P, Ditchfield D, Klepacka DH, Shangaris P, Urbani L, Loukogeorgakis SP, Eaton S, De Coppi P (2014) Isolation of esophageal stem cells with potential for therapy. Pediatr Surg Int 30(12):1249–1256. https://doi.org/10.1007/s00383-014-3615-6
155. Emmerson E, Knox SM (2018) Salivary gland stem cells: a review of development, regeneration and cancer. Genesis 56(5):e23211. https://doi.org/10.1002/dvg.23211
156. Maimets M, Rocchi C, Bron R, Pringle S, Kuipers J, Giepmans BN, Vries RG, Clevers H, de Haan G, van Os R, Coppes RP (2016) Long-term in vitro expansion of salivary gland stem cells driven by Wnt signals. Stem Cell Rep 6(1):150–162. https://doi.org/10.1016/j.stemcr.2015.11.009
157. Denny PC, Chai Y, Klauser DK, Denny PA (1993) Parenchymal cell proliferation and mechanisms for maintenance of granular duct and acinar cell populations in adult male mouse submandibular gland. Anat Rec 235(3):475–485
158. Hisatomi Y, Okumura K, Nakamura K, Matsumoto S, Satoh A, Nagano K, Yamamoto T, Endo F (2004) Flow cytometric isolation of endodermal progenitors from mouse salivary gland differentiate into hepatic and pancreatic lineages. Hepatology 39(3):667–675
159. Lombaert IM, Brunsting JF, Wierenga PK, Faber H, Stokman MA, Kok T, Visser WH, Kampinga HH, de Haan G, Coppes RP (2008) Rescue of salivary gland function after stem cell transplantation in irradiated glands. PLoS One 3(4):e2063. https://doi.org/10.1371/journal.pone.0002063
160. Nanduri LS, Maimets M, Pringle SA, van der Zwaag M, van Os RP, Coppes RP (2011) Regeneration of irradiated salivary glands with stem cell marker expressing cells. Radiother Oncol 99(3):367–372. https://doi.org/10.1016/j.radonc.2011.05.085
161. Pringle S, Nanduri LS, van der Zwaag M, van Os R, Coppes RP (2011) Isolation of mouse salivary gland stem cells. J Vis Exp 48:2484. https://doi.org/10.3791/2484
162. Shin HS, Hong HJ, Koh WG, Lim JY (2018) Organotypic 3D culture in nanoscaffold microwells supports salivary gland stem-cell-based organization. ACS Biomater Sci Eng 4 (12):4311–4320. https://doi.org/10.1021/acsbiomaterials.8b00894
163. Barker N, Bartfeld S, Clevers H (2010) Tissue-resident adult stem cell populations of rapidly self-renewing organs. Cell Stem Cell 7(6):656–670. https://doi.org/10.1016/j.stem.2010.11.016
164. Barker N (2014) Adult intestinal stem cells: critical drivers of epithelial homeostasis and regeneration. Nat Rev Mol Cell Biol 15(1):19–33. https://doi.org/10.1038/nrm3721
165. Heath JP (1996) Epithelial cell migration in the intestine. Cell Biol Int 20(2):139–146
166. Sakthianandeswaren A, Christie M, D'Andreti C, Tsui C, Jorissen RN, Li S, Fleming NI, Gibbs P, Lipton L, Malaterre J, Ramsay RG, Phesse TJ, Ernst M, Jeffery RE, Poulsom R, Leedham SJ, Segditsas S, Tomlinson IP, Bernhard OK, Simpson RJ, Walker F, Faux MC, Church N, Catimel B, Flanagan DJ, Vincan E, Sieber OM (2011) PHLDA1 expression marks the putative epithelial stem cells and contributes to intestinal tumorigenesis. Cancer Res 71 (10):3709–3719. https://doi.org/10.1158/0008-5472.CAN-10-2342
167. Barker N, van Es JH, Kuipers J, Kujala P, van den Born M, Cozijnsen M, Haegebarth A, Korving J, Begthel H, Peters PJ, Clevers H (2007) Identification of stem cells in small intestine and colon by marker gene Lgr5. Nature 449(7165):1003–1007
168. Wang F, Scoville D, He XC, Mahe MM, Box A, Perry JM, Smith NR, Lei NY, Davies PS, Fuller MK, Haug JS, McClain M, Gracz AD, Ding S, Stelzner M, Dunn JC, Magness ST, Wong MH, Martin MG, Helmrath M, Li L (2013) Isolation and characterization of intestinal stem cells based on surface marker combinations and colony-formation assay. Gastroenterology 145(2):383–395.e1-21. https://doi.org/10.1053/j.gastro.2013.04.050
169. Gjorevski N, Sachs N, Manfrin A, Giger S, Bragina ME, Ordóñez-Morán P, Clevers H, Lutolf MP (2016) Designer matrices for intestinalstem cell and organoid culture. Nature 539 (7630):560–564. https://doi.org/10.1038/nature20168
170. Sugimoto S, Sato T (2017) Establishment of 3D intestinal organoid cultures from intestinal stem cells. Methods Mol Biol 1612:97–105. https://doi.org/10.1007/978-1-4939-7021-6_7

171. Kohli L, Passegué E (2014) Surviving change: the metabolic journey of hematopoietic stem cells. Trends Cell Biol 24(8):479–487. https://doi.org/10.1016/j.tcb.2014.04.001
172. Eaves CJ (2015) Hematopoietic stem cells: concepts, definitions, and the new reality. Blood 125(17):2605–2613. https://doi.org/10.1182/blood-2014-12-570200
173. Wang LD, Wagers AJ (2011) Dynamic niches in the origination and differentiation of haematopoietic stem cells. Nat Rev Mol Cell Biol 12(10):643–655. https://doi.org/10.1038/nrm3184. Erratum in: Nat Rev Mol Cell Biol. 2011;13(1):12
174. MacLean AL, Lo Celso C, Stumpf MP (2017) Concise review: stem cell population biology: insights from hematopoiesis. Stem Cells 35(1):80–88. https://doi.org/10.1002/stem.2508
175. Szade K, Gulati GS, Chan CKF, Kao KS, Miyanishi M, Marjon KD, Sinha R, George BM, Chen JY, Weissman IL (2018) Where hematopoietic stem cells live: the bone marrow niche. Antioxid Redox Signal 29(2):191–204. https://doi.org/10.1089/ars.2017.7419
176. Cooper B (2011) The origins of bone marrow as the seedbed of our blood: from antiquity to the time of Osler. Proc (BaylUniv Med Cent) 24(2):115–118
177. Neumann E (1882) DasGesetz der Verbreitung des Gelben und rotten Knochenmaarkes. Zentralbl Med Wissensch 20:321–323
178. Neumann E (1912) Hamatologische student II. Die Variabilitat der Leukamie. Virch Arch 207:379–412
179. Seita J, Weissman IL (2010) Hematopoietic stem cell: self-renewal versus differentiation. Wiley Interdiscip Rev Syst Biol Med 2(6):640–653. https://doi.org/10.1002/wsbm.86
180. Kumar S, Geiger H (2017) HSC niche biology and HSC expansion ex vivo. Trends Mol Med 23(9):799–819. https://doi.org/10.1016/j.molmed.2017.07.003
181. Boitano AE, Wang J, Romeo R, Bouchez LC, Parker AE, Sutton SE, Walker JR, Flaveny CA, Perdew GH, Denison MS, Schultz PG, Cooke MP (2010) Aryl hydrocarbon receptor antagonists promote the expansion of human hematopoietic stem cells. Science 329 (5997):1345–1348. https://doi.org/10.1126/science.1191536
182. Vo LT, Daley GQ (2015) De novo generation of HSCs from somatic and pluripotent stem cell sources. Blood 125(17):2641–2648. https://doi.org/10.1182/blood-2014-10-570234
183. Flores-Guzmán P, Fernández-Sánchez V, Mayani H (2013) Concise review: ex vivo expansion of cord blood-derived hematopoietic stem and progenitor cells: basic principles, experimental approaches, and impact in regenerative medicine. Stem Cells Transl Med 2 (11):830–838. https://doi.org/10.5966/sctm.2013-0071
184. Pineault N, Abu-Khader A (2015) Advances in umbilical cord blood stem cell expansion and clinical translation. Exp Hematol 43(7):498–513. https://doi.org/10.1016/j.exphem.2015.04.011
185. Ivanovs A, Rybtsov S, Ng ES, Stanley EG, Elefanty AG, Medvinsky A (2017) Human haematopoietic stem cell development: from the embryo to the dish. Development 144 (13):2323–2337. https://doi.org/10.1242/dev.134866
186. Daniel MG, Pereira CF, Lemischka IR, Moore KA (2016) Making a hematopoietic stem cell. Trends Cell Biol 26(3):202–214. https://doi.org/10.1016/j.tcb.2015.10.002
187. Civin CI, Strauss LC, Brovall C, Fackler MJ, Schwartz JF, Shaper JH (1984) Antigenic analysis of hematopoiesis. III. A hematopoietic progenitor cell surface antigen defined by a monoclonal antibody raised against KG-1a cells. J Immunol 133(1):157–165
188. Tindle RW, Nichols RA, Chan L, Campana D, Catovsky D, Birnie GD (1985) A novel monoclonal antibody BI-3C5 recognises myeloblasts and non-B non-T lymphoblasts in acute leukaemias and CGL blast crises, and reacts with immature cells in normal bone marrow. Leuk Res 9(1):1–9
189. Holden JT, Geller RB, Farhi DC, Holland HK, Stempora LL, Phillips CN, Bray RA (1995) Characterization of Thy-1 (CDw90) expression in CD34+ acute leukemia. Blood 86(1):60–65
190. Craig W, Kay R, Cutler RL, Lansdorp PM (1993) Expression of Thy-1 on human hematopoietic progenitor cells. J Exp Med 177(5):1331–1342
191. Ashman LK, Cambareri AC, To LB, Levinsky RJ, Juttner CA (1991) Expression of the YB5.B8 antigen (c-kit proto-oncogene product) in normal human bone marrow. Blood 78(1):30–37

192. Yin AH, Miraglia S, Zanjani ED, Almeida-Porada G, Ogawa M, Leary AG, Olweus J, Kearney J, Buck DW (1997) AC133, a novel marker for human hematopoietic stem and progenitor cells. Blood 90(12):5002–5012
193. Bonanno G, Perillo A, Rutella S, De Ritis DG, Mariotti A, Marone M, Meoni F, Scambia G, Leone G, Mancuso S, Pierelli L (2004) Clinical isolation and functional characterization of cord blood CD133+ hematopoietic progenitor cells. Transfusion 44(7):1087–1097
194. Kastan MB, Schlaffer E, Russo JE, Colvin OM, Civin CI, Hilton J (1990) Direct demonstration of elevated aldehyde dehydrogenase in human hematopoietic progenitor cells. Blood 75 (10):1947–1950
195. Jones RJ, Barber JP, Vala MS, Collector MI, Kaufmann SH, Ludeman SM, Colvin OM, Hilton J (1995) Assessment of aldehyde dehydrogenase in viable cells. Blood 85(10):2742–2746
196. Chowdhury S, Chandra S, Mandal C (2014) 9-O-acetylated sialic acids differentiating normal haematopoietic precursors from leukemic stem cells with high aldehyde dehydrogenase activity in children with acute lymphoblastic leukaemia. Glycoconj J 31(6-7):523–535. https://doi.org/10.1007/s10719-014-9550-x
197. Cicuttini FM, Welch KL, Boyd AW (1994) The effect of cytokines on CD34+Rh-123high and low progenitor cells from human umbilical cord blood. Exp Hematol 22:1244–1251
198. Cardoso AA, Li ML, Batard P, Hatzfeld A, Brown EL, Levesque JP, Sookdeo H, Panterne B, Sansilvestri P, Clark SC et al (1993) Release from quiescence of CD34+ CD38- human umbilical cord blood cells reveals their potentiality to engraft adults. Proc Natl Acad Sci U S A 90(18):8707–8711
199. de Wynter EA, Nadali G, Coutinho LH, Testa NG (1996) Extensive amplification of single cells from CD34+ subpopulations in umbilical cord blood and identification of long-term culture-initiating cells present in two subsets. Stem Cells 14(5):566–576
200. Mayani H, Lansdorp PM (1994) Thy-1 expression is linked to functional properties of primitive hematopoietic progenitor cells from human umbilical cord blood. Blood 83 (9):2410–2417
201. Szilvassy SJ (2003) The biology of hematopoietic stem cells. Arch Med Res 34:446–460
202. Bhatia M, Wang JC, Kapp U, Bonnet D, Dick JE (1997) Purification of primitive human hematopoietic cells capable of repopulating immune-deficient mice. Proc Natl Acad Sci U S A 94(10):5320–5325
203. Kollet O, Peled A, Byk T, Ben-Hur H, Greiner D, Shultz L, Lapidot T (2000) beta2 microglobulin-deficient (B2m(null)) NOD/SCID mice are excellent recipients for studying human stem cell function. Blood 95(10):3102–3105
204. Douay L (2001) Experimental culture conditions are critical for the ex-vivo expansion of hematopoietic cells. J Hematother Stem Cell Res 10:341–346
205. Bai T, Li J, Sinclair A, Imren S, Merriam F, Sun F, O'Kelly MB, Nourigat C, Jain P, Delrow JJ, Basom RS, Hung HC, Zhang P, Li B, Heimfeld S, Jiang S, Delaney C (2019) Expansion of primitive human hematopoietic stem cells by culture in a zwitterionic hydrogel. Nat Med 25 (10):1566–1575. https://doi.org/10.1038/s41591-019-0601-5
206. Manohar R, Komori J, Guzik L, Stolz DB, Chandran UR, LaFramboise WA, Lagasse E (2011) Identification and expansion of a unique stem cell population from adult mouse gallbladder. Hepatology 54(5):1830–1841. https://doi.org/10.1002/hep.24568
207. Lugli N, Kamileri I, Keogh A, Malinka T, Sarris ME, Talianidis I, Schaad O, Candinas D, Stroka D, Halazonetis TD (2016) R-spondin 1 and noggin facilitate expansion of resident stem cells from non-damaged gallbladders. EMBO Rep 17(5):769–779. https://doi.org/10.15252/embr.201642169
208. Manohar R, Li Y, Fohrer H, Guzik L, Stolz DB, Chandran UR, LaFramboise WA, Lagasse E (2015) Identification of a candidate stem cell in human gallbladder. Stem Cell Res 14 (3):258–269. https://doi.org/10.1016/j.scr.2014.12.003
209. Zhang L, Theise N, Chua M, Reid LM (2008) The stem cell niche of human livers: symmetry between development and regeneration. Hepatology 48(5):1598–1607. https://doi.org/10.1002/hep.22516

210. Theise ND, Saxena R, Portmann BC, Thung SN, Yee H, Chiriboga L, Kumar A, Crawford JM (1999) The canals of Hering and hepatic stem cells in humans. Hepatology 30(6):1425–1433
211. Kuwahara R, Kofman AV, Landis CS, Swenson ES, Barendswaard E, Theise ND (2008) The hepatic stem cell niche: identification by label-retaining cell assay. Hepatology 47 (6):1994–2002. https://doi.org/10.1002/hep.22218
212. Overturf K, Al-Dhalimy M, Ou CN, Finegold M, Grompe M (1997) Serial transplantation reveals the stem-cell-like regenerative potential of adult mouse hepatocytes. Am J Pathol 151 (5):1273–1280
213. Yamazaki T, Enosawa S, Kasahara M, Fukuda A, Sakamoto S, Shigeta T, Nakazawa A, Tokiwa T (2012) Isolation of hepatic progenitor cells from human liver with cirrhosis secondary to biliary atresia using EpCAM or Thy-1 markers. Cell Med 3(1-3):121–126. https://doi.org/10.3727/215517912X639441
214. Castell JV, Gómez-Lechón MJ (2009) Liver cell culture techniques. Methods Mol Biol 481:35–46. https://doi.org/10.1007/978-1-59745-201-4_4
215. Fukuda T, Takayama K, Hirata M, Liu YJ, Yanagihara K, Suga M, Mizuguchi H, Furue MK (2017) Isolation and expansion of human pluripotent stem cell-derived hepatic progenitor cells by growth factor defined serum-free culture conditions. Exp Cell Res 352(2):333–345. https://doi.org/10.1016/j.yexcr.2017.02.022
216. Berika M, Elgayyar ME, El-Hashash AH (2014) Asymmetric cell division of stem cells in the lung and other systems. Front Cell Dev Biol 2:33. https://doi.org/10.3389/fcell.2014.00033
217. Li F, He J, Wei J, Cho WC, Liu X (2015) Diversity of epithelial stem cell types in adult lung. Stem Cells Int 2015:728307. https://doi.org/10.1155/2015/728307
218. Bishop AE (2004) Pulmonary epithelial stem cells. Cell Prolif 37(1):89–96
219. Gilpin SE, Charest JM, Ren X, Tapias LF, Wu T, Evangelista-Leite D, Mathisen DJ, Ott HC (2016) Regenerative potential of human airway stem cells in lung epithelial engineering. Biomaterials 108:111–119. https://doi.org/10.1016/j.biomaterials.2016.08.055
220. Karp PH, Moninger TO, Weber SP, Nesselhauf TS, Launspach JL, Zabner J, Welsh MJ (2002) An in vitro model of differentiated human airway epithelia. Methods for establishing primary cultures. Methods Mol Biol 188:115–137
221. Barkauskas CE, Cronce MJ, Rackley CR, Bowie EJ, Keene DR, Stripp BR, Randell SH, Noble PW, Hogan BL (2013) Type 2 alveolar cells are stem cells in adult lung. J Clin Invest 123(7):3025–3036. https://doi.org/10.1172/JCI68782
222. Lloyd-Lewis B, Harris OB, Watson CJ, Davis FM (2017) Mammary stem cells: premise, properties, and perspectives. Trends Cell Biol 27(8):556–567. https://doi.org/10.1016/j.tcb.2017.04.001
223. Tsai YC, Lu Y, Nichols PW, Zlotnikov G, Jones PA, Smith HS (1996) Contiguous patches of normal human mammary epithelium derived from a single stem cell: implications for breast carcinogenesis. Cancer Res 56(2):402–404
224. Shi AP, Fan ZM, Ma KW, Jiang YF, Wang L, Zhang KW, Fu SB, Xu N, Zhang ZR (2017) Isolation and characterization of adult mammary stem cells from breast cancer-adjacent tissues. Oncol Lett 14(3):2894–2902. https://doi.org/10.3892/ol.2017.6485
225. Wang D, Cai C, Dong X, Yu QC, Zhang XO, Yang L, Zeng YA (2015) Identification of multipotent mammary stem cells by protein C receptor expression. Nature 517(7532):81–84. https://doi.org/10.1038/nature13851
226. Stingl J, Eaves CJ, Kuusk U, Emerman JT (1998) Phenotypic and functional characterization in vitro of a multipotent epithelial cell present in the normal adult human breast. Differentiation 63(4):201–213
227. Péchoux C, Gudjonsson T, Ronnov-Jessen L, Bissell MJ, Petersen OW (1999) Human mammary luminal epithelial cells contain progenitors to myoepithelial cells. Dev Biol 206 (1):88–99
228. Emerman JT, Wilkinson DA (1990) Routine culturing of normal, dysplastic and malignant human mammary epithelial cells from small tissue samples. In Vitro Cell Dev Biol 26 (12):1186–1194

229. Santamaria X, Mas A, Cervelló I, Taylor H, Simon C (2018) Uterine stem cells: from basic research to advanced cell therapies. Hum Reprod Update 24(6):673–693. https://doi.org/10.1093/humupd/dmy028
230. Taylor HS (2004) Endometrial cells derived from donor stem cells in bone marrow transplant recipients. JAMA 292(1):81–85
231. Teixeira J, Rueda BR, Pru JK (2008) Uterine stem cells. In: StemBook [internet]. Harvard Stem Cell Institute, Cambridge, MA
232. Tanaka M, Kyo S, Kanaya T, Yatabe N, Nakamura M, Maida Y, Okabe M, Inoue M (2003) Evidence of the monoclonal composition of human endometrial epithelial glands and mosaic pattern of clonal distribution in luminal epithelium. Am J Pathol 163(1):295–301
233. Leong KG, Wang BE, Johnson L, Gao WQ (2008) Generation of a prostate from a single adult stem cell. Nature 456(7223):804–808. https://doi.org/10.1038/nature07427
234. Hu WY, Hu DP, Xie L, Li Y, Majumdar S, Nonn L, Hu H, Shioda T, Prins GS (2017) Isolation and functional interrogation of adult human prostate epithelial stem cells at single cell resolution. Stem Cell Res 23:1–12. https://doi.org/10.1016/j.scr.2017.06.009
235. Komeya M, Ogawa T (2015) Spermatogonial stem cells: Progress and prospects. Asian J Androl 17(5):771–775. https://doi.org/10.4103/1008-682X.154995
236. de Rooij DG (2017) The nature and dynamics of spermatogonial stem cells. Development 144 (17):3022–3030. https://doi.org/10.1242/dev.146571
237. Aloisio GM, Nakada Y, Saatcioglu HD, Peña CG, Baker MD, Tarnawa ED, Mukherjee J, Manjunath H, Bugde A, Sengupta AL, Amatruda JF, Cuevas I, Hamra FK, Castrillon DH (2014) PAX7 expression defines germline stem cells in the adult testis. J Clin Invest 124 (9):3929–3944. https://doi.org/10.1172/JCI75943
238. Komai Y, Tanaka T, Tokuyama Y, Yanai H, Ohe S, Omachi T, Atsumi N, Yoshida N, Kumano K, Hisha H, Matsuda T, Ueno H (2014) Bmi1 expression in long-term germ stem cells. Sci Rep 4:6175. https://doi.org/10.1038/srep06175
239. Goodyear S, Brinster R (2017) Isolation of the spermatogonial stem cell-containing fraction from testes. Cold Spring Harb Protoc 2017(4):pdb.prot094185. https://doi.org/10.1101/pdb.prot094185
240. Goodyear S, Brinster R (2017) Culture and expansion of primary undifferentiated spermatogonial stem cells. Cold Spring Harb Protoc 2017(4):pdb.prot094193. https://doi.org/10.1101/pdb.prot094193
241. Mushahary D, Spittler A, Kasper C, Weber V, Charwat V (2018) Isolation, cultivation, and characterization of human mesenchymal stem cells. Cytometry A 93(1):19–31. https://doi.org/10.1002/cyto.a.23242
242. Ullah I, Subbarao RB, Rho GJ (2015) Human mesenchymal stem cells-current trends and future prospective. Biosci Rep 35:e00191
243. Friedenstein AJ, Gorskaja JF, Kulagina NN (1976) Fibroblast precursors in normal and irradiated mouse hematopoietic organs. Exp Hematol 4(5):267–274
244. Li H, Ghazanfari R, Zacharaki D, Lim HC, Scheding S (2016) Isolation and characterization of primary bone marrow mesenchymal stromal cells. Ann N Y Acad Sci 1370:109–118
245. Pittenger MF, Mackay AM, Beck SC, Jaiswal RK, Douglas R, Mosca JD, Moorman MA, Simonetti DW, Craig S, Marshak DR (1999) Multilineage potential of adult human mesenchymal stem cells. Science 284:143–147
246. Ab Kadir R, Zainal Ariffin SH, Megat Abdul Wahab R, Kermani S, Senafi S (2012) Characterization of mononucleated human peripheral blood cells. Sci World J 2012:843843. https://doi.org/10.1100/2012/843843
247. Wagner W, Wein F, Seckinger A, Frankhauser M, Wirkner U, Krause U, Blake J, Schwager C, Eckstein V, Ansorge W (2005) Comparative characteristics of mesenchymal stem cells from human bone marrow, adipose tissue, and umbilical cord blood. Exp Hematol 33:1402–1416
248. Zhang X, Yang M, Lin L, Chen P, Ma K, Zhou C, Ao Y (2006) Runx2 overexpression enhances osteoblastic differentiation and mineralization in adipose-derived stem cells in vitro and in vivo. Calcif Tissue Int 79:169–178

249. Huang GT, Gronthos S, Shi S (2009) Mesenchymal stem cells derived from dental tissues vs. those from other sources: their biology and role in regenerative medicine. J Dent Res 88:792–806
250. Riekstina U, Muceniece R, Cakstina I, Muiznieks I, Ancans J (2008) Characterization of human skin-derived mesenchymal stem cell proliferation rate in different growth conditions. Cytotechnology 58(3):153–162. https://doi.org/10.1007/s10616-009-9183-2
251. Bartsch G, Yoo JJ, De Coppi P, Siddiqui MM, Schuch G, Pohl HG, Fuhr J, Perin L, Soker S, Atala A (2005) Propagation, expansion, and multilineage differentiation of human somatic stem cells from dermal progenitors. Stem Cells Dev 14(3):337–348
252. Rotter N, Oder J, Schlenke P, Lindner U, Böhrnsen F, Kramer J, Rohwedel J, Huss R, Brandau S, Wollenberg B, Lang S (2008) Isolation and characterization of adult stem cells from human salivary glands. Stem Cells Dev 17(3):509–518. https://doi.org/10.1089/scd.2007.0180
253. Jiao F, Wang J, Dong ZL, Wu MJ, Zhao TB, Li DD, Wang X (2012) Human mesenchymal stem cells derived from limb bud can differentiate into all three embryonic germ layers lineages. Cell Reprogram 14(4):324–333. https://doi.org/10.1089/cell.2012.0004
254. Morito T, Muneta T, Hara K, Ju YJ, Mochizuki T, Makino H, Umezawa A, Sekiya I (2008) Synovial fluid-derived mesenchymal stem cells increase after intra-articular ligament injury in humans. Rheumatology (Oxford) 47(8):1137–1143. https://doi.org/10.1093/rheumatology/ken114
255. Allickson JG, Sanchez A, Yefimenko N, Borlongan CV, Sanberg PR (2011) Recent studies assessing the proliferative capability of a novel adult stem cell identified in menstrual blood. Open Stem Cell J 3(2011):4–10
256. Tsai MS, Lee JL, Chang YJ, Hwang SM (2004 Jun) Isolation of human multipotent mesenchymal stem cells from second-trimester amniotic fluid using a novel two-stage culture protocol. Hum Reprod 19(6):1450–1456
257. Cai J, Li W, Su H, Qin D, Yang J, Zhu F, Xu J, He W, Guo X, Labuda K, Peterbauer A, Wolbank S, Zhong M, Li Z, Wu W, So KF, Redl H, Zeng L, Esteban MA, Pei D (2010) Generation of human induced pluripotent stem cells from umbilical cord matrix and amniotic membrane mesenchymal cells. J Biol Chem 285(15):11227–11234. https://doi.org/10.1074/jbc.M109.086389
258. Raynaud C, Maleki M, Lis R, Ahmed B, Al-Azwani I, Malek J, Safadi F, Rafii A (2012) Comprehensive characterization of mesenchymal stem cells from human placenta and fetal membrane and their response to osteoactivin stimulation. Stem Cells Int 2012:658356
259. Hou T, Xu J, Wu X, Xie Z, Luo F, Zhang Z, Zeng L (2009) Umbilical cord Wharton's jelly: a new potential cell source of mesenchymal stromal cells for bone tissue engineering. Tissue Eng Part A 15:2325–2334
260. Kita K, Gauglitz GG, Phan TT, Herndon DN, Jeschke MG (2010) Isolation and characterizationof mesenchymal stem cells from the sub-amniotic human umbilical cord lining membrane. Stem Cells Dev 19:491–502
261. Wang HS, Hung SC, Peng ST, Huang CC, Wei HM, Guo YJ, Fu YS, Lai MC, Chen CC (2004) Mesenchymal stem cells in the Wharton's jelly of the human umbilical cord. Stem Cells 22:1330–1337
262. Bieback K, Kern S, Kluter H, Eichler H (2004) Critical parameters for the isolation of mesenchymal stem cells from umbilical cord blood. Stem Cells 22:625–634
263. La Rocca G, Anzalone R, Corrao S, Magno F, Loria T, Iacono ML, Di Stefano A, Giannuzzi P, Marasà L, Cappello F (2009) Isolation and characterization of Oct-41/HLAG1 mesenchymal stem cells from human umbilical cord matrix: differentiation potential and detection of new markers. Histochem Cell Biol 131:267–282
264. Miao Z, Jin J, Chen L, Zhu J, Huang W, Zhao J, Qian H, Zhang X (2006) Isolation of mesenchymal stem cells from human placenta: comparison with human bone marrow mesenchymal stem cells. Cell Biol Int 30:681–687

265. Schüring AN, Schulte N, Kelsch R, Röpke A, Kiesel L, Götte M (2011) Characterization of endometrial mesenchymal stem-like cells obtained by endometrial biopsy during routine diagnostics. Fertil Steril 95(1):423–426. https://doi.org/10.1016/j.fertnstert.2010.08.035
266. Kita K, Gauglitz GG, Phan TT, Herndon DN, Jeschke MG (2010) Isolation and characterization of mesenchymal stem cells from the sub-amniotic human umbilical cord lining membrane. Stem Cells Dev 19(4):491–502. https://doi.org/10.1089/scd.2009.0192
267. Pendleton C, Li Q, Chesler DA, Yuan K, Guerrero-Cazares H, Quinones-Hinojosa A (2013) Mesenchymal stem cells derived from adipose tissue vs bone marrow: in vitro comparison of their tropism towards gliomas. PLoS One 8:e58198
268. Otsuru S, Hofmann TJ, Olson TS, Dominici M, Horwitz EM (2013) Improved isolation and expansion of bone marrow mesenchymal stromal cells using a novel marrow filter device. Cytotherapy 15(2):146–153. https://doi.org/10.1016/j.jcyt.2012.10.012
269. Wu XB, Tao R (2012) Hepatocyte differentiation of mesenchymal stem cells. Hepatobiliary Pancreat Dis Int 11(4):360–371
270. Ranera B, Remacha AR, Álvarez-Arguedas S, Castiella T, Vázquez FJ, Romero A, Zaragoza P, Martín-Burriel I, Rodellar C (2013) Expansion under hypoxic conditions enhances the chondrogenic potential of equine bone marrow-derived mesenchymal stem cells. Vet J 195(2):248–251. https://doi.org/10.1016/j.tvjl.2012.06.008
271. Zhang Y, Mignone J, MacLellan WR (2015) Cardiac regeneration and stem cells. Physiol Rev 95(4):1189–1204. https://doi.org/10.1152/physrev.00021.2014
272. Beltrami AP, Barlucchi L, Torella D, Baker M, Limana F, Chimenti S, Kasahara H, Rota M, Musso E, Urbanek K, Leri A, Kajstura J, Nadal-Ginard B, Anversa P (2003) Adult cardiac stem cells are multipotent and support myocardial regeneration. Cell 114(6):763–776
273. Bearzi C, Rota M, Hosoda T, Tillmanns J, Nascimbene A, De Angelis A, Yasuzawa-Amano S, Trofimova I, Siggins RW, Lecapitaine N, Cascapera S, Beltrami AP, D'Alessandro DA, Zias E, Quaini F, Urbanek K, Michler RE, Bolli R, Kajstura J, Leri A, Anversa P (2007) Human cardiac stem cells. Proc Natl Acad Sci U S A 104(35):14068–14073
274. Oh H, Bradfute SB, Gallardo TD, Nakamura T, Gaussin V, Mishina Y, Pocius J, Michael LH, Behringer RR, Garry DJ, Entman ML, Schneider MD (2003) Cardiac progenitor cells from adult myocardium: homing, differentiation, and fusion after infarction. Proc Natl Acad Sci U S A 100(21):12313–12318
275. Pfister O, Oikonomopoulos A, Sereti KI, Sohn RL, Cullen D, Fine GC, Mouquet F, Westerman K, Liao R (2008) Role of the ATP-binding cassette transporter Abcg2 in the phenotype and function of cardiac side population cells. Circ Res 103(8):825–835. https://doi.org/10.1161/CIRCRESAHA.108.174615
276. Feige P, Brun CE, Ritso M, Rudnicki MA (2018) Orienting muscle stem cells for regeneration in homeostasis, aging, and disease. Cell Stem Cell 23(5):653–664. https://doi.org/10.1016/j.stem.2018.10.006
277. Partridge TA, Morgan JE, Coulton GR, Hoffman EP, Kunkel LM (1989) Conversion of mdx myofibres from dystrophin-negative to -positive by injection of normal myoblasts. Nature 337 (6203):176–179
278. Morgan JE, Partridge TA (2003) Muscle satellite cells. Int J Biochem Cell Biol 35 (8):1151–1156
279. Fukada S, Yamaguchi M, Kokubo H, Ogawa R, Uezumi A, Yoneda T, Matev MM, Motohashi N, Ito T, Zolkiewska A, Johnson RL, Saga Y, Miyagoe-Suzuki Y, Tsujikawa K, Takeda S, Yamamoto H (2011) Hesr1 and Hesr3 are essential to generate undifferentiated quiescent satellite cells and to maintain satellite cell numbers. Development 138 (21):4609–4619. https://doi.org/10.1242/dev.067165
280. Gnocchi VF, White RB, Ono Y, Ellis JA, Zammit PS (2009) Further characterisation of the molecular signature of quiescent and activated mouse muscle satellite cells. PLoS One 4(4): e5205. https://doi.org/10.1371/journal.pone.0005205
281. Wang YX, Dumont NA, Rudnicki MA (2014) Muscle stem cells at a glance. J Cell Sci 127 (Pt 21):4543–4548. https://doi.org/10.1242/jcs.151209

282. Hüttner SS, Ahrens HE, Schmidt M, Henze H, Jung MJ, Schüler SC, von Maltzahn J (2019) Isolation and culture of individual myofibers and their adjacent muscle stem cells from aged and adult skeletal muscle. Methods Mol Biol 2045:25–36. https://doi.org/10.1007/7651_2019_209
283. Gentile P, Piccinno MS, Calabrese C (2019) Characteristics and potentiality of human adipose-derived stem cells (hASCs) obtained from enzymatic digestion of fat graft. Cells 8 (3):E282. https://doi.org/10.3390/cells8030282
284. Takahashi K, Tanabe K, Ohnuki M, Narita M, Ichisaka T, Tomoda K, Yamanaka S (2007) Induction of pluripotent stem cells from adult human fibroblasts by defined factors. Cell 131 (5):861–872
285. Yu J, Vodyanik MA, Smuga-Otto K, Antosiewicz-Bourget J, Frane JL, Tian S, Nie J, Jonsdottir GA, Ruotti V, Stewart R, Slukvin II, Thomson JA (2007) Induced pluripotent stem cell lines derived from human somatic cells. Science 318(5858):1917–1920
286. Bragança J, Lopes JA, Mendes-Silva L, Almeida Santos JM (2019) Induced pluripotent stem cells, a giant leap for mankind therapeutic applications. World J Stem Cells 11(7):421–430. https://doi.org/10.4252/wjsc.v11.i7.421
287. Edmondson R, Broglie JJ, Adcock AF, Yang L (2014) Three-dimensional cell culture systems and their applications in drug discovery and cell-based biosensors. Assay Drug Dev Technol 12(4):207–218. https://doi.org/10.1089/adt.2014.573
288. Langhans SA (2018) Three-dimensional in vitro cell culture models in drug discovery and drug repositioning. Front Pharmacol 9:6. https://doi.org/10.3389/fphar.2018.00006

Cryopreservation of Stem Cells

3

Contents

Abstract

Stem cells are imperative in regenerative medicine. However, their availability cannot be ensured all the time. Here comes the role of cryopreservation where the stem cells are preserved for indefinite period of time at low temperature. Freezing is lethal to the cells. Additionally, the properties of the stem cells are also affected due to freezing. Therefore, a number of techniques of cryopreservation have solved this issue while some are yet to be explored. The chapter deals with the multiple aspects of cryopreservation of the stem cells.

Keywords

Cryopreservation · Cryoprotectants · Stem cells · Intracellular freezing · Vitrification · Dimethyl sulfoxide

S. Chowdhury, S. Ghosh, *Stem Cells*, https://doi.org/10.1007/978-981-16-1638-9_3

Abbreviations

AP	Alkaline phosphatase
ASCs	Adult stem cells
BM	Bone marrow
CAS	Cells Alive System
CESCs	Corneal epithelial stem cells
CFC	Colony-forming cells
CPA	Cryoprotective agents
CPLL	Carboxylated e-poly-L-lysine
CQAs	Critical quality attributes
CSSCs	Corneal stromal stem cells
DFPCs	Dental follicle progenitor cells
DMSO	Dimethyl sulfoxide
DPSCs	Dental pulp stem cells
EG	Ethylene glycol
ESC	Embryonic stem cells
FACS	Fluorescence-activated cell sorting
FBS	Foetal bovine serum
HAP	Hair follicle-associated pluripotent stem cells
HES	Hydroxyethyl starch
ICC	Immunocytochemical
iPSCs	Induced pluripotent stem cells
MSCs	Mesenchymal stem cells
PB	Peripheral blood
PDLSCs	Periodontal ligament stem cells
PEG	Polyethylene glycol
RT-PCR	Reverse transcription polymerase chain reaction
SCAP	Stem cells from apical papilla
SCK	StemCell Keep™
SHED	Stem cells from human exfoliated deciduous teeth
SKPs	Skin-derived precursors
SNP	Single nucleotide polymorphism
STR	Single tandem repeat
UCB	Umbilical cord blood
WJ	Wharton's jelly

3.1 Introduction

Stem cell transplantation finds importance in treating many malignant and non-malignant diseases. Thus, cryopreservation of marrow cells as a source of the stem cells is of utmost significance. This method finds importance in cord blood transplantation [1]. Similar process of cryopreservation exists for other stem cell types. Cryopreservation is the process of preserving the architecturally intact living cells and the tissues which is re-used later for therapeutic or research purpose [2]. Different cryopreservatives find application with numerous freezing and thawing techniques. The efficiency of freezing is critical for repopulating ability of long-term stored marrow [1].

3.2 Cryopreservation and Cryoprotectants

Cryopreservation is the process of preserving architecturally intact viable cells and tissues, using sub-zero temperatures and cryoprotectants. It is applied for the enduring storage of these biological materials containing live cells, with the aim of using it in future. However, freezing is a lethal process. When any living cell is cooled to <0 °C, the biological properties are dictated by freezing of the water, which archetypally consists of about 80% of the frozen mass. Freezing is the formation of crystalline ice from water and results in the increased concentration of dissolved solutes in the residual liquid phase and precipitation of any solutes which surpass their solubility limit [2–4].

Cryoprotectants or cryoprotective agents (CPA) are a group of compounds, usually fluid, extensively used in cryopreservation, for minimizing freezing injury and maximizing cryoprotection. These compounds are biologically acceptable, less toxic compounds, extremely soluble in water even at lower temperatures and produce profound depression of the freezing temperature. Thus, they have the ability to pierce the cells and reduce formation of ice at any specified temperature by increasing the overall concentration of all solutes in the system and include compounds like dimethyl sulfoxide (DMSO), glycerol, propanediol, and ethanediol [2, 5].

3.3 History of Cryopreservation

The first method of freezing animal cells was discovered in 1948 with consequent restoration of the structure and function. Polge, Smith, and Parkes for the first time used 10–20% of glycerol during freezing in 1949. This addition permitted the spermatozoa of the cock to endure continued freezing at −80 °C. The existing concepts of injury due to freezing, that was prevelant in those early days considered ice crystals destroying the cells by direct mechanical action, teasing apart and intracellular structures [6].

Glycerol reduces the quantity of formed ice by increasing the concentration of the solute. However, freezing an aqueous solution increases the concentration of the solutes in diminishing volume of the remnant solution, which is a central origin of injury. In the 1950s, James Lovelock proved that the concentration of salt, rather than the ice, is the main cause of freezing injury to the cells, and that glycerol helps to protect against this damage by modulating the increase in salt concentration at the time of freezing [7]. The rate of change of temperature checks the transport of water through the cell membrane, and meanderingly the probability of intracellular freezing was discovered by Mazur in 1963 [8].

3.4 Factors Affecting Cryopreservation

3.4.1 Temperature

As introduced earlier, the rate of cooling, i.e., degree of conversion of water to ice determines the rate at which the concentration of the solution surrounding the cells changes. Thus, when the temperature changes slowly, it controls the osmolality of the surrounding fluid, indirectly affecting the degree at which water is transported out of the cells at the time of cooling. If the thermodynamic equilibrium can be maintained throughout the cell membrane by rapid external movement of water, the cytoplasm will not cool beyond its freezing point and there would be no intracellular ice formation. The contrary is also true [2].

3.4.2 Solution Effect

Lovelock experimentally demonstrated that haemolysis of erythrocytes occurred when they were cooled to and thawed from a particular sub-zero temperature in saline suspension. Comparable haemolysis was detected in a cell suspension exposed to sodium chloride concentration, produced by freezing to that particular temperature and then returning it to isotonic saline. He further established that in the presence of glycerol, such haemolysis commenced at a lesser temperature at which the critical concentration of salt was produced. These observations ascertained that freezing injury is due to indirect effects of ice formation and extracellular ice is harmless to cells [2]. The salt present in the suspending medium as well as the cryoprotectant is concentrated to the same level. Pegg and Diaper demonstrated that erythrocytes suffer more damage in a salt concentration containing glycerol. Thus, cryoprotectants are equally damaging [9].

3.4.3 Intracellular Freezing

Unswerving experimental evidences suggest that slight intracellular ice is suitable for recovery and warming rate has a reflective consequence on the cell. The behaviour of tiny intracellular ice crystals varies between slow and rapid warming. While slow warming permits the recrystallization of the crystals, during rapid warming the ice merely melts. Since the cooling rate impacts the intracellular ice formation and the warming rate controls the fate of that ice afterward, followed by differential behaviour of cells with respect to water permeability and predisposition to intracellular ice, it shadows that cells will vary in their cooling and warming requirements and cooling rate will interact with the warming rate [10].

3.4.4 The Packing Volume of Cells

The cryopreservation of cells which are packed in a very dense volume imposes problem in the recovery. It has been observed that there is a reduction in the recovery rate with amplified cell packing. The most probable clarification for this may be that densely packed cells are damaged due to mechanical stress when the channels within which they are impounded change shape. This is an outcome of recrystallization of the ice that defines their boundaries [11].

3.5 Cryoprotection

Cryoprotection is the process of treating the tissues or cells with high concentration of cryoprotectants. This handling generates a driving force for the movement of water and solutes through osmosis and diffusion, respectively. The process of freezing induces such deviations as well. The cell membrane is a selectively permeable membrane and there are many factors affecting this diffusion, excess of which can damage the cells. The methods of osmosis and diffusion are of great significance in cryopreservation. Auspiciously, there is sophisticated quantifiable description of mass transfer processes [2, 12].

3.5.1 Introducing the Cryoprotectant

Osmotic dehydration occurs when cells are introduced to increased concentration of the cryoprotectant. The cell volume increases due to entrance of cryoprotective compound and water till it reaches the final volume. The final equilibrium volume should be identical to the normal volume and depends on the concentration of the impermeant solutes in the solution. This is because the cryoprotectant occupies space inside the cells and therefore the volume of water must be less than the physiological water content, if the entire volume is to be normal. The degree of modification of the volume, especially the equilibrium volume, are vital and must be optimized in cryopreservation procedures [13].

3.5.2 Removing the Cryoprotectant

The cells tend to swell beyond their original volume due to osmotic uptake of water, when a pervading cryoprotectant is removed from them, by exposing them to lesser concentration of the compound. As the cryoprotectant moves out, shrinkage of cells occur. This is balanced by water intake in the cells to maintain the osmotic balance. Since cells are over sensitive to swelling than to the shrinkage, therefore, removal of the cryoprotectants is more perilous than their addition. Care should be taken when designing procedures for the recovery of the cryopreserved cells so that both proportion of change of volume and the final volume is taken into consideration [2].

3.5.3 Freezing and Thawing of Cells

The shrinkage of cells depends on the rate of ice formation. The solution surrounding the cells often concentrates due to freezing, causing shrinkage of these cells. This phenomenon is an enormously imperative determinant of intracellular freezing. The final extent of shrinkage depends on the concentration of the cryoprotectant [2].

3.5.4 Exposure of Cells to Nonpermeating Solutes

When cells are dipped in a solution containing nonpermeating solute, they attain an equilibrium volume which is the inverse function of osmolality of that solution i.e., more the osmolality of solution, less will be the equilibrium value of the cell [13].

3.6 Cryopreservation: The Process

The process of cryopreservation varies with different cell types in different mammalian species with respect to their cryobiological response and the cryosurvival during the freezing cycle [14, 15]. The process of cryopreservation is grouped into the following four types, namely, slow freezing, vitrification, sub-zero non-freezing storage and preservation in the dry state [5].

In slow freezing protocol or the conventional cryopreservation, cells are cooled to below freezing point. As a result, the formation of ice mass containing pure crystalline water is observed while the unfrozen fraction contains all cells and solutes. The concentrations of salts, sugars, and cryoprotectant (e.g. glycerol) increase, while the volume of the unfrozen fraction declines. The upsurge of osmotic strength causes efflux of water from the cells. Slow cooling is required to allow adequate outflow of water to diminish the chance of the formation of intracellular ice. As the cooling continues, the viscosity of the unfrozen fraction becomes too high to crystallize. The residual unfrozen fraction gets converted to an amorphous solid which contains no ice crystal [16, 17] (Fig. 3.1).

Vitrification or the rapid freezing method of cryopreservation is a process by which cell suspensions are transformed directly from the aqueous phase to a glass state, after direct exposure to liquid nitrogen [18]. Rapid freezing leads to the formation of amorphous ice; the transition of water to the glassy state can be done directly [3, 4]. Two types of vitrification processes are used in cryopreservation. The first one is equilibrium vitrification where formulation of cryoprotectant mixtures and their introduction to the cell suspensions is done in a step-by-step fashion, to help reduce chemical toxicity before the cells are flash frozen with liquid nitrogen. The danger of imposing osmotic damage to the cells remains [19]. Nonequilibrium vitrification is the second method which uses tremendously high freezing rates in combination with lower concentrations of cryoprotectant mixture for the preserving of cells. In this approach, vitrification systems are separated into carrier-based systems and carrier-free systems [17, 19].

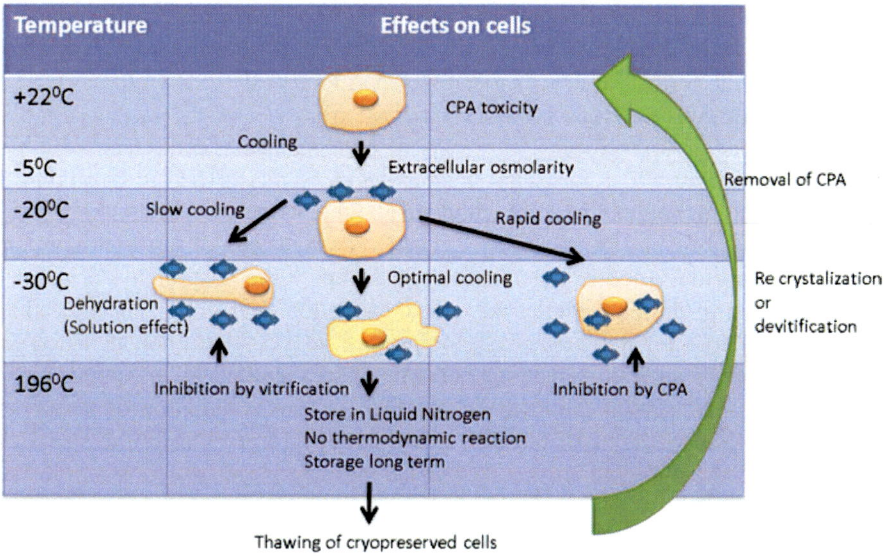

Fig. 3.1 The effect of agents causing cryoinjury of cells while freezing and thawing. Cryoinjuries are caused, partly, by the solution effect (leading to osmotic shock) and intracellular ice formation (leading to breakdown of intracellular structures). *CPA* cryoprotective agent

In sub-zero non-freezing storage, originally reported by Matsuda et al, the cells are stored at temperatures between 0 °C and the freezing point of the cells. The cells stored at −4°Care kept in a computer-programmed incubator which averts ice formation in the −4 °C range. Thus, cells preserved in this way showed better functional status upon rewarming [20].

In nature, some organisms can survive in in a dehydrated state. Trying to preserve cells in this condition requires a complex phenotypic response to environmental changes involving intracellular accumulation of cytoprotectants. Generally, the storage of mammalian cells in dry state is not effectively possible because of the problems in introducing the disaccharide trehalose (disaccharide of glucose, 342 Da) and amino acids (used as preservatives in plants) into the intracellular region [5, 20].

The major steps in cryopreservation are [1]: the mixing of CPAs with cells or tissues prior to cooling; [2] cooling of the cells or tissues to a lower temperature followed by its storage; [3] warming of cells or tissues; and [4] removal of CPAs from cells or tissues after thawing. The appropriate use of CPAs is therefore vital in improving the viability of the sample to be cryopreserved [5, 21, 22].

3.7 Different Types of Cryoprotectants

As discussed earlier, cryoprotectants are indispensable for cryopreservation. Different types of cryoprotectants have been developed to reduce ice formation within the cryopreserved cells. The cryoprotectants are divided into two groups. The first group includes cell membrane-permeating cryoprotectants, with low molecular weight like DMSO, glycerol, and 1,2-propanediol. They can enter the cell cytoplasmic membrane, stop the formation of ice crystal nuclei, and also slow down the ice crystal growth inside cells [5, 23, 24].

The second group includes non-membrane permeating cryoprotectants with high molecular weight like dextran, hydroxyethyl starch (HES), 2-methyl-2,4-pentanediol, and polymer like polyvinylpyrrolidone, numerous sugars like glucose, sucrose, galactose, and trehalose [15, 25] and a separate class of anti-freeze proteins [26]. They reside in the extracellular space, allowing the dehydration of cell and minimization of the intracellular ice crystal formation, and stabilizing the cellular membrane [24]. Certain biomaterials, for instance, polyvinyl alcohol, alginates, and chitosan can be used to hinder the growth of ice crystals [27]. Application of antioxidants and similar compounds also help in direct inhibition of ice crystal formation [28, 29]. Some of the CPAs used in cryopreservation of the stem cells are discussed below.

3.7.1 Glycerol

Glycerol was the first discovered cryopreservative [6]. It is a non-electrolyte that reduces the concentration of electrolytes in the residual unfrozen solution in and around the cell at any particular temperature [5]. Glycerol has been extensively used in combination with other CPAs for cryopreservation of the human ESCs and induced pluripotent stem cells (iPSCs) [30].

3.7.2 DMSO

DMSO was synthesized by the Russian scientist Alexander Zaytsevin 1866 and has since been used for the cryopreservation of cultured mammalian cells primarily because it is less expensive and has low cytotoxicity [31, 32]. Similar to glycerol, DMSO functions by reducing the electrolyte concentration in residual unfrozen solution in and around a cell at a temperature. Nevertheless, DMSO is toxic and using it as a CPA reduces survival rate and initiation of the cell differentiation [33]. This is primarily because DMSO is toxic at temperatures more than 4°C, even at low concentrations (about 1%), and with a rise in temperature it decomposes into a mixture of toxic products with the characteristic odour of dimethyl sulphide. Furthermore, it is not possible to remove it by washing, even using a specialized system for washing cell transplants [25]. Thus, these adverse effects of DMSO impose restriction in its regular usage as a CPA [5].

3.7.3 Polymers

The cell suspension to be cryopreserved is often entrapped within a capsule with a CPA. Amongst the encapsulating materials, the synthetic non-penetrating polymers can provide cryoprotection of the cells within framework, thereby bypassing the limitations of diffusion in higher-dimensional cryopreservation. Vinyl-derived polymers, like polyethylene glycol (PEG), polyvinyl alcohol, and HES have the capacity to decrease the size of formed ice crystals [5, 27, 34].

3.7.4 Trehalose

Sugars are non-toxic and have cryoprotective properties. Trehalose, a non-reducing disaccharide, not cytotoxic in nature, is present in high concentration in several organisms like bacteria and yeast, who can survive in complete dehydration [35]. It is an alternative CPA due to its stability upon freezing [36]. Mammals do not produce trehalose but it is an operative cryoprotectant, which decreases the amount of cell injury by ice crystallization [37]. Its defensive role is connected both to the osmotic effect and specific interactions with phospholipids and labile proteins of cell membrane, preventing their damage and denaturation due to desiccation and oxidative stress [30, 38].

3.7.5 HES

HES is a complex carbohydrate, used in cryopreservation. It is a synthetic polymer based on the purified starch of potatoes or corn that has been modified by hydroxyethylation at carbon position C2, C3 or C6 [39]. It is extensively used in the cryopreservation of peripheral blood (PB) and bone marrow (BM) derived stem cells [40, 41].

3.7.6 Proteins

A water-soluble sticky protein sericin, of molecular weight ~30 kDa, isolated from the cocoon of silkworm, has been established as a DMSO-replacing CPA for human adipose tissue-derived stem or progenitor cells or the hepatocytes [32, 33]. Small anti-freeze proteins derived from the marine teleosts or fishes have also been included in the list of CPAs [5, 42].

3.7.7 Cell Banker Series

In order to attain improved survival rates subsequent to freezing and thawing, a Cell Banker series (Nippon Zenyaku Kogyo Co., Ltd., Fukushima, Japan) has been developed for rapid cell cryopreservation at −80 °C [5, 43]. This Cell Banker series of cryopreservation media contains glucose, 10% DMSO, a prescribed high polymer, and pH adjustors [44]. The serum-containing Cell Bankers is suitable enough to be used for cryopreservation of almost all mammalian cells. However, conventional cryopreservation media, containing foetal bovine serum (FBS) and other xenogeneic substances, prohibits its usage for homogenous cryopreservation protocol for clinical use in humans [45]. Thus, the non-serum-type Cell Banker 2 is optimum for the cryopreservation of cells in serum-free culture conditions. Cell Banker3 (or Stem cell Banker) is composed of 10% DMSO and other inorganic compounds, and gratifies the criterion of a chemically defined known ingredient that is xenofree and is thus suitable for the preservation of somatic stem cells and iPSCs [5].

3.8 Cryopreservation of Stem Cells

Both embryonic and the adult stem cells (ASCs) are considered as treasure-troves for their potential application in regenerative medicine. The fields of tissue engineering, gene therapy, regenerative medicine, and cell transplantation are principally dependent on the ability to preserve, store, and transport these stem cells without modifying their cellular and genetic contents. Thus, cryopreservation holds immense importance in taking forward the applications of various stem cell types [5, 43]. Table 3.1 gives a brief illustration of some stem cell types and their methods of cryopreservation, the details of which are mentioned in the later part of the chapter.

3.8.1 Cryopreservation of Embryonic Stem Cells

The main objective of cryopreservation of human embryonic stem cells (ESCs) is to preserve their integrity and functional properties from any impairment induced by ice crystal formation during cooling to very low temperatures [76]. The human ESCs can be cryopreserved successfully by vitrification. It was first devised by Fujioka et al. who established a simple yet large-scale vitrification method for the primate ESCs in a solution of 1 M acetamide, 2 M DMSO, and 3 M propylene glycol [77].

The human ESCs cannot be maintained efficiently in an undifferentiated state with DMSO [46]. Several studies have established conditions of human ESC cryopreservation without using DMSO [78, 79]. Matsumura et al. explored the consequence of adding a devitrification inhibitor compound, carboxylated e-poly-L-lysine (CPLL) to ethylene glycol (EG)-based vitrification solutions. CPLL was devised as a highly efficient, low-toxicity, novel cryoprotective alternative to DMSO [47, 48]. The use of a CPA, StemCell Keep™ (SCK) containing 6.5 M EG, 10% CPLL, and 0.7 M sucrose without serum, was found to exhibit high efficiency for the cryopreservation and banking of the human ESCs [4].

Table 3.1 Some of the different types of cryoprotectant in different stem cell types

Stem cell type		Cryoprotectant	Effect	Reference
Human embryonic stem cells		DMSO	Cannot maintain undifferentiated state	[46]
		Carboxylated e-poly-L-lysine (CPLL) and ethylene glycol	Exhibit high efficacy for the cryopreservation	[47, 48]
		6%HES, 5%DMSO, and 5% EG in saline	Beneficial for large-scale cryopreservation	[49]
Umbilical cord blood stem cells		DMSO and dextran	No change from control cells	[50]
		DMSO and sucrose	No change from control cells	[51]
		DMSO and HES	No change from control cells	[52]
Haematopoietic stem cells (HSCs)		DMSO	Associated toxicity of DMSO	[1, 53]
		HES and DMSO	Better cryopreservation of PB stem cells	[54]
		Lower DMSO concentration with pentastarch		[55]
		DMSO and trehalose		[56]
		0.3 M sucrose with 5% DMSO		[57]
		Only trehalose	Enhanced cell survival of PB stem cells	[58]
Neural stem cells		EG and sucrose	No change from control cells	[59]
Stem cells in the eye	Corneal epithelial stem cells (CESCs)	DMSO, propylene glycol or ethylene glycol	No change from control cells	[60]
	Corneal stromal stem cells (CSSCs)	2% human serum, 10% DMSO	No change from control cells	[61]
Dental stem cells	Dental pulp stem cells (DPSCs)	90% FBS and 10% DMSO	Viability reduced	[62]
		10% DMSO and 90% FBS	No effect on cell proliferation	[63]
		10% DMSO + 90% serum-free medium	No effect on cell proliferation	[64]
	Stem cells from exfoliating deciduous teeth (SHED)	90% FBS and 10% DMSO	Decreased viability after cryopreservation	[65]
		RPMI 1640 medium, 10% DMSO and 10% human albumin	No effect on viability and cell proliferation	[66]
	Stem cells from apical	90% FCS with 10% DMSO,	No effect in cell proliferation	[67]

(continued)

Table 3.1 (continued)

Stem cell type		Cryoprotectant	Effect	Reference
	papilla (SCAP)	90% FCS with 10% glycerol or 90% FCS with 10% ethylene glycol		
	Periodontal ligament stem cells (PDLSCs)	90% FBS + 10% DMSO	Ineffective on cell proliferation	[68]
		Programmed freezer with magnetic field, termed cells alive system (CAS)	Increased cell survival	[69]
	Dental follicle progenitor cells (DFPCs)	0.05 M glucose, 0.05 M sucrose and 1.5 M ethylene glycol	No effect in differentiation potential	[70]
Stem cells from the epidermis (skin)	Hair follicle-associated-pluripotent (HAP) stem cells	DMSO with medium	Less growth	[71]
		10% glycerol with 90% FBS	Extensive growth	[72]
		10% DMSO with 90% FBS	Less growth	[72]
	Skin-derived precursors (SKPs)	5% DMSO without FBS	Better differentiation potential	[73]
Mesenchymal stem cells (MSCs)		10% polyvinylpyrrolidone in DMEM	Better cell viability and adipogenic and osteogenic differentiability	[74]
		0.05 M glucose, 0.05 M sucrose, and 1.5 M ethylene glycol in phosphate buffer saline	Successful isolation and characterization of MSCs after 3 months of cryopreservation	[70]
Induced pluripotent stem cells (iPSCs)		Euro-Collins solution as base with 40% (v/v) EG and 10% (w/v) PEG	Higher recovery rate of iPSCs	[75]
		6%HES, 5%DMSO, and 5% EG in saline	Beneficial for large-scale banking	[49]
		Trehalose alone or in combination with EG or glycerol		[30]

The conventional slow freezing method uses a combination of Pronase/EDTA for Stem™ and CP-5E™ (6% HES, 5%DMSO, and 5% EG in saline). CP-5E™ is highly effective for the cryopreservation of human ESCs [49].

3.8.2 Cryopreservation of Adult Stem Cells (ASCs)

As discussed in Chap. 2, the adult stem cells (ASCs) encompass all stem cell types from umbilical cord blood (UCB) stem cells to the organ/tissue specific stem cells. The stem cells which can be easily isolated and have the ability to transdifferentiate are usually cryopreserved for future usage in autologous transplantation or reconstitution. In this section, we will deliberate the different aspects of cryopreservation of some of these cell types.

3.8.2.1 Cryopreservation of Umbilical Cord Blood (UCB) Stem Cells

Umbilical cord blood (UCB) haematopoietic stem cell transplantation is a dynamic development in the treatment for a variety of diseases including haematopoietic and non-haematopoietic malignancies, haemoglobinopathies, BM failure syndromes, and metabolic and immunodeficiency disorders. Successful cryopreservation of these stem cells is the prelude to such application. The UCB stem cells can be cryopreserved using DMSO and dextran [50]. The Wharton's jelly (WJ) from the umbilical cord stroma can be cryopreserved for future isolation of mesenchymal stem cells (MSCs) using 0.1 M sucrose and 1.5 M DMSO by slow freezing [51]. Most of the cryopreservation protocols employ DMSO as the main cryoprotectant. DMSO, having a low molecular weight (78.13 g/mol), can penetrate into the cell via the plasma membrane, preventing ice crystals formation by steady hydrogen bonding. However, due to its toxic nature DMSO is being substituted with DMSO-free systems for cryopreservation using non-penetrating cryoprotectants like (e.g., sucrose, glucose, galactose, or trehalose) and intracellular cryoprotectants (e.g., EG, propylene glycol, glycerol, methanol, formamide, and butanediol) [25, 34]. A combination of DMSO and HES (molecular weight—450,000 Dalton) is also an appropriate option for cryopreserving cord derived stem cells [52].

3.8.2.2 Cryopreservation of Tissue Specific Adult Stem Cells

In many cryopreservation protocols rigorous contemplation of sterility and viability is neglected. Therefore, these protocols include serum and protein additives, which introduce many contaminants, and slow cooling with DMSO/glycerol-based cryopreservation solutions, which damages cell survival.

Cryopreservation of Haematopoietic Stem Cells (HSC)

The haematopoietic stem cells (HSCs) are used for transplantation in patients requiring treatment of haematologic and non-haematologic malignancies. Three chief sources of HSCs are known which include the BM harvested from the cavity of the hipbone, PB obtained through leukapheresis, and UCB collected from placenta subsequent to childbirth [80].

If cells can be transferred from the donor to recipient within 72 h, protocols for initial storage at supra freezing temperatures are in position. However, the present therapeutic strategies demand that progenitor cells must be cryopreserved, for the autologous and allogeneic transplants. This strategy is harmless and not associated with any hostile outcomes regarding graft failure or graft versus host disease [1].

Although differences exist and is followed differently in the different transplantation protocols, a process of cryopreservation and its successful application includes steps of (1) collection and culture of the healthy donor cells, (2) use of cryopreservatives to such cultured healthy stem cells, (3) use of appropriate protocol for actual freezing, (4) detection of viable frozen cells after about 72 h of freezing, (5) thawing of cells, (6) thawed cells to be washed, free of cryopreservative, conditioned of the donor and then used for transplantation [1].

There are several differences in practices and use of cryopreservative, procedure of freezing, freezing temperature, rate of freezing, durability, thawing temperature, and rate of thawing. In the last one and half decades the temperature has varied from −196, −156, or −80 °C, depending on the storage in the liquid and vapour phase nitrogen and in cryopreservation mechanical freezers, respectively [1] and has been standardized to −196 to −80 °C. Due to reports of spreading infectious agents like Aspergillus and virus through nitrogen tanks (liquid phase), storage in the vapour nitrogen phase, at −156 °C finds importance [1].

A standardized NIH protocol used by Roger Williams Medical Center, Bone Marrow Transplant Unit, East Wing, Providence, Rhode Island included preservation of allogeneic and autologous peripheral HSC and BM transplant specimens by collecting cells and centrifuging to develop the cell rich pellet. Donor plasma finds use in autologous transplants. The supernatant is used for the re-liquidification of the pelleted cells, after adding heparinized Plasmalyte solution and 10% DMSO as cryopreservative which is stored initially at −4 °C and then frozen to −156 °C in the vapour phase and to −196 °C when stored in the liquid phase, depending on the location in the storage container. Prior to myeloablative therapy, live cells are detected by Trypan Blue assay or propidium iodide (PI) assay. Cells are thawed by treatment at 37 °C before the actual stem cell infusion [1, 53] (Fig. 3.2).

Although, DMSO is considered to be the golden standard in the cryopreservation of the HSCs, several other methods of cryopreservation have been devised by several investigators. In 2010, McCullough et al. discovered the application of HES mixed with DMSO at varying ratios to augment cryopreservation of PB stem cells [54]. Hayakawa et al. reported that lower DMSO concentration with pentastarch improves cryopreservation process over 10% DMSO and can serve as an alternate freezing medium [55]. A combination of DMSO and trehalose also improves cryopreservation [56]. The use of 5% DMSO with 0.3 M sucrose led to better functional capacity of HSCs than those cryopreserved in 10% DMSO with 10% FBS [57]. Cryopreservation of PB stem cells with trehalose exhibited enhanced survival of cells compared to those cryopreserved in 10% DMSO and 90% FBS [58, 81].

Cryopreservation of Neural Stem Cells

For the cryopreservation of neural stem cells, a serum- and protein-free vitrification process was developed. The procedure involved short step-wise exposure to vitrification solution composed of EG and sucrose and removal of vitrification solution at room temperature. To guard the samples from the introduction of other contaminants during handling and cryostorage, an original "straw-in-straw" method

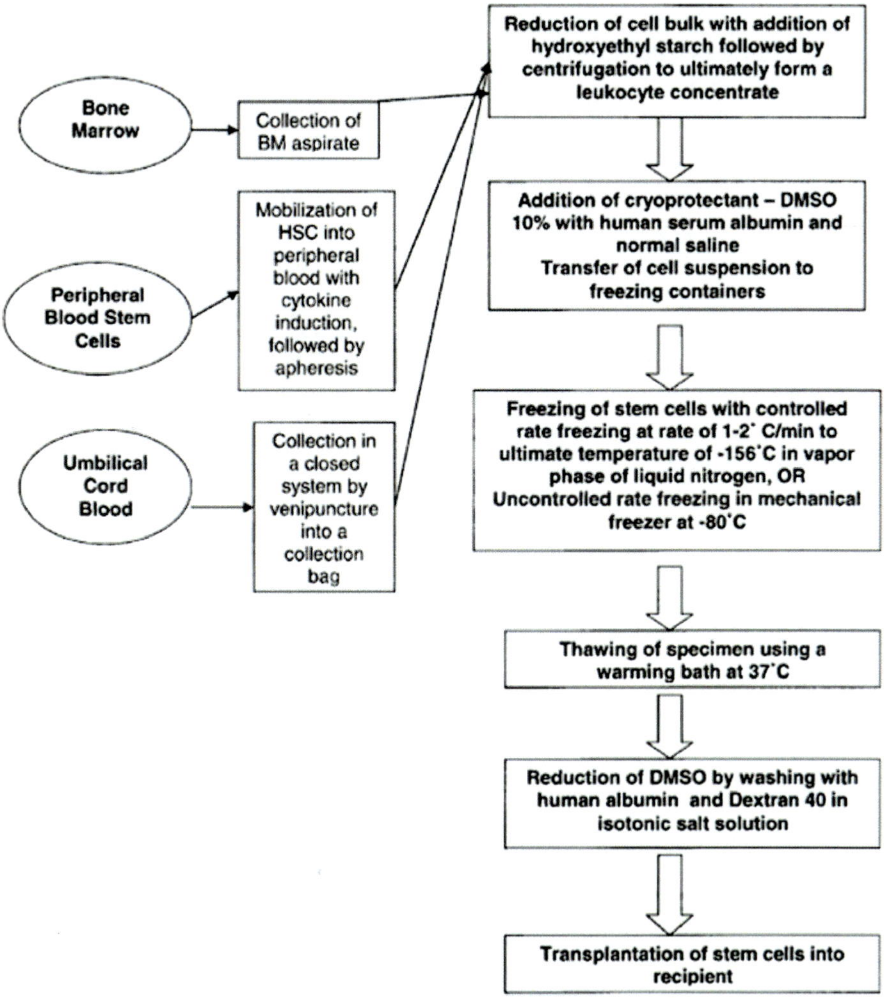

Fig. 3.2 Cryopreservation process for the bone marrow, peripheral blood, and UCB stem cells. Image adapted from Berz D, McCormack EM, Winer ES, Colvin GA, and Quesenberry PJ, Cryopreservation of Hematopoietic Stem Cells Am J Hematol. 2007 June; 82(6): 463–472 [1] with permission

(250 μL sterile straw placed in 500 μL straw) for direct immersion into liquid nitrogen and storing the samples was also introduced. It was found that this sterilized method for xeno-free cryopreservation served as a preparatory phase for the development of cryopreservation protocols for human neural stem and progenitor cells to be used later for clinical use [59].

Neurospheres cultures of 3–4 days, not exceeding 100–150 μm diameter can be cryopreserved as complete neurospheres, or single-cell populations. The cells are suspended in pre-warmed TrypLE followed by keeping them in water bath at 37 °C for 3 min. This is followed by addition of complete human NSC proliferation medium. Cell are spun, supernatant removed and resuspended in cryopreservation media. They are cooled to room temperature, transferred to −80 °C freezer overnight and then in the liquid nitrogen [82].

Cryopreservation of the Stem Cells in the Eye: Corneal Epithelial (CESCs) and Stromal Stem Cells (CSSCs)

The limbus is the fine ring of tissue bordering the cornea and is the home of the corneal epithelial stem cells, also known as limbal stem cells. Suspension cultures obtained from cadaveric corneoscleral rims containing these progenitor cells can be cryopreserved using DMSO, propylene glycol or EG. Cells cryopreserved using propylene glycol yielded more functionally potent cells post-cryopreservation while concentration of cryoprotectant above 5% was detrimental for the cells [60].

The corneal stromal stem cells (CSSCs) can be cryopreserved using 2% human serum, 10% DMSO and culture medium consisting of DMEM, low glucose, and other necessary supplements. Cells are initially cooled at a rate of −1 °C/min using an isopropyl alcohol in −80 °C freezer followed by storing in liquid nitrogen [61].

Cryopreservation of Dental Stem Cells (DSCs)

The purpose of short or prolonged cryopreservation of dental stem cells (DSCs) is the cryostorage of dental stem cell lineages after their successful isolation and expansion from the dental related tissues. The entire tooth either deciduous or permanent can be cryopreserved to delay the expensive process of isolation and cryopreservation of the DSCs. In spite of its toxicity, DMSO is used for cryopreservation [24]. A combination of DMSO and HES can also be used [41].

The dental pulp stem cells (DPSCs) can be cryopreserved using FBS and DMSO in varying concentration cooling them to 4 °C for 1 h, then at −20 °C for 2 h and −80 °C overnight with reduced viability after thawing [62, 63]. A serum-free method of cryopreservation, followed by cryopreserving in isopropanol filled cryobox has been devised by Ducret et al, exhibits higher viability [64].

Stem cells from the human exfoliated deciduous teeth (SHED) can similarly be cryopreserved using 90% FBS and 10% DMSO [65]. A combination of RPMI-1640 medium, 10% DMSO, and 10% human albumin used by Gioventu et al. showed no effect on viability and proliferation [66].

The periodontal ligament stem cells (PDLSCs) can be cryopreserved using 90% FBS and 10% DMSO, with no effect on cell proliferation [68]. A non-conventional way using programmed freezer and a magnetic field, known as Cells Alive System (CAS) increased the cell survival post-thawing, in comparison to conventional freezing method [69].

The stem cells from apical papilla (SCAP) are cryopreserved using 90% FCS with 10% DMSO, 90% FCS with 10% glycerol or 90% FCS with 10% EG. There was no effect in cell proliferation [67].

Dental follicle progenitor cells (DFPCs) can be cryopreserved using 0.05 M sucrose, 0.05 M glucose, and 1.5 M EG. The differentiation potential of these cryopreserved cells remained unaffected [70, 71, 83].

Cryopreservation of Stem Cells from the Epidermis (Skin)

The entire hair follicle from mouse can be cryopreserved by slow-rate cooling in DMSO-containing medium or TC-Protector medium and stored in liquid nitrogen or at −80 °C experiments. The cryopreserved follicles can be thawed at 37 °C in water bath for about 90 s (slow recovery) with mild shaking, and separated in three parts (upper, middle, and lower). The upper part of hair follicle can be isolated and cultured in DMEM with 10% FBS [71, 84].

Cryopreservation of whisker hair follicles in either 10% DMSO or glycerol with 90% FBS added as the freezing medium is also achievable. For short duration, cryovials with hair follicles were stored in an −80 °C freezer. It was observed that the hair follicle-associated pluripotent (HAP) stem cells were more potent in glycerol-cryopreserved follicles [72, 84]. The HAP cells can be isolated from anyone and cryopreserved without loss of differentiation potential [85].

Skin-derived precursors (SKPs) are a possibly autologous and an accessible source of neural precursor cells which can be used for drug screening or cell-based treatments, in neurological disorders. They can be cryopreserved with improved survival and proliferation potential compared to other working preparations, in 5% DMSO without FBS [73].

Cryopreservation of Mesenchymal Stem Cells (MSCs)

Human mesenchymal stem cells (MSCs) are non-haematopoietic cells, multipotent stem cells which differentiate into mesodermal and endodermal lineages. They were first derived from the BM [86], and since then have been isolated from practically every tissue including the perivascular area. They hold immense potential in regenerative biology [22, 87]. DMSO is usually avoided in the cryopreservation of MSCs because due to infusion of MSCs frozen in DMSO, patients grow mild complications like nausea, headache, vomiting, hypertension, diarrhoea, and hypotension [88].

The MSCs can be cryopreserved in serum-free media. They have been cryopreserved using a freezing medium containing high molecular weight polymers, like polyvinylpyrrolidone and/or methylcellulose as CPA instead of DMSO in DMEM and stored in liquid nitrogen [74]. In another new method of cryopreservation, a mixture of 0.05 M sucrose, 0.05 M glucose, and 1.5 M EG in phosphate buffer saline was used, which enabled effective isolation and characterization of MSCs after 3 months of cryopreservation [70]. The second imperative factor in cryopreservation of MSCs is the freezing temperature rate; slow freezing at the rate of 1 °C/min is the optimum rate [22, 89].

3.8.3 Cryopreservation of Induced Pluripotent Stem Cells (iPSCs)

Human induced pluripotent stem cells (iPSCs) have immense possibility in regenerative medicine and discovery of drug. Therefore, it is indispensable to establish extremely efficient cryopreservation methods for these cells. A vitrification method was devised by Nishigaki et al, using a DMSO-free, serum-free medium, VS2E, that uses Euro-Collins solution as a base with 40% EG and 10% PEG as cryoprotectants.

This amalgamation of vitrification and cryoprotectants resulted in a greater recovery rate of iPSCs cells. After the vitrification and warming, they could be cultured effortlessly. After several sub-culturing steps, cells expressed undifferentiated cell markers like Oct-3/4 and SSEA-4 and also exhibited alkaline phosphatase (AP) activity. The pluripotent nature of human iPSCs was sustained, as confirmed by teratoma formation upon iPSC transplantation into severe combined immunodeficient mice [75].

The conventional slow freezing method using a combination 6% HES, 5% DMSO, and 5% EG in saline is extremely effective for cryopreservation of the small cell clumps produced by iPSCs. This novel method is beneficial for large-scale iPSC banking for experimental usage [49]. Trehalose alone or used in combination with EG or glycerol as extracellular CPAs, in the absence of animal serum or album can also be used for cryopreservation of iPSC [30].

3.9 Quality Assessment of Stem Cells Recovered from Cryopreservation

Stem cells must have the skill of surviving prolonged storage and retain their functions after recovery. Cell viability testing after thawing is important in evaluating the consequences of cryopreservation on the cells. Quality assessment which includes viability tests, functional assays, stability measurements, and authentication must be applied to analyse the genetic integrity, metabolic functions, and totipotency of the post-thaw stem cells after recovery [90–92].

To determine the post-thaw viability of stem cells, Trypan Blue exclusion assay is done [33]. Stability assessment for stem cells includes the evaluation of totipotency by cell differentiation potential assays and morphological characteristics, studies of cytological and karyotype testing, phenotypes, confirmation of functional biochemistry, omics analyses, biomarker studies, investigations of the structure of genome, DNA damage and repair, chromatin analysis, analysis of DNA adducts, and confirmation of genetic modifications [93]. Furthermore, since these cells have been detained in long-term storage, they may have been cross-contaminated with other samples. Therefore, authentication is important and quality control needs to be executed to ensure sample identity following the retrieval of the cells from storage [92, 94].

After cryopreservation, cultured human ESCs can be evaluated for their cell count, viability, ability for attachment, and characterized by AP activity, immunocytochemical (ICC) staining, fluorescence-activated cell sorting (FACS), reverse transcription polymerase chain reaction (RT-PCR), and karyotyping. AP staining is a simple assay to assess pluripotency and is used to identify the formation of undifferentiated colonies. In successful cryopreservation, cryopreserved human ESCs retain their pluripotency, continually express stem cell markers as revealed by the information from FACS, ICC, and RT-PCR. The teratoma assay displays the ability of the ESCs to differentiate into three germ layers [4]. Similarly, the cryopreserved neural stem or progenitor cells are checked for cell markers, proliferation or multipotent differentiation [59].

In order to evaluate the role of cryopreserved HSCs, cell viability can be measured by trypan blue exclusion test and the numbers of colony-forming cells (CFC) and CD34 cells can successfully predict engraftment kinetics and haematopoietic recovery [1, 92, 95].

The critical quality attributes (CQAs) for iPSCs, to be used for therapeutic purpose after cryopreservation comprise of viability, identity, genetic fidelity and stability, microbiological sterility, characterization, and potency. Single tandem repeat (STR) genotyping of iPSC seed and master cell banks help in identity of these cells [96, 97]. The iPSCs cannot be sterilized and therefore, microbiological sterility is most important in delivering therapies. Microbiological sterility is done by testing for mycoplasma, bacteria, virus, and endotoxin. Residual vector testing is mandatory for checking genetic fidelity and stability followed by karyotyping, single nucleotide polymorphism (SNP) arrays, and genetic and disease marker analysis. Viability analysis is done by standard testing using trypan blue while FACS and ICC help in characterization of these cells. Potency is the qualitative measure of the biological activity of the iPSCs and is estimated by phenotypic and molecular pluripotency assays [98].

3.10 Discussion

The fruitful utilization of stem cells in research, translational or commercial reasons requires the successful implementation of cryopreservation method [30]. The advantages of cryopreservation are many. It provides an equipped source of stem cells for autologous transplantation. However, freezing is lethal to cells. Even if freezing is acceptable (conventional cryopreservation) or prohibited (vitrification), the cryoprotectant has to gain access to all parts of the system. There are frequent barriers to the unrestricted diffusion of solutes which result in change in compartment volume. Hence, the processes of diffusion and osmosis have significant effects during the introduction of cryoprotectants, removal of cryoprotectants, freezing process, and during thawing [2]. In many cases, the CPA is toxic and is also difficult to wash away during clinical use. Post-thawing of the stem cells also affects their properties. For example, the therapeutic potential of MSCs substantially reduces during cryostorage (the so-called "cryo stun effect") [25].

To conclude, cryopreservation is only the initiating process of producing stem cells and their derivatives for therapy. While the process is routine for the HSCs and MSCs, at least for autologous use, cryopreservation of ESCs and iPSCs cells are in preliminary state and needs to be further explored [19].

References

1. Berz D, McCormack EM, Winer ES, Colvin GA, Quesenberry PJ (2007) Cryopreservation of hematopoietic stem cells. Am J Hematol 82(6):463–472
2. Pegg DE (2007) Principles of cryopreservation. Methods Mol Biol 368:39–57. https://doi.org/10.1007/978-1-59745-362-2_3
3. Zhmakin AI (2009) Fundamentals of cryobiology. Physical phenomena and mathematical models. Springer-Verlag, Berlin
4. Ota A, Matsumura K, Lee JJ, Sumi S, Hyon SH (2017) StemCell Keep™ is effective for cryopreservation of human embryonic stem cells by vitrification. Cell Transplant 26 (5):773–787. https://doi.org/10.3727/096368916X692654
5. Jang TH, Park SC, Yang JH, Kim JY, Seok JH, Park US, Choi CW, Lee SR, Han J (2017) Cryopreservation and its clinical applications. Integr Med Res 6(1):12–18. https://doi.org/10.1016/j.imr.2016.12.001
6. Polge C, Smith AU, Parkes AS (1949) Revival of spermatozoa after vitrification and dehydration at low temperatures. Nature 164(4172):666. https://doi.org/10.1038/164666a0
7. Lovelock JE (1953) The mechanism of the protective action of glycerol against haemolysis by freezing and thawing. Biochim Biophys Acta 11(1):28–36. https://doi.org/10.1016/0006-3002(53)90005-5
8. Mazur P (1963) Kinetics of water loss from cells at subzero temperatures and the likelihood of intracellular freezing. J Gen Physiol 47:347–369
9. Pegg DE, Diaper MP (1988) On the mechanism of injury to slowly frozen erythrocytes. Biophys J 54(3):471–488
10. Leibo SP (1977) Fundamental cryobiology of mouse ova and embryos. Ciba Found Symp 52:69–96
11. Pegg DE, Diaper MP, Skaer HL, Hunt CJ (1984) The effect of cooling rate and warming rate on the packing effect in human erythrocytes frozen and thawed in the presence of 2 M glycerol. Cryobiology 21(5):491–502
12. Kleinhans FW (1998) Membrane permeability modeling: Kedem-Katchalsky vs a two-parameter formalism. Cryobiology 37(4):271–289
13. Pegg DE, Hunt CJ, Fong LP (1987) Osmotic properties of the rabbit corneal endothelium and their relevance to cryopreservation. Cell Biophys 10(2):169–189
14. Gao D, Critser JK (2000) Mechanisms of cryoinjury in living cells. ILAR J 41(4):187–196
15. Karlsson JO, Toner M (1996) Long-term storage of tissues by cryopreservation: critical issues. Biomaterials 17(3):243–256
16. Mandawala AA, Harvey SC, Roy TK, Fowler KE (2016) Cryopreservation of animal oocytes and embryos: current progress and future prospects. Theriogenology 86(7):1637–1644. https://doi.org/10.1016/j.theriogenology.2016.07.018
17. Yong KW, Wan Safwani WK, Xu F, Wan Abas WA, Choi JR, Pingguan-Murphy B (2015) Cryopreservation of human mesenchymal stem cells for clinical applications: current methods and challenges. Biopreserv Biobank 13(4):231–239. https://doi.org/10.1089/bio.2014.0104
18. Moon JH, Lee JR, Jee BC, Suh CS, Kim SH, Lim HJ, Kim HK (2008) Successful vitrification of human amnion-derived mesenchymal stem cells. Hum Reprod 23(8):1760–1770. https://doi.org/10.1093/humrep/den202
19. Hunt CJ (2011) Cryopreservation of human stem cells for clinical application: a review. Transfus Med Hemother 38(2):107–123
20. Matsuda H, Yagi T, Matsuoka J, Yamamura H, Fujisawa K, Endo A, Okada Y, Matsukawa H, Nakao A, Oishi M, Matsuno T, Tanaka N (1999) Effects of subzero nonfreezing storage in UW solution on isolated hepatocytes and cultured sinusoidal endothelial cells of rat liver. Transplant Proc 31(7):2913–2917
21. Gao HH, Li ZP, Wang HP, Zhang LF, Zhang JM (2016) Cryopreservation of whole bovine ovaries: comparisons of different thawing protocols. Eur J Obstet Gynecol Reprod Biol 204:104–107. https://doi.org/10.1016/j.ejogrb.2016.06.016

22. Ullah I, Subbarao RB, Rho GJ (2015) Human mesenchymal stem cells - current trends and future prospective. Biosci Rep 35(2):e00191. https://doi.org/10.1042/BSR20150025
23. Zambelli A, Poggi G, Da Prada G, Pedrazzoli P, Cuomo A, Miotti D, Perotti C, Preti P, Robustellidella CG (1998) Clinical toxicity of cryopreserved circulating progenitor cells infusion. Anticancer Res 18(6B):4705–4708
24. Pilbauerová N, Suchánek J (2018) Cryopreservation of dental stem cells. Acta Med (Hradec Kralove) 61(1):1–7. https://doi.org/10.14712/18059694.2018.16
25. Arutyunyan I, Fatkhudinov T, Sukhikh G (2018) Umbilical cord tissue cryopreservation: a short review. Stem Cell Res Ther 9(1):236. https://doi.org/10.1186/s13287-018-0992-0
26. Chao H, Davies PL, Carpenter JF (1996) Effects of antifreeze proteins on red blood cell survival during cryopreservation. J Exp Biol 199(Pt 9):2071–2076
27. Sambu S (2015) A Bayesian approach to optimizing cryopreservation protocols. PeerJ 3:e1039. https://doi.org/10.7717/peerj.1039
28. Safa S, Moghaddam G, Jozani RJ, Daghigh Kia H, Janmohammadi H (2016) Effect of vitamin E and selenium nanoparticles on post-thaw variables and oxidative status of rooster semen. Anim Reprod Sci 174:100–106. https://doi.org/10.1016/j.anireprosci.2016.09.011
29. Krausková Ľ, Procházková J, Klašková M, Filipová L, Chaloupková R, Malý S, Damborský J, Heger D (2016) Suppression of protein inactivation during freezing by minimizing pH changes using ionic cryoprotectants. Int J Pharm 509(1–2):41–49. https://doi.org/10.1016/j.ijpharm.2016.05.031
30. Ntai A, La Spada A, De Blasio P, Biunno I (2018) Trehalose to cryopreserve human pluripotent stem cells. Stem Cell Res 31:102–112. https://doi.org/10.1016/j.scr.2018.07.021
31. Lovelock JE, Bishop MW (1959) Prevention of freezing damage to living cells by dimethyl sulphoxide. Nature 183(4672):1394–1395
32. Slichter SJ, Jones M, Ransom J, Gettinger I, Jones MK, Christoffel T, Pellham E, Bailey SL, Corson J, Bolgiano D (2014) Review of in vivo studies of dimethyl sulfoxide cryopreserved platelets. Transfus Med Rev 28(4):212–225. https://doi.org/10.1016/j.tmrv.2014.09.001
33. Yong KW, Pingguan-Murphy B, Xu F, Abas WA, Choi JR, Omar SZ, Azmi MA, Chua KH, Wan Safwani WK (2015) Phenotypic and functional characterization of long-term cryopreserved human adipose-derived stem cells. Sci Rep 5:9596. https://doi.org/10.1038/srep09596
34. Best BP (2015) Cryoprotectant toxicity: facts, issues, and questions. Rejuvenation Res 18 (5):422–436. https://doi.org/10.1089/rej.2014.1656. Erratum in: Rejuvenation Res. 2018 Feb;21(1):87
35. Crowe JH, Crowe LM, Oliver AE, Tsvetkova N, Wolkers W, Tablin F (2001) The trehalose myth revisited: introduction to a symposium on stabilization of cells in the dry state. Cryobiology 43(2):89–105
36. Buchanan SS, Gross SA, Acker JP, Toner M, Carpenter JF, Pyatt DW (2004) Cryopreservation of stem cells using trehalose: evaluation of the method using a human hematopoietic cell line. Stem Cells Dev 13(3):295–305
37. Crowe JH, Crowe LM (2000) Preservation of mammalian cells-learning nature's tricks. Nat Biotechnol 18(2):145–146
38. Benaroudj N, Lee DH, Goldberg AL (2001) Trehalose accumulation during cellular stress protects cells and cellular proteins from damage by oxygen radicals. J Biol Chem 276 (26):24261–24267
39. Boldt J (2009) Modern rapidly degradable hydroxyethyl starches: current concepts. Anesth Analg 108(5):1574–1582. https://doi.org/10.1213/ane.0b013e31819e9e6c
40. Luo K, Wu G, Wang Q, Sun Y, Liu H (1994) Effect of dimethylsulfoxide and hydroxyethyl starch in the preservation of fractionated human marrow cells. Cryobiology 31(4):349–354
41. Stolzing A, Naaldijk Y, Fedorova V, Sethe S (2012) Hydroxyethylstarch in cryopreservation - mechanisms, benefits and problems. Transfus Apher Sci 46(2):137–147. https://doi.org/10.1016/j.transci.2012.01.007

42. Fletcher GL, Hew CL, Davies PL (2001) Antifreeze proteins of teleost fishes. Annu Rev Physiol 63:359–390
43. Miyagi-Shiohira C, Kurima K, Kobayashi N, Saitoh I, Watanabe M, Noguchi Y, Matsushita M, Noguchi H (2015) Cryopreservation of adipose-derived mesenchymal stem cells. Cell Med 8 (1–2):3–7. https://doi.org/10.3727/215517915X689100
44. Holm F, Ström S, Inzunza J, Baker D, Strömberg AM, Rozell B, Feki A, Bergström R, Hovatta O (2010) An effective serum- and xeno-free chemically defined freezing procedure for human embryonic and induced pluripotent stem cells. Hum Reprod 25(5):1271–1279. https://doi.org/10.1093/humrep/deq040
45. Li Y, Tan JC, Li LS (2010) Comparison of three methods for cryopreservation of human embryonic stem cells. Fertil Steril 93(3):999–1005. https://doi.org/10.1016/j.fertnstert.2008.10.052
46. Katkov II, Kim MS, Bajpai R, Altman YS, Mercola M, Loring JF, Terskikh AV, Snyder EY, Levine F (2006) Cryopreservation by slow cooling with DMSO diminished production of Oct-4 pluripotency marker in human embryonic stem cells. Cryobiology 53(2):194–205
47. Matsumura K, Bae JY, Hyon SH (2010) Polyampholytes as cryoprotective agents for mammalian cell cryopreservation. Cell Transplant 19(6):691–699. https://doi.org/10.3727/096368910X508780
48. Matsumura K, Bae JY, Kim HH, Hyon SH (2011) Effective vitrification of human induced pluripotent stem cells using carboxylated ε-poly-l-lysine. Cryobiology 63(2):76–83. https://doi.org/10.1016/j.cryobiol.2011.05.003
49. Imaizumi K, Iha M, Nishishita N, Kawamata S, Nishikawa S, Akuta T (2016) A simple and efficient method of slow freezing for human embryonic stem cells and induced pluripotent stem cells. Methods Mol Biol 1341:15–24. https://doi.org/10.1007/7651_2015_211
50. Elmoazzen H, Holovati JL (2015) Cord blood clinical processing, cryopreservation, and storage. Methods Mol Biol 1257:369–379. https://doi.org/10.1007/978-1-4939-2193-5_18
51. Da-Croce L, Gambarini-Paiva GH, Angelo PC, Bambirra EA, Cabral AC, Godard AL (2013) Comparison of vitrification and slow cooling for umbilical tissues. Cell Tissue Bank 14 (1):65–76. https://doi.org/10.1007/s10561-012-9301-9
52. Strobel J, Hohensee F, Kuta P, Eckstein R, Zingsem J (2017) Comparison of six different cryoprotective agents used for deep freezing and storage of CD34+ cells derived from cord blood and peripheral blood stem cell concentrates. Clin Lab 63(3):543–550. https://doi.org/10.7754/Clin.Lab.2016.160928
53. Lecchi L, Giovanelli S, Gagliardi B, Pezzali I, Ratti I, Marconi M (2016) An update on methods for cryopreservation and thawing of hemopoietic stem cells. Transfus Apher Sci 54(3):324–336. https://doi.org/10.1016/j.transci.2016.05.009
54. McCullough J, Haley R, Clay M, Hubel A, Lindgren B, Moroff G (2010) Long-term storage of peripheral blood stem cells frozen and stored with a conventional liquid nitrogen technique compared with cells frozen and stored in a mechanical freezer. Transfusion 50(4):808–819. https://doi.org/10.1111/j.1537-2995.2009.02482.x. Erratum in: Transfusion. 2010 Oct;50 (10):2285
55. Hayakawa J, Joyal EG, Gildner JF, Washington KN, Phang OA, Uchida N, Hsieh MM, Tisdale JF (2010) 5% dimethyl sulfoxide (DMSO) and pentastarch improves cryopreservation of cord blood cells over 10% DMSO. Transfusion 50(10):2158–2166. https://doi.org/10.1111/j.1537-2995.2010.02684.x
56. Chen G, Yue A, Ruan Z, Yin Y, Wang R, Ren Y, Zhu L (2016) Comparison of the effects of different cryoprotectants on stem cells from umbilical cord blood. Stem Cells Int 2016:1396783. https://doi.org/10.1155/2016/1396783
57. Petrenko YA, Jones DR, Petrenko AY (2008) Cryopreservation of human fetal liver hematopoietic stem/progenitor cells using sucrose as an additive to the cryoprotective medium. Cryobiology 57(3):195–200. https://doi.org/10.1016/j.cryobiol.2008.08.003

58. Martinetti D, Colarossi C, Buccheri S, Denti G, Memeo L, Vicari L (2017) Effect of trehalose on cryopreservation of pure peripheral blood stem cells. Biomed Rep 6(3):314–318. https://doi.org/10.3892/br.2017.859
59. Kuleshova LL, Tan FCK, Magalhães R, Gouk SS, Lee KH, Dawe GS (2009) Effective cryopreservation of neural stem or progenitor cells without serum or proteins by vitrification. Cell Transplant 18(2):135–144. https://doi.org/10.3727/096368909788341298
60. Osei-Bempong C, Ghareeb AE, Lako M, Figueiredo FC, Armitage WJ (2018) Defining the optimal cryoprotectant and concentration for cryopreservation of limbal stem cells. Cryobiology 84:98–102. https://doi.org/10.1016/j.cryobiol.2018.07.008
61. Sun Y, Dos Santos A, Balayan A, Deng SX (2020) Evaluation of cryopreservation media for the preservation of human corneal stromal stem cells. Tissue Eng Part C Methods 26(1):37–43. https://doi.org/10.1089/ten.TEC.2019.0195
62. Davies OG, Smith AJ, Cooper PR, Shelton RM, Scheven BA (2014) The effects of cryopreservation on cells isolated from adipose, bone marrow and dental pulp tissues. Cryobiology 69 (2):342–347. https://doi.org/10.1016/j.cryobiol.2014.08.003
63. Papaccio G, Graziano A, d'Aquino R, Graziano MF, Pirozzi G, Menditti D, De Rosa A, Carinci F, Laino G (2006) Long-term cryopreservation of dental pulp stem cells (SBP-DPSCs) and their differentiated osteoblasts: a cell source for tissue repair. J Cell Physiol 208(2):319–325
64. Ducret M, Fabre H, Farges JC, Degoul O, Atzeni G, McGuckin C, Forraz N, Mallein-Gerin F, Perrier-Groult E (2015) Production of human dental pulp cells with a medicinal manufacturing approach. J Endod 41(9):1492–1499. https://doi.org/10.1016/j.joen.2015.05.017
65. Ji EH, Song JS, Kim SO, Jeon M, Choi BJ, Lee JH (2014) Viability of pulp stromal cells in cryopreserved deciduous teeth. Cell Tissue Bank 15(1):67–74. https://doi.org/10.1007/s10561-013-9375-z
66. Gioventù S, Andriolo G, Bonino F, Frasca S, Lazzari L, Montelatici E, Santoro F, Rebulla P (2012) A novel method for banking dental pulp stem cells. Transfus Apher Sci 47(2):199–206. https://doi.org/10.1016/j.transci.2012.06.005
67. Ding G, Wang W, Liu Y, An Y, Zhang C, Shi S, Wang S (2010) Effect of cryopreservation on biological and immunological properties of stem cells from apical papilla. J Cell Physiol 223 (2):415–422. https://doi.org/10.1002/jcp.22050
68. Vasconcelos RG, Ribeiro RA, Vasconcelos MG, Lima KC, Barboza CA (2012) In vitro comparative analysis of cryopreservation of undifferentiated mesenchymal cells derived from human periodontal ligament. Cell Tissue Bank 13(3):461–469. https://doi.org/10.1007/s10561-011-9271-3
69. Kaku M, Kamada H, Kawata T, Koseki H, Abedini S, Kojima S, Motokawa M, Fujita T, Ohtani J, Tsuka N, Matsuda Y, Sunagawa H, Hernandes RA, Ohwada N, Tanne K (2010) Cryopreservation of periodontal ligament cells with magnetic field for tooth banking. Cryobiology 61(1):73–78. https://doi.org/10.1016/j.cryobiol.2010.05.003
70. Park BW, Jang SJ, Byun JH, Kang YH, Choi MJ, Park WU, Lee WJ, Rho GJ (2017) Cryopreservation of human dental follicle tissue for use as a resource of autologous mesenchymal stem cells. J Tissue Eng Regen Med 11(2):489–500. https://doi.org/10.1002/term.1945
71. Kajiura S, Mii S, Aki R, Hamada Y, Arakawa N, Kawahara K, Li L, Katsuoka K, Hoffman RM, Amoh Y (2015) Cryopreservation of the hair follicle maintains pluripotency of nestin-expressing hair follicle-associated pluripotent stem cells. Tissue Eng Part C Methods 21 (8):825–831. https://doi.org/10.1089/ten.TEC.2014.0500
72. Cao W, Li L, Tran B, Kajiura S, Amoh Y, Liu F, Hoffman RM (2015) Extensive hair shaft growth after mouse whisker follicle isolation, cryopreservation and transplantation in nude mice. PLoS One 10(12):e0145997. https://doi.org/10.1371/journal.pone.0145997
73. Bakhtiari M, Mansouri K, Sadeghi Y, Mostafaie A (2012) Proliferation and differentiation potential of cryopreserved human skin-derived precursors. Cell Prolif 45(2):148–157. https://doi.org/10.1111/j.1365-2184.2011.00803.x

74. Thirumala S, Gimble JM, Devireddy RV (2010) Cryopreservation of stromal vascular fraction of adipose tissue in a serum-free freezing medium. J Tissue Eng Regen Med 4(3):224–232. https://doi.org/10.1002/term.232
75. Nishigaki T, Teramura Y, Nasu A, Takada K, Toguchida J, Iwata H (2011) Highly efficient cryopreservation of human induced pluripotent stem cells using a dimethyl sulfoxide-free solution. Int J Dev Biol 55(3):305–311. https://doi.org/10.1387/ijdb.103145tn
76. Serdarogullari M, Coban O, Boynukalin FK, Bilgin EM, Findikli N, Bahceci M (2019) Successful application of a single warming protocol for embryos cryopreserved by either slow freezing or vitrification techniques. Syst Biol Reprod Med 65(1):12–19. https://doi.org/10.1080/19396368.2018.1487477
77. Fujioka T, Yasuchika K, Nakamura Y, Nakatsuji N, Suemori H (2004) A simple and efficient cryopreservation method for primate embryonic stem cells. Int J Dev Biol 48(10):1149–1154
78. Kuleshova L, Gianaroli L, Magli C, Ferraretti A, Trounson A (1999) Birth following vitrification of a small number of human oocytes: case report. Hum Reprod 14(12):3077–3079
79. Nishigaki T, Teramura Y, Suemori H, Iwata H (2010) Cryopreservation of primate embryonic stem cells with chemically-defined solution without Me2SO. Cryobiology 60(2):159–164. https://doi.org/10.1016/j.cryobiol.2009.10.007
80. Barriga F, Ramírez P, Wietstruck A, Rojas N (2012) Hematopoietic stem cell transplantation: clinical use and perspectives. Biol Res 45(3):307–316. https://doi.org/10.4067/S0716-97602012000300012
81. Hornberger K, Yu G, McKenna D, Hubel A (2019) Cryopreservation of hematopoietic stem cells: emerging assays, cryoprotectant agents, and technology to improve outcomes. Transfus Med Hemother 46(3):188–196. https://doi.org/10.1159/000496068
82. Crook JM, Tomaskovic-Crook E (2017) Culturing and cryobanking human neural stem cells. Methods Mol Biol 1590:199–206. https://doi.org/10.1007/978-1-4939-6921-0_15
83. Hilkens P, Driesen RB, Wolfs E, Gervois P, Vangansewinkel T, Ratajczak J, Dillen Y, Bronckaers A, Lambrichts I (2016) Cryopreservation and banking of dental stem cells. Adv Exp Med Biol 951:199–235
84. Hoffman RM, Kajiura S, Cao W, Liu F, Amoh Y (2016) Cryopreservation of hair-follicle associated pluripotent (HAP) stem cells maintains differentiation and hair-growth potential. Adv Exp Med Biol 951:191–198
85. Amoh Y, Hoffman RM (2017) Hair follicle-associated-pluripotent (HAP) stem cells. Cell Cycle 16(22):2169–2175. https://doi.org/10.1080/15384101.2017.1356513
86. Pittenger MF, Mackay AM, Beck SC, Jaiswal RK, Douglas R, Mosca JD, Moorman MA, Simonetti DW, Craig S, Marshak DR (1999) Multilineage potential of adult human mesenchymal stem cells. Science 284(5411):143–147
87. Crisan M, Yap S, Casteilla L, Chen CW, Corselli M, Park TS, Andriolo G, Sun B, Zheng B, Zhang L, Norotte C, Teng PN, Traas J, Schugar R, Deasy BM, Badylak S, Buhring HJ, Giacobino JP, Lazzari L, Huard J, Péault B (2008) A perivascular origin for mesenchymal stem cells in multiple human organs. Cell Stem Cell 3(3):301–313. https://doi.org/10.1016/j.stem.2008.07.003
88. Rodríguez L, Velasco B, García J, Martín-Henao GA (2005) Evaluation of an automated cell processing device to reduce the dimethyl sulfoxide from hematopoietic grafts after thawing. Transfusion 45(8):1391–1397
89. Thirumala S, Zvonic S, Floyd E, Gimble JM, Devireddy RV (2005) Effect of various freezing parameters on the immediate post-thaw membrane integrity of adipose tissue derived adult stem cells. Biotechnol Prog 21(5):1511–1524
90. Stacey G (2004) Fundamental issues for cell-line banks in biotechnology and regulatory affairs. In: Fuller BJ, Lane N, Benson EE (eds) Life in the frozen state. CRC Press, Boca Raton, pp 437–452
91. Day JG, Stacey G (2007) Cryopreservation and freeze-drying protocols. Springer, Dordrecht
92. Sun C, Yue J, He N, Liu Y, Zhang X, Zhang Y (2016) Fundamental principles of stem cell banking. Adv Exp Med Biol 951:31–45

93. Chen YK, Liu QH, Li J, Xiao ZZ, Xu SH, Shi XH, Ma DY (2010) Effect of long-term cryopreservation on physiological characteristics, antioxidant activities and lipid peroxidation of red seabream (Pagrus major) sperm. Cryobiology 61(2):189–193. https://doi.org/10.1016/j.cryobiol.2010.07.003
94. Benson E, Betson F, Fuller BJ, Harding K, Kofanova O (2013) Translating cryobiology principles into trans-disciplinary storage guidelines for biorepositories and biobanks: a concept paper. Cryo Letters 34(3):277–312
95. Winter JM, Jacobson P, Bullough B, Christensen AP, Boyer M, Reems JA (2014) Long-term effects of cryopreservation on clinically prepared hematopoietic progenitor cell products. Cytotherapy 16(7):965–975. https://doi.org/10.1016/j.jcyt.2014.02.005
96. MacLeod RA, Dirks WG, Matsuo Y, Kaufmann M, Milch H, Drexler HG (1999) Widespread intraspecies cross-contamination of human tumor cell lines arising at source. Int J Cancer 83 (4):555–563
97. Gymrek M, McGuire AL, Golan D, Halperin E, Erlich Y (2013) Identifying personal genomes by surname inference. Science 339(6117):321–324. https://doi.org/10.1126/science.1229566
98. Sullivan S, Stacey GN, Akazawa C, Aoyama N, Baptista R, Bedford P, BennaceurGriscelli A, Chandra A, Elwood N, Girard M, Kawamata S, Hanatani T, Latsis T, Lin S, Ludwig TE, Malygina T, Mack A, Mountford JC, Noggle S, Pereira LV, Price J, Sheldon M, Srivastava A, Stachelscheid H, Velayudhan SR, Ward NJ, Turner ML, Barry J, Song J (2018) Quality control guidelines for clinical-grade human induced pluripotent stem cell lines. Regen Med 13 (7):859–866. https://doi.org/10.2217/rme-2018-0095

Glycosylation of Stem Cells

4

Contents

Abstract

Glycosylation is a posttranslational modification during biosynthesis of biological macromolecules including protein and lipid. Glycosylated proteins and lipids play vital role in cell–cell interaction, signalling, recognition of ligand and finds importance in maintaining the cellular physiology and normal function of cell and its proper development. Deviation in glycosylation of cells can lead to diseases. Stem cells have been known to express markers that are glycosylated. In this chapter we discuss glycosylation in the context of stem cells. Although, the glycosylation has been reported in stem cells, their exact function in maintaining stemness, self-renewal, and process of differentiation is yet to be studied. We also discuss the glycosylation reported from cancer stem cells.

Keywords

Glycosylation · Stem cells · iPSc · MSC · *N*-glycosylation · N–glycans

Abbreviations

AML	Acute myeloid leukaemia
ASC	Adult stem cells

S. Chowdhury, S. Ghosh, *Stem Cells*, https://doi.org/10.1007/978-981-16-1638-9_4

Asn	Asparagine
B4GALNT3	β1,4-N-acetylgalactosaminyltransferase III
BMP	Bone morphogenetic proteins
CDG	Disorders of glycosylation
COG	Oligomeric Golgi
CSC	Cancer stem cell
Dol-P	Dolichol phosphate
DS	Dermatan sulphate
ECM	Extracellular matrix
EGF	Epidermal growth factors
EMT	Epithelial mesenchymal transition
EpCAM	Epithelial cell adhesion molecule
ER	Endoplasmic reticulum
ESA	Epithelial surface antigen
ESCs	Embryonic stem cells
FGF2	Fibroblast growth factor 2
Gal	Galactose
GalNAc	*N*-acetylgalactosamine
GALNTs	GalNAc-transferases
GalT	Galactosyltransferases
GCS	Glucosylceramide synthase
GF	Glycosyltransferases
GlcNAc	*N*-acetylglucosamine
GSLs	Glycosphingolipids
HA	Hyaluronan
HS	Heparin sulphate
HSCs	Haematopoietic stem cells
iPSC	Induced pluripotent stem cells
LeX	Lewis X
LIF	Leukaemia inhibitory factor
LIFR	LIF receptor
Man	Mannose
Man-6-P	Man-6-phosphate
MD	Muscular Dystrophy
MSC	Mesenchymal stromal cells
MUC1	Mucin1
NSC	Neural stem cells
OGAs	*O*-GlcNAcases
O-GlcNAc	*O*-linked GlcNAc
OGTs	*O*-linked GlcNAc (*O*-GlcNAc) transferases
OST	Oligosaccharyltransferase
PMM2	Phosphomannomutase 2
Sialyl-T	Sialyltransferase
SSEA	Stage-specific embryonic antigen
ST6GAL1	β-galactosideα2,6-sialyltransferase
STAT	Signal transducer and activator of transcription

4.1 Introduction

Glycosylation is a co-translational and posttranslational event in biosynthesis of macromolecules involving ubiquitous modification of biomolecules like proteins and lipids in all living cells by incorporation of monosaccharides or oligosaccharide moieties forming highly complex glycoproteins and glycolipids. They are expressed on the mammalian cell surface varying in their diversity both in the glycan arrangement and the underlying protein or lipid and varies with the cell lineage.

Glycosylation of macromolecules have been known to be associated in regulating protein quality control and recognition has been implicated in development and in the biology of disease.

N-glycosylation of macromolecules at asparagine starts in the cell organelle, endoplasmic reticulum (ER) and undergoes maturation in the Golgi apparatus involving enzymes glycosyltransferases, glycosidases, and nucleotide sugar transporters. The process in order to be successful needs efficient processing, quality control of the developing sugar chain coordinated with protein trafficking and sorting within the secretory pathway within the Golgi cisternae within the Golgi apparatus. Golgi stacks and Golgi stacking factors GRASP55 and GRASP65 and oligomeric Golgi (COG) complex containing enzymes play role in accurate glycosylation [1–3].

In the last two decades considerable studies have been done on glycosylated moeties and the process of glycosylation is better understood. Pluripotent stem cells have also revealed glycosylation and finds importance in normal development. Induced pluripotent stem cells (iPSC) derived from patients have been used to model human diseases including congenital disorders associated with glycosylation [4].

4.2 Glycosylation

The biosynthesis and association of glycans, on secreted proteins and cell surface glycoproteins, occurs in the ER and Golgi apparatus and synthesize and arrangement of *N*-glycans is mediated by enzymes including nucleotide sugar transporters, glycosidases, and glycosyltransferases as it moves through the secretory pathway [1–3].

N-glycans are added on nascent polypeptides in the ER by the oligosaccharyl transferase (OST) complex to initiate the process of glycosylation (Fig. 4.1). Nucleotide sugar transporters transport sugar donors from the cytosol into the luminal compartments of the Golgi apparatus. Specific glycosyltransferases which are type-II membrane proteins found in the lumens of ER and Golgi apparatus, transfers the sugars from a sugar nucleotide donor to an acceptor substrate to produce structurally diverse glycosylated proteins.

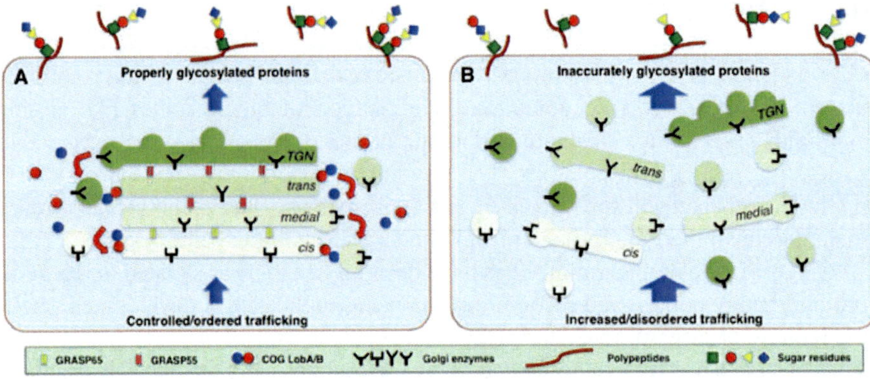

Fig. 4.1 Glycosylation of proteins and lipids. Image adapted with permission from Zhang and Wang [5]

Around 50% of the cellular proteins that undergo glycosylation are secreted and plays role in protein folding and quality control of synthesized and secreted protein. Cell surface glycoproteins and glycolipids act as receptors for ligands, mediate protein–protein or cell–cell interactions, downstream signalling pathways, rate of internalization and degradation, cell function, impact developmental processes, and affect the disease onset and progression in diseases like diabetes and cancer. Defective glycosylation can lead to diseases, congenital disorders of glycosylation (CDGs), affecting to human health, immunity, susceptibility to infectious diseases, attachment and bacterial entry into host cell, entry of fungi and viruses into cells [1–3].

4.2.1 *N*-glycosylation

N-glycans are formed when *N*-acetylglucosamine (GlcNAc) is attached to the Asn (N) side chain with a β-1N linkage wherein the Asn-linked glycoconjugates contain a GlcNAc2 mannose (Man)3 core. Different numbers of other monosaccharides can be added or removed to form high-mannose *N*-glycan.

N-glycosylation involves the formation of a lipid precursor in which N-acetylglucosamine (GlcNAc) and Mannose (Man) in a branched carbohydrate architecture is attached to dolichol phosphate (Dol-P) on the ER cytoplasmic side, which is later turned to the ER lumen, following addition of Man and glucose to form a 14 sugar complex structure , Glc3Man9GlcNAc2 (Fig. 4.2) [6, 7].

An oligosaccharyl transferase enables addition of carbohydrate chain to Asn-X-Ser/Thr site of protein wherein X may be any amino acid excepting Proline. This carbohydrate-protein conjugate is processed further in ER. Here the glucose residues are removed, followed by its transit to the cis- Golgi where it undergoes maturation by specific mannosidases acting on carbohydrate structures on its way to

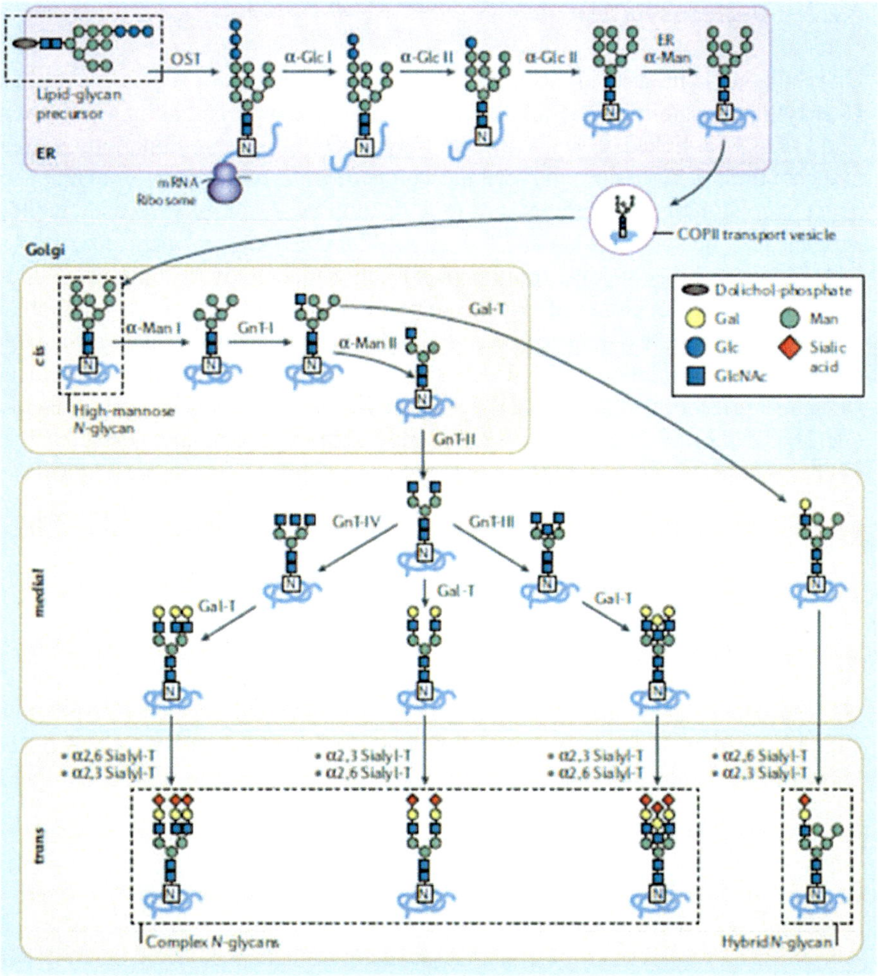

Fig. 4.2 *N*-glycan synthesis starts in the ER by transfer of a lipid-glycan precursor (glucose $(Glc)_3$ mannose $(Man)_9$ *N*-acetylglucosamine $(GlcNAc)_2$. This is followed by removal of glucose residues by two α-glucosidases (α-Glc I–II) and an initial Man residue removed by the ER α-mannosidase (ER α-Man). This is followed by entry of glycoprotein into Golgi apparatus to undergo modification by α-mannosidase I and II (α-Man I–II). Glycosidases and transferases including GlcNAc-transferase I–IV (GnT-I–IV), β1,4-galactosyltransferases (Gal-T), α2,3-sialyltransferase (α2,3-Sialyl-T), and α2,6-sialyltransferase (α2,6-Sialyl-T) leads to generation of diverse forms of *N*-glycans with linked terminal sialic acids. Image and legend adapted with permission from Reily et al. [6]

medial-Golgi followed by its transit to *trans*-Golgi producing complex *N*-glycans by adding GlcNAc, galactose, fucose sugars and sialic acid [6].

4.2.2 *O*-glycosylation

Glycosylation in humans involves the most common addition of sugars GlcNAc and *N*-acetylgalactosamine (GalNAc) linked to Serine (Ser) or Threonine (Thr) amino acids. GalNAc linked glycans or mucin-type *O*-glycans are found in nuclear, mitochondrial, cytoplasmic intracellular compartments, and mucins containing six major core structures including core of 1–4, terminal GalNAc (Tn) and sialyl-Tn antigens and the synthesis is initially mediated by GalNAc transferases (GALNTs), with amino acid specificities. Addition of glycans sequentially to the initial GalNAc leads to formation of clustered glycoproteins, in mucins and human immunoglobulin A1 (IgA1) during its transportation through the *cis*-, *medial*-, and *trans*-Golgi compartments but unlike *N*-glycans, pre and post-processing trimming of existing sugar residues does not occur. *O*-glycan chains can be modified with galactose, GlcNAc, sialic acid, fucose by distinct glycosyltransferases and is regulated by *O*-GlcNAcases (OGAs) and *O*-linked GlcNAc (*O*-GlcNAc) transferases (OGTs).

4.2.3 O-GlcNAcyl Modifications

GlcNAc is linked to serine or threonine amino acid residue. GlcNAc is added by *O*-GlcNAc transferase while *O*-GlcNAcase causes its removal [9] (Fig 4.1). *O*-GlcNAcylation has been found to occur in proteins shuttling between the cell's cytoplasm and nucleus playing roles in regulation of diverse functions. *O*-GlcNAc is known to play role in regulation of stem cell pluripotency and reprogramming in embryonic stem cells (ESCs).

4.2.4 Glycosphingolipids (GSLs)

GSLs, abundant in expression in human's cell membranes lipid bilayer are composed of sphingolipid, to which a carbohydrate, glucose or galactose is attached at the C1 hydroxyl (–OH) position of a ceramide at the ER or golgi cytoplasmic side. This is further processed by flipping to the luminal side by lipid specific enzymes initially followed by glycosyltransferases and the diversity in composition affects their properties.

4.2.5 Proteoglycans and Glycosaminoglycan

Proteoglycans [6, 11, 12] are glycoproteins expressed in the extracellular matrix (ECM) that, with *N*-glycans and *O*-glycans, and long sugar repeats linked by O-linked glycosylation motifs called as glycosaminoglycans. In contrast to *N*-glycans with 5–12 monosaccharides, a glycosaminoglycan motif includes greater than 80 sugar residues.

Keratan sulphate is sulphated glucosaminoglycans, composed of a poly-*N*-acetyllactosamine chain of disaccharide repeats of GlcNAc or GalNAc, with an uronic acid like galactose, glucuronic or iduronic acid. Glycosaminoglycans including chondroitin sulphate, heparan sulphate, hyaluronan, and keratin sulphate are found in glycocalyx.

4.2.6 Events of Glycosylation

Extrinsic glycosylation involves the role of glycosyltransferases by ageing glycans to acceptor molecules of galactose (Gal)(β3)-GlcNAc, Gal(β3)-GalNAc, Gal(β4)-GlcNAc, and GalNAc; β3 and β4. These processes can form Lewis and sialyl Lewis antigens, sialyl-Tn and Tn antigens, constituting of *O*-glycans with a terminal residue of GalNAc or sialylated GalNAc. This process plays role in leukocyte trafficking in inflammatory responses, haematopoietic cells, and stem cells interaction in the bone marrow [6, 13–17].

4.3 Glycosylation and Stem Cells

Stem cells are progenitor cells, undifferentiated in characters with key properties of pluripotency and self-renewal expressing specific markers. Glycosylation has been studied to play major role in the embryonic development.

O-fucosylation of Notch receptors has been reported to control differentiation of haematopoietic cells. O-fucosylation of apolipoprotein B plays major role in proper midline pattern development in zebrafish by the sonic hedgehog signalling pathway [18].

In mammals different cell types express diversity in their *N*-glycome profiles indicating that cell differentiation is associated with alterations in the *N*-glycan profile. The role of glycosylation and differentiation is stem cells to differentiated lineages are not very well studied. Although some preliminary reports suggest a possible role of changes of glycosylation in cell differentiation, the exact step of differentiation of cells into lineages from stem cells that such differentiation of glycosylation takes place, is unknown and forms a new area of study [6, 18].

A study on the *N*-glycan profiles of two separate clonal populations of mesenchymal stromal cells (MSCs) including clone (Y101), clone (Y202), although revealed changes in *N*-glycome profile [18, 19], upon differentiation did not affect the process of differentiation. Thus *N*-glycans may play role in self-renewal rather than only for cell fate determination [19].

Cell surface glycans, of stem cells interact with several ligands including bone morphogenetic proteins (BMP), epidermal growth factors (EGF), fibroblast growth factors (FGF), Wnt, Hedgehog mediating cell–cell interactions and interactions between cell and extracellular matrix and participates in cell signalling events contributing in maintenance of the stem cell properties stemness, and self-renewal and mediate differentiation into different lineages [18, 20].

Intracellular *O*-linked N-acetylglucosamine has been reported to regulate transcription factors of stemness and phosphorylation of downstream cell signalling components [18, 20]. It is also observed that different glycans involved in maintaining the stem cell property are conserved in mammals through *Drosophila* in evolution and their understanding plays important role in basic research and applied research in understanding of disease initiation and progression with its clinical applications [18, 20]. Glycosylation in stem cells have been extensively studied in Drosophila while reports on the study of glycosylation of stem cells in humans are on the rise. Carbohydrate containing markers have been reported from mammalian pluripotent stem cells including embryonic stem (ES) cells and induced pluripotent cells.

Mouse ES cells are known to express SSEA (Stage-specific embryonic antigen)-1, or Lewis X (LeX) carbohydrate antigen, Galβ1,4(Fucα1,3) GlcNAc, on glycoproteins and glycolipids, but human stem cells lack expression of SSEA-1 [4].

Human ES and iPS cells express markers including SSEA-3, SSEA-48, TRA-1-60, and TRA-1-81, in which SSEA-3 and SSEA-4, reveals structures of R-3GalNAcβ1, 3Galα1,4R′ and NeuAcα2,3Galβ1,3GalNAcβ1,3Galα1,4R′ respectively, on globo-series glycolipids. TRA-1-60 and TRA-1-81 antigens are known to express type 1 lactosamine (Galβ1,3GlcNAcβ1,3Galβ1,4GlcNAc) on *O*-glycan type mucin [4, 21].

SSEA-5 and Globo H on stem cells have revealed expression of H type 1 (Fucα1,2Galβ1,3GlcNAc) or H type 3 (Fucα1,2Galβ1,3GalNAc) on a glycolipid or a *O*-glycan type mucin. However, keratan sulphate has been reported to lack hypersulphated structures [18, 20, 22].

ES cells called as naive cells are known to express β-galactoside α2,6-sialyltransferase (ST6GAL1), globotriaosylceramide Gb3, LacdiNAc, heparan sulphate, and *O*-GlcNAc that are required for the maintenance of the naive state or pluripotent state of the stem cells. LacdiNAc expression is conserved from *Drosophila*, belonging to Class Insecta under an invertebrate Phylum of Phylum Arthropoda to the humans belonging to class Mammalia under SubPhylum Vertebrata, under Phylum Chordata and glycoprotein 130 (gp130) and LacdiNAc on LIF receptor (LIFR) enables maintenance of the naive state by LIF/Stat3signalling [18, 20, 22].

LacdiNAc has been reported to decrease during differentiation in mouse stem cells. In naive mouse ES cells, heparin sulphate (HS) playing role in BMP4/Smad signalling is reported to be expressed more as compared to other glycosaminoglycans including hyaluronan (HA), CS/dermatan sulphate (DS), and glycosaminoglycan expression is reported to increase during differentiation [18, 23].

GlcNAc has been reported to be vital for viability of naive ES cell. Transcription factors, Oct4 and Sox2, are reported to be *O*-GlcNAcylated and their *O*-GlcNAc modifications undergo rapid removal during differentiation [18, 24].

4.3.1 Glycosylation and Embryonic Stem Cells (ESC) Biomarkers

Glycosylated ESCs markers TRA-1-60 and TRA-1-81 and, SSEA 3 and 4 have been associated with pluripotency and stemness. Together with regulating pluripotency, glycosylation has been reported in playing role in regulating reprogramming of stem cell transcription factors in mouse ESCs [18, 25]. *O*-GlcNAc residues have been reported in transcription factors Oct4 and Sox2 in the undifferentiated ESCs of mouse and have been associated with enhancement of transcriptional activity of pluripotency-related genes, leading to pluripotency and reprogramming of embryonic fibroblasts in mouse which is lost on cell differentiation and lineage commitment [18, 24–26].

4.3.2 Glycosylation and Adult Stem Cell (ASC) Biomarkers

ASC maintain tissue homeostasis by replacing damaged worn out cells and tissues. LeX antigen expressed on glycolipid and glycoproteins is reported to regulate activity of neural precursors cells. LeX (SSEA-1) has been reported specifically from neural stem cells (NSCs) in adult mouse [27–29]. SLeX glycans on CD44, a stem cell marker on MSCs has been reported to facilitate their trafficking to bone. CD133, an haematopoietic stem cells (HSCs) marker, and marker for progenitor cells, NSCs, and prostate stem cells [30–34] also reveal glycosylation.

4.3.3 Glycosylation and Self-Renewal of Stem Cell

Stem cells have the dual property of pluripotency and differentiation into lineages. Self-renewal is reported to involve several signalling pathways like leukaemia inhibitory factor (LIF)/signal transducer and activator of transcription (STAT3), bone morphogenic protein 4 (BMP4)/Smad, Wnt/β-catenin, fibroblast growth factor 2 (FGF2) pathway, and activin/nodal in mouse ESCs and human ESCs. Glycosylation has been reported to regulate such pathways in ESCs. Specific cell surface glycan LacdiNAc (GalNAcβ1- 4GlcNAc) has been reported to lead to self-renewal of ESCs in mouse by regulating LIF/STAT3 signalling pathway. Heparin sulphate (HS) has been reported to enable self-renewal and pluripotency of mouse ESCs and regulate Nanog expression through auto or paracrine signalling pathways like the Wnt/β-catenin signalling pathways and play role in modulating pluripotency and lineage commitment of ESCs [23, 35–37].

4.4 Defective Glycosylation and Congenital Disorders

Role of glycans in development finds importance as genetic defects affecting glycosylation process can be lethal for the development. Genetic disorders called as congenital disorders of glycosylation (CDG) reveal affected glycosylation events and has been associated with inflammatory responses, cancer cell metastasis, viral immune escape, regulation of apoptosis and known to affect kidney function.

CDGs are involved in muscular, developmental, and neurological functions, in paediatric population leading to multi-system disorders and could be of two types:

type I and type II CDG [16, 38–46]. Type I CDGs involves structural abnormalities in oligosaccharide on the glycolipid precursor prior to the attachment to the proteins amino acid residue. Type II CDGs involve dysregulation in the *N*-linked branching structure on the newly formed glycoprotein. Under-glycosylated serum transferrin has been reported in children.

4.4.1 CDGs Associated with N-glycan

2 (PMM2)-CDG, the most common CDG manifesting neurological or multi-system phenotype is caused by PMM2 mutations, encoding phosphor mannomutase2 (PMM2) that converts Man-6-phosphate (Man-6-P) to Man-1-P, a precursor for GDP-(Dol-P-Man) synthesis. Severe forms of mutation leading to expression of completely inactive form can be embryonic lethal. Milder forms could lead to disorders with no therapeutic options

MPI-CDG, another common CDG, is caused by mutations in *MPI*, encoding Man-6-P isomerase that converts Man-6-P and fructose-6-phosphate and dietary supplementation of mannose (Man) is an effective treatment for MPI-CDG [16, 38–50].

4.4.2 CDGs Associated with O-glycan

Abnormal Man *O*-glycosylation of α-dystroglycan can lead to congenital muscular dystrophies, called as Walker–Warburg syndrome (WWS) and muscle eye brain disease (MEBD).

O-Mannosylation of α-dystroglycan starts in the ER by the attachment of Man to a Ser or Thr residue, catalysed by the enzyme *O*-mannosyltransferase complex containing protein *O*-mannosyltransferase or POMT1 and POMT2 enzymes [51]. The Man-Ser/Thr is acted upon by multiple enzymes in the Golgi apparatus to generate a long polymeric glycan chain. However, dysregulation in this process affects the cell membranes and causes muscular dystrophy (MD), disease of the central nervous system (CNS) and ocular diseases.

4.4.3 CDG Associated with N-glycanase 1 (NGLY1) Deficiency

NGLY1 deficiency [52–56] due to nonsense mutation is an autosomal recessive disorder of the ER associated protein degradation pathway leading to loss of Peptide: N-glycosidase F or PNGase, affecting translocation of misfolded proteins across the ER membrane into the cytosol for subsequent degradation by the proteasome thereby leading to dysfunction of the nervous system, abnormal tear production, and liver disease.

4.4.4 CDG and Therapies

Therapies of CDC lie in supplementation in diet, such as Man therapy in human phosphomannose isomerise (MPI) congenital disorders of glycosylation (MPI-CDG) which is an autosomal recessive congenital disorder of N-linked glycosylation.

Carbamoyl-Phosphate Synthetase 2, Aspartate Transcarbamylase, and Dihydroorotase are also termed as CAD. CAD-CDG is a rare congenital disorder of glycosylation caused by mutations in the CAD gene with symptoms of anisopoikilocytosis, epileptic encephalopathy, global developmental delay, and normocytic anaemia.

UDP-N-acetylglucosamine-2-epimerase/N-acetylmannosamine kinase is coded by the GNE gene. GNE-CDG or hereditary inclusion body myopathy affecting skeletal muscles other than the quadriceps muscles due to defective UDP-GlcNAc-2-epimerase/Man-NAc kinase.

PGM1-CDG or Phosphoglucomutase 1 deficiency is a rare congenital disorder characterized with orofacial malformations, many patients reveal cleft palate and multi-system abnormalities including cardiomyopathy, sudden death in childhood, coagulopathy, endocrinopathies, and abnormal liver functions due to abnormal protein glycosylation.

Solute carrier family 35 Member C1 or SLC35C1 gene encodes for a GDP-fucose transmembrane transporter (FucT1) expressed in the Golgi apparatus. SLC35C1-CDG or Leukocyte adhesion deficiency type 2 affects the expression of a GDP-fucose transporter, thereby preventing the synthesis of fucosylated glycans, like sialyl Lewis X (sLeX), manifested by clinical symptoms of intellectual disability, short height, a broad flat nasal bridge, frequent infections.

Identification of the type of genetic mutations in individual patients is thus critical and personalized medicine finds importance [57].

4.5 Glycosylation and Cancer Stem Cells (CSC)

Cancer Stem Cells (CSCs) are a set of cells within tumours expressing most common biomarkers of CD24, CD44, and CD133 and revealing stem cell like properties of self-renewal, differentiation, and tumorigenicity properties when transplanted into an animal host and are known to act as sites of initiation of metastasis. Epithelial mesenchymal transition (EMT) can increase role of CSCs. Different regulatory and signalling pathways including Hedgehog, Notch, Wnt/β-catenin, and microRNAs (miRNAs) are known to regulate CSC properties. CSC reveal resistance to chemo and radiation therapy [58].

4.5.1 Glycosylation and Epithelial Mesenchymal Transition (EMT) in Cancer

EMT has been reported to play role in embryonic development and organ formation associated with loss in cells polarity from apical to basal polarity, altered fibroblastic nature, downregulation of epithelial markers, and upregulated expression of mesenchymal markers. EMT has been reported to be associated with pathogenesis, in tumour metastasis [59].

Glycosylation and its aberrant behaviour have been reported to regulate EMT and cell migration. Downregulation of glycosphingolipids (GSLs), gangliotetraosylceramide Gg4, and ganglioside (GM2), and Gg4 synthase (Gg4S) has been reported from induced EMT induced by TGFβ in epithelial cells of mouse and human origin. *O*-glycosylation has been reported to regulate EMT process in prostate epithelial cells in human. Polypeptide N-Acetylgalactosaminyltransferase 14 gene codes for GALNT14, which belongs to subfamily of glycosyltransferases located within the Golgi that catalyses *O*-glycosylation of mammalian proteins by mediating transfer of N-acetyl-D-galactosamine (GalNAc) to its peptide substrate and plays role in regulating pathways of cell proliferation, migration, and invasion by inducing mesenchymal EMT gene expression and by stimulating *matrix metalloproteinase-2* (MMP-2) a type IV collagenase called as activity in breast carcinoma [60].

O-GlcNAcylation of Guanine nucleotide-binding protein subunit beta-2-like 1 protein (GNB2L1) is known to play role in regulation of metastasis by modulating the translation of EMT proteins. This has been associated with the chemoresistance observed of gastric cancer [61].

EMT and altered glycosylation in cancer cells have been thought to be related by activation of hexosamine biosynthetic pathway as cells take up more glucose during EMT through this pathway leading to aberrant expression of cell surface glycosylation (sialylation α2-6,polyLacNAc and fucosylation) and *O*-GlcNAcylation [62–71]. Core 2 O-glycans has been reported to enable tumour metastasis evading immune surveillance by natural killer (NK) cells while Core 3 *O*-glycans or *O*-mannosyl receptors has been reported to repress tumour metastasis by modulating integrin mediated signalling [62–71].

4.5.2 CSC Markers and Glycosylation

Markers of CSC are overexpressed on CSCs from tumours and their glycosylation is reported from CSC [72, 73].

4.5.2.1 CD44

CD44, with several isoforms CD44s and CD44v is a transmembrane glycoprotein and reveals expression in different haematopoietic cells. CD44 is known to play role in lymphocyte homing and glycosaminoglycan, hyaluronic acid (HA) mediated cell adhesion, maintenance of stem cells and stem cell niche (SCN), maintenance of quiescence and resistance to apoptosis.

CD44 has been reported in many tumours playing role in the development of tumour. It has also been reported together with CD24 and epithelial surface antigen (ESA). Both *N*- and *O*-linked glycan modification has been reported in CD44. Differential glycosylation has been reported to differ the molecular weight of the CD44 variants.

CD44 *N*-linked glycosylation sites enable CD44 mediated adhesion to HA in Chinese Hamster Ovary (CHO) cell line ldl-D [60] and ovarian tumours [74–80]. While different modifications of *N*-linked *N*-acetylglucosamine residue, *O*-linked glycans on *N*-deglycosylated cD44, are known to affect CD44 mediated HA binding, incorporation of *N*-acetylgalactosamine into non–N-linked glycans has been reported to increase HA binding by cell surface CD44. Incorporation of α2,3-linked sialic acid on N-linked glycans has been known to inhibit CDD44 binding to HA [79].

4.5.2.2 CD133

Cell surface marker CD133 (Prominin-1) is expressed in HSCs and progenitor cells but lacks expression in adult tissues and its downregulation has been associated with malignant haematopoietic diseases and in myelodysplastic syndrome (MDS). CD133 has been reported to play role in activation of WNT signalling pathway in ESCs, adult, and CSCs. Aberrant expression of glycosylation of CD133 has been associated with stem cell maintenance and α2,3-sialylation of CD133 has been associated with conferring stability in NSCs and glioma initiating cells [81–88].

4.5.2.3 CD24

CD 24, expressed in human B cells, tumours, and mice, is a cell surface glycoprotein linked to glycosylphosphatidylinositol (GPI). It is used as a marker to separate haematopoietic cells from neuronal cells. CD24 with a core of 27 amino acids has been reported to be heavily glycosylated with *N*- and *O*-linked glycans. Overexpression of CD24 has been reported from cancers associated with cell adhesion to collagen, fibronectin, and laminin, tumour cell proliferation, progression, and metastasis. Aberrant glycosylation and hyperglycosylation have been associated with cancer [89–103]. CD 24 modified by sialyl lewis x or sLeX has been known to mediate P-selectin-mediated rolling in breast carcinoma and P-selectin-mediated rolling and lung colonization of tumours.

ESA or epithelial specific antigen or epithelial cell adhesion molecule (*EpCAM*) is a known CSC marker. Co expression of CD44, CD24 and ESA/EpCAM as been reported in pancreatic CSC. Increased *N*-linked glycosylation at Asn198 on ESA is known to regulate its property of surface retention and stability. CD24 expressing cells have been reported from ovarian and colorectal cancers [89–103].

4.5.3 Epithelial Cell Adhesion Molecule (EpCAM), CSCs Maintenance and Differential Glycosylation

Cell surface glycoprotein, EpCAM or ESA is known to upregulate the fatty acid binding protein E-FABP thereby controlling cellular metabolism in tumours. It is reported to be overexpressed in different tumours and CSCs and play role in promoting cell cycle and proliferation of cells by upregulating the proto-oncogene c-myc and cyclin A or E proteins thereby contributing to carcinogenesis. EpCAM plays role in maintenance of human ESCs in the undifferentiated phenotype by regulating genes and transcription factors including c-MYC, KLF4, NANOG, OCT-4, and SOX2.

EpCAM in epithelial cells have been known to be *N*-glycosylated at positions of Asn-74, Asn-111, and Asn-198 in human cells and at Asn-88 and Asn-51 in insect cells and is hyperglycosylated in head and neck cancer, as compared to normal epithelia. *N*-glycosylation of EpCAM has been associated with regulation of apoptosis in breast cancer cells [104–112].

4.5.4 CSCs and Mucins

Mucins predominantly expressed on epithelial cells reveal heavy glycosylation with *O*-linkages at serine, threonine, and proline amino acid-rich regions of variable tandem repeat regions and *N*-linked glycans and is known to confer antimicrobial functions. MUC1, MUC4, MUC5AC, and MUC16 have been implicated in tumour-promoting potential. Aberrant glycosylation is associated with cancer. *N*-terminal cleaved mucin 1 (MUC1) has been reported from undifferentiated human pluripotent stem cells to act as growth factor receptor causing its growth. MUC1 overexpression is observed in human stem cells, cord blood cells and from acute myeloid leukaemia (AML). Aberrant forms of Mucin 4 are reported from many cancers and act as a diagnostic cancer marker. It is known to maintain CSC population in ovarian cancer and is overexpressed in CD133+ CSCs population of pancreatic cancer with gemcitabine resistance. MUC16 (CA125) reveals a tumour-promoting role in ovarian and pancreatic and is a source of CSCs in ovarian cancer [113–128].

4.5.5 Glycosyltransferases (GF) in Stemness of CSCs

The different cell membrane glycans play role in different cellular signalling. Glucosylceramide synthase (GCS) catalysing ceramide glycosylation has been known to play role in conferring stemness of CSCs [129–132]. Overexpression of GCS is reported from breast, colon, leukaemia, and drug-resistant cancer cells. In breast CSCs, ceramide glycosylation and globotriosylceramide (Gb3) and glucosylceramide synthase (GCS) are reported to be overexpressed. Overexpressed Gb3 has been reported to upregulate fibroblast growth factors (FGF-2), CD44 and transcription factor Oct4 and enable maintenance of stemness of breast CSCs by the c-Src/β-catenin signalling pathway. Suppression of GCS has been known to disrupt Gb3 and destroy the breast CSCs [132] and therefore could be a potential strategy in targeting the CSC. Overexpression of β1,4-N-acetylgalactosaminyltransferase III (B4GALNT3), involved in synthesis of LacdiNAc structures has been reported in colon CSCs. In colon CSCs, MGAT5 gene coding for α-1,6-mannosylglycoprotein 6-β-*N*-acetylglucosaminyltransferase A can affect Wnt signalling by modifying Frizzled-7 (FZD-7) and Wnt receptor with β-(1,6)-branched *N*-glycan moiety, thus promoting tumour progression. Fucosyltransferases (Fut1-4), GDP-Fucose transporters, GDP-fucose synthetic enzymes (FX, GMDS) are overexpressed in pancreatic cancer.

In pancreatic and ovarian cancer, the enzyme ST6Gal-1 contributes to cancer stem cell (CSC) phenotypes by regulating the stem cell transcription factors like Sox9 and Slug. In bladder CSCs, GALNT1 is known to regulate the property of self-renewal and plays role in maintenance of bladder CSCs (BCMab1+CD44+).

4.6 Discussion

Although it has been possible to identify the glycans associated with stem cell markers in humans, their exact role in stem cells has yet not been ascertained. Glycosylation of CSCs has been implicated in cancer cell maintenance, stemness, pluripotency, tumour development, cell adhesion, EMT transition and metastasis, cell survival, invasiveness, drug resistance, and cell death or apoptosis. However, the role played by aberration in glycosylation in stem cell maintenance needs further research. Altered glycosylation on CSCs finds importance in the synthesis of glycopeptide based vaccine which may be a step towards eradication of CSCs and cancer cells and inhibition of tumour recurrence.

References

1. Roth Z, Yehezkel G, Khalaila I (2012) Identification and quantification of protein glycosylation. Int J Carbohydr Chem 2012:640923
2. Shental-BechorD LY (2008) Effect of glycosylation on protein folding: a close look at thermodynamic stabilization. Proc Natl Acad Sci 105(24):8256–8261. https://doi.org/10.1073/pnas.0801340105
3. Ghosh S (2020) Sialic acids and sialoglycoconjugates in the biology of life. In: Health and disease. Elsevier, London
4. Berger RP, Dookwah M, Steet R, Dalton S (2016) Glycosylation and stem cells: regulatory roles and application of iPSCs in the study of glycosylation-related disorders. Bioessays 38 (12):1255–1265
5. Zhang X, Wang Y (2016) Glycosylation quality control by the golgi structure. J Mol Biol 428 (16):3183–3193
6. Reily C, Stewart TJ, Renfrow MB et al (2019) Glycosylation in health and disease. Nat Rev Nephrol 15:346–366. https://doi.org/10.1038/s41581-019-0129-4
7. Stanley P, Taniguchi N, Aebi M (2017) Essentials of glycobiology. Spring Harbor Laboratory Press, Cold Spring Harbor, pp 99–111

9. Hart GW, Akimoto Y (2009) The O-GlcNAc modification. In: Varki A, Cummings RD, Esko JD et al (eds) Essentials of glycobiology, 2nd edn. Cold Spring Harbor Laboratory Press, Cold Spring Harbor
11. Lozzo RV, Schaefer L (2015) Proteoglycan form and function: a comprehensive nomenclature of proteoglycans. Matrix Biol 42:11–55
12. Lindahl U, Couchman J, Kimata K, Esko JD (2017) Essentials of glycobiology. Cold Spring Harbor Laboratory Press, Cold Spring Harbor, pp 207–221
13. Lee MM et al (2014) Platelets support extracellular sialylation by supplying the sugar donor substrate. J Biol Chem 289:8742–8748
14. Lee-Sundlov MM et al (2017) Circulating blood and platelets supply glycosyltransferases that enable extrinsic extracellular glycosylation. Glycobiology 27:188–198
15. Manhardt CT, Punch PR, Dougher CWL, Lau JTY (2017) Extrinsic sialylation is dynamically regulated by systemic triggers in vivo. J Biol Chem 292:13514–13520
16. Sadat MA et al (2014) Glycosylation, hypogammaglobulinemia, and resistance to viral infections. N Engl J Med 370:1615–1625
17. Wandall HH et al (2012) The origin and function of platelet glycosyltransferases. Blood 120:626–635
18. Barkeer S, Chugh S, Batra SK, Ponnusamy MP (2018) Glycosylation of cancer stem cells: function in stemness, tumorigenesis, and metastasis. Neoplasia 20(8):813–825
19. Wilson KM, Thomas-Oates JE, Genever PG, Ungar D (2016) Glycan profiling shows unvaried N-Glycomes in MSC clones with distinct differentiation potentials. Front Cell Dev Biol 4:52. https://doi.org/10.3389/fcell.2016.00052
20. Nishihara S (2018) Glycans in stem cell regulation: from Drosophila tissue stem cells to mammalian pluripotent stem cells. FEBS Lett 592(23):3773–3790
21. Vossaert L, Scheerlinck E, Deforce D (2016) Embryonic stem cells: keeping track of the pluripotent status, pluripotent stem cells - from the bench to the clinic. IntechOpen, London. https://doi.org/10.5772/62651
22. Tateno H, Toyota M, Saito S et al (2011) Glycome diagnosis of human induced pluripotent stem cells using lectin microarray. J Biol Chem 286(23):20345–20353
23. Kraushaar DC, Dalton S, Wang L (2013) Heparan sulfate: a key regulator of embryonic stem cell fate. Biol Chem 394(6):741–751
24. Myers SA, Peddada S, Chatterjee N et al (2016) SOX2 O-GlcNAcylation alters its protein-protein interactions and genomic occupancy to modulate gene expression in pluripotent cells. elife 5:e10647
25. Zhao W, Ji X, Zhang F, Li L, Ma L (2012) Embryonic stem cell markers. Molecules 17 (6):6196–6236
26. Shi G, Jin Y (2010) Role of Oct4 in maintaining and regaining stem cell pluripotency. Stem Cell Res Ther 1(5):39
27. Lanctot PM, Gage FH, Varki AP (2007) The glycans of stem cells. Curr Opin Chem Biol 11 (4):373–380. https://doi.org/10.1016/j.cbpa.2007.05.032
28. Yagi H, Kato K (2017) Functional roles of glycoconjugates in the maintenance of stemness and differentiation process of neural stem cells. Glycoconj J 34:757–763
29. Hamouda H, Ullah M, Berger M, Sittinger M, Tauber R, Ringe J, Blanchard V (2013) N-glycosylation profile of undifferentiated and adipogenically differentiated human bone marrow mesenchymal stem cells: towards a next generation of stem cell markers. Stem Cells Dev 22:3100–3113
30. Miraglia S, Godfrey W, Yin AH, Atkins K, Warnke R, Holden JT, Bray RA, Waller EK, Buck DW (1997) A novel five-transmembrane hematopoietic stem cell antigen: isolation, characterization, and molecular cloning. Blood 90:5013–5021
31. Sackstein R, Merzaban JS, Cain DW, Dagia NM, Spencer JA, Lin CP, Wohlgemuth R (2008) Ex vivo glycan engineering of CD44 programs human multipotent mesenchymal stromal cell trafficking to bone. Nat Med 14:181–187

32. Vander Griend DJ, Karthaus WL, Dalrymple S, Meeker A, DeMarzo AM, Isaacs JT (2008) The role of CD133 in normal human prostate stem cells and malignant cancer-initiating cells. Cancer Res 68:9703–9711
33. Zhou F, Cui C, Ge Y, Chen H, Li Q, Yang Z, Wu G, Sun S, Chen K, Gu J (2010) Alpha2,3-Sialylation regulates the stability of stem cell marker CD133. J Biochem 148:273–280
34. Zoller M (2015) CD44, hyaluronan, the hematopoietic stem cell, and leukemia-initiating cells. Front Immunol 6:235
35. He S, Nakada D, Morrison SJ (2009) Mechanisms of stem cell self-renewal. Annu Rev Cell Dev Biol 25:377–406
36. Sasaki N, Okishio K, Ui-Tei K, Saigo K, Kinoshita-Toyoda A, Toyoda H, Nishimura T, Suda Y, Hayasaka M, Hanaoka K (2008) Heparan sulfate regulates self-renewal and pluripotency of embryonic stem cells. J Biol Chem 283:3594–3606
37. Sasaki N, Shinomi M, Hirano K, Ui-Tei K, Nishihara S (2011) LacdiNAc (GalNAcbeta1-4GlcNAc) contributes to self-renewal of mouse embryonic stem cells by regulating leukemia inhibitory factor/STAT3 signaling. Stem Cells 29:641–650
38. Al Teneiji A et al (2017) Phenotypic and genotypic spectrum of congenital disorders of glycosylation type I and type II. Mol Genet Metab 120:235–242
39. Freeze HH, Eklund EA, Ng BG, Patterson MC (2015) Neurological aspects of human glycosylation disorders. Annu Rev Neurosci 38:105–125
40. Freeze HH, Schachter H, Kinoshita T (2017) Essentials of glycobiology. Spring Harbor Laboratory Press, Cold Spring Harbor, pp 569–582
41. Hansen L et al (2015) A glycogene mutation map for discovery of diseases of glycosylation. Glycobiology 25:211–224
42. Hennet T, Cabalzar J (2015) Congenital disorders of glycosylation: a concise chart of glycocalyx dysfunction. Trends Biochem Sci 40:377–384
43. Monticelli M, Ferro T, Jaeken J, Dos Reis Ferreira V, Videira PA (2016) Immunological aspects of congenital disorders of glycosylation (CDG): a review. J Inherit Metab Dis 39:765–780
44. Ng BG, Freeze HH (2018) Perspectives on glycosylation and its congenital disorders. Trends Genet 34:466–476
45. Peanne R et al (2017) Congenital disorders of glycosylation (CDG): quo vadis? Eur J Med Genet 61:643–663
46. Witters P, Cassiman D, Morava E (2017) Nutritional therapies in congenital disorders of glycosylation (CDG). Nutrients 9:E1222
47. Stibler H, Borg S, Allgulander C (1979) Clinical significance of abnormal heterogeneity of transferrin in relation to alcohol consumption. Acta Med Scand 206:275–281
48. Yamashita K et al (1993) Sugar chains of serum transferrin from patients with carbohydrate deficient glycoprotein syndrome. Evidence of asparagine-N-linked oligosaccharide transfer deficiency. J Biol Chem 268:5783–5789
49. Witters P et al (2018) Long-term follow-up in PMM2-CDG: are we ready to start treatment trials? Genet Med. https://doi.org/10.1038/s41436-018-0301-4
50. Yuste-Checa P et al (2017) Pharmacological chaperoning: a potential treatment for PMM2-CDG. Hum Mutat 38:160–168
51. Beltran-Valero de Bernabe D et al (2002) Mutations in the O-mannosyltransferase gene POMT1 give rise to the severe neuronal migration disorder Walker-Warburg syndrome. Am J Hum Genet 71:1033–1043
52. Enns GM et al (2014) Mutations in NGLY1 cause an inherited disorder of the endoplasmic reticulum associated degradation pathway. Genet Med 16:751–758
53. Galeone A et al (2017) Tissue-specific regulation of BMP signaling by Drosophila N-glycanase 1. elife 6:e27612
54. Kong J et al (2018) Mitochondrial function requires NGLY1. Mitochondrion 38:6–16

55. Owings KG, Lowry JB, Bi Y, Might M, Chow CY (2018) Transcriptome and functional analysis in a Drosophila model of NGLY1 deficiency provides insight into therapeutic approaches. Hum Mol Genet 27:1055–1066
56. Zhang L, Ten Hagen KG (2017) Enzymatic insights into an inherited genetic disorder. elife 6: e31127
57. Brasil S et al (2018) CDG therapies: from bench to bedside. Int J Mol Sci 19:E1304
58. Yu Z, Pestell TG, Lisanti MP, Pestell RG (2012) Cancer stem cells. Int J Biochem Cell Biol 44 (12):2144–2151
59. Lamouille S, Xu J, Derynck R (2014) Molecular mechanisms of epithelial-mesenchymal transition. Nat Rev Mol Cell Biol 15(3):178–196. https://doi.org/10.1038/nrm3758
60. Li X, Wang X, Tan Z, Chen S, Guan F (2016) Role of glycans in cancer cells undergoing epithelial-mesenchymal transition. Front Oncol 6:33
61. Cheng S, Mao Q, Dong Y et al (2017) GNB2L1 and its O-GlcNAcylation regulates metastasis via modulating epithelial-mesenchymal transition in the chemoresistance of gastric cancer. PLoS One 12(8):e0182696
62. Cheng S, Mao Q, Dong Y, Ren J, Su L, Liu J, Liu Q, Zhou J, Ye X, Zheng S (2017) GNB2L1 and its O-GlcNAcylation regulates metastasis via modulating epithelial-mesenchymal transition in the chemoresistance of gastric cancer. PLoS One 12:e0182696
63. Freire-de-Lima L, Gelfenbeyn K, Ding Y, Mandel U, Clausen H, Handa K, Hakomori SI (2011) Involvement of O-glycosylation defining oncofetal fibronectin in epithelial-mesenchymal transition process. Proc Natl Acad Sci U S A 108:17690–17695
64. Freire-de-Lima L (2014) Sweet and sour: the impact of differential glycosylation in cancer cells undergoing epithelial-mesenchymal transition. Front Oncol 4:59
65. Guan F, Handa K, Hakomori SI (2009) Specific glycosphingolipids mediate epithelial-to-mesenchymal transition of human and mouse epithelial cell lines. Proc Natl Acad Sci U S A 106:7461–7466
66. Guan F, Schaffer L, Handa K, Hakomori SI (2010) Functional role of gangliotetraosylceramide in epithelial-to-mesenchymal transition process induced by hypoxia and by TGF-{beta}. FASEB J 24:4889–4903
67. Huanna T, Tao Z, Xiangfei W, Longfei A, Yuanyuan X, Jianhua W, Cuifang Z, Manjing J, Wenjing C, Shaochuan Q (2015) GALNT14 mediates tumor invasion and migration in breast cancer cell MCF-7. Mol Carcinog 54:1159–1171
68. Hugo H, Ackland ML, Blick T, Lawrence MG, Clements JA, Williams ED, Thompson EW (2007) Epithelial-mesenchymal and mesenchymal-epithelial transitions in carcinoma progression. J Cell Physiol 213:374–383
69. Lee JM, Dedhar S, Kalluri R, Thompson EW (2006) The epithelial-mesenchymal transition: new insights in signaling, development, and disease. J Cell Biol 172:973–981
70. Oliveira IA, de Queiroz RM, Marinho-Carvalho MM, Sola-Penna M, de Paula IF, Gondim KC, McComb ME (2016) Epithelial mesenchymal transition induces aberrant glycosylation through hexosamine biosynthetic pathway activation. J Biol Chem 291:12917–12929
71. Thiery JP, Acloque H, Huang RY, Nieto MA (2009) Epithelial-mesenchymal transitions in development and disease. Cell 139:871–890
72. Medema JP (2013) Cancer stem cells: the challenges ahead. Nat Cell Biol 15:338–344
73. Mallard BW, Tiralongo J (2017) Cancer stem cell marker glycosylation: nature, function and significance. Glycoconj J 34:441–452
74. Bartolazzi A, Nocks A, Aruffo A, Spring F, Stamenkovic I (1996) Glycosylation of CD44 is implicated in CD44-mediated cell adhesion to hyaluronan. J Cell Biol 132:1199–1208
75. Camp RL, Kraus TA, Pure E (1991) Variations in the cytoskeletal interaction and posttranslational modification of the CD44 homing receptor in macrophages. J Cell Biol 115:1283–1292
76. Catterall JB, Jones LM, Turner GA (1999) Membrane protein glycosylation and CD44 content in the adhesion of human ovarian cancer cells to hyaluronan. Clin Exp Metastasis 17:583–591
77. Hakomori S (1996) Tumor malignancy defined by aberrant glycosylation and sphingo(glyco)-lipid metabolism. Cancer Res 56:5309–5318

78. Rodgers AK, Nair A, Binkley PA, Tekmal R, Schenken RS (2011) Inhibition of CD44 N- and O-linked glycosylation decreases endometrial cell lines attachment to peritoneal mesothelial cells. Fertil Steril 95:823–825
79. Skelton TP, Zeng C, Nocks A, Stamenkovic I (1998) Glycosylation provides both stimulatory and inhibitory effects on cell surface and soluble CD44 binding to hyaluronan. J Cell Biol 140:431–446
80. Yan Y, Zuo X, Wei D (2015) Concise review: emerging role of CD44 in cancer stem cells: a promising biomarker and therapeutic target. Stem Cells Transl Med 4:1033–1043
81. Bidlingmaier S, Zhu X, Liu B (2008) The utility and limitations of glycosylated human CD133 epitopes in defining cancer stem cells. J Mol Med 86:1025–1032
82. Hemmoranta H, Satomaa T, Blomqvist M, Heiskanen A, Aitio O, Saarinen J, Natunen J, Partanen J, Laine J, Jaatinen T (2007) N-glycan structures and associated gene expression reflect the characteristic N-glycosylation pattern of human hematopoietic stem and progenitor cells. Exp Hematol 35:1279–1292
83. Iida H, Suzuki M, Goitsuka R, Ueno H (2012) Hypoxia induces CD133 expression in human lung cancer cells by up-regulation of OCT3/4 and SOX2. Int J Oncol 40:71–79
84. Kato Y, Katoh M (2007) Comparative genomics on PROM1 gene encoding stem cell marker CD133. Int J Mol Med 19:967–970
85. Kemper K, Sprick MR, de Bree M, Scopelliti A, Vermeulen L, Hoek M, Zeilstra J, Pals ST, Mehmet H, Stassi G (2010) The AC133 epitope, but not the CD133 protein, is lost upon cancer stem cell differentiation. Cancer Res 70:719–729
86. Lehnus KS, Donovan LK, Huang X, Zhao N, Warr TJ, Pilkington GJ, An Q (2013) CD133 glycosylation is enhanced by hypoxia in cultured glioma stem cells. Int J Oncol 42:1011–1017
87. Liu Y, Ren S, Xie L, Cui C, Xing Y, Liu C, Cao B, Yang F, Li Y, Chen X (2015) Mutation of N-linked glycosylation at Asn548 in CD133 decreases its ability to promote hepatoma cell growth. Oncotarget 6:20650–20660
88. Osmond TL, Broadley KW, McConnell MJ (2010) Glioblastoma cells negative for the anti-CD133 antibody AC133 express a truncated variant of the CD133 protein. Int J Mol Med 25:883–888
89. Aigner S, Ramos CL, Hafezi-Moghadam A, Lawrence MB, Friederichs J, Altevogt P, Ley K (1998) CD24 mediates rolling of breast carcinoma cells on P-selectin. FASEB J 12:1241–1251
90. Fang X, Zheng P, Tang J, Liu Y (2010) CD24: from A to Z. Cell Mol Immunol 7:100–103
91. Friederichs J, Zeller Y, Hafezi-Moghadam A, Grone HJ, Ley K, Altevogt P (2000) The CD24/P-selectin binding pathway initiates lung arrest of human A125 adenocarcinoma cells. Cancer Res 60:6714–6722
92. Hurt EM, Kawasaki BT, Klarmann GJ, Thomas SB, Farrar WL (2008) CD44+ CD24(-) prostate cells are early cancer progenitor/stem cells that provide a model for patients with poor prognosis. Br J Cancer 98:756–765
93. Kay R, Rosten PM, Humphries RK (1991) CD24, a signal transducer modulating B cell activation responses, is a very short peptide with a glycosyl phosphatidylinositol membrane anchor. J Immunol 147:1412–1416
94. Kristiansen G, Sammar M, Altevogt P (2004) Tumour biological aspects of CD24, a mucin-like adhesion molecule. J Mol Histol 35:255–262
95. Kristiansen G, Winzer KJ, Mayordomo E, Bellach J, Schluns K, Denkert C, Dahl E, Pilarsky C, Altevogt P, Guski H (2003) CD24 expression is a new prognostic marker in breast cancer. Clin Cancer Res 9:4906–4913
96. Lee CJ, Dosch J, Simeone DM (2008) Pancreatic cancer stem cells. J Clin Oncol 26:2806–2812
97. Lee HJ, Choe G, Jheon S, Sung SW, Lee CT, Chung JH (2010) CD24, a novel cancer biomarker, predicting disease-free survival of non-small cell lung carcinomas: a retrospective study of prognostic factor analysis from the viewpoint of forthcoming (seventh) new TNM classification. J Thorac Oncol 5:649–657

98. Ricardo S, Vieira AF, Gerhard R, Leitao D, Pinto R, Cameselle-Teijeiro JF, Milanezi F, Schmitt F, Paredes J (2011) Breast cancer stem cell markers CD44, CD24 and ALDH1: expression distribution within intrinsic molecular subtype. J Clin Pathol 64:937–946
99. Springer T, Galfre G, Secher DS, Milstein C (1978) Monoclonal xenogeneic antibodies to murine cell surface antigens: identification of novel leukocyte differentiation antigens. Eur J Immunol 8:539–551
100. Su D, Deng H, Zhao X, Zhang X, Chen L, Chen X, Li Z, Bai Y, Wang Y, Zhong Q (2009) Targeting CD24 for treatment of ovarian cancer by short hairpin RNA. Cytotherapy 11:642–652
101. Visvader JE, Lindeman GJ (2008) Cancer stem cells in solid tumours: accumulating evidence and unresolved questions. Nat Rev Cancer 8:755–768
102. Yang CH, Wang HL, Lin YS, Kumar KP, Lin HC, Chang CJ, Lu CC, Huang TT, Martel J, Ojcius DM (2014) Identification of CD24 as a cancer stem cell marker in human nasopharyngeal carcinoma. PLoS One 9:e099412
103. Zheng J, Li Y, Yang J, Liu Q, Shi M, Zhang R, Shi H, Ren Q, Ma J, Guo H (2011) NDRG2 inhibits hepatocellular carcinoma adhesion, migration and invasion by regulating CD24 expression. BMC Cancer 11(251):251–259
104. Baeuerle PA, Gires O (2007) EpCAM (CD326) finding its role in cancer. Br J Cancer 96:417–423
105. Munz M, Kieu C, Mack B, Schmitt B, Zeidler R, Gires O (2004) The carcinoma-associated antigen EpCAM upregulates c-myc and induces cell proliferation. Oncogene 23:5748–5758
106. Munz M, Zeidler R, Gires O (2005) The tumour-associated antigen EpCAM upregulates the fatty acid binding protein E-FABP. Cancer Lett 225:151–157
107. Lu TY, Lu RM, Liao MY, Yu J, Chung CH, Kao CF, Wu HC (2010) Epithelial cell adhesion molecule regulation is associated with the maintenance of the undifferentiated phenotype of human embryonic stem cells. J Biol Chem 285:8719–8732
108. Ng VY, Ang SN, Chan JX, Choo AB (2010) Characterization of epithelial cell adhesion molecule as a surface marker on undifferentiated human embryonic stem cells. Stem Cells 28:29–35
109. Chong JM, Speicher DW (2001) Determination of disulfide bond assignments and N-glycosylation sites of the human gastrointestinal carcinoma antigen GA733-2 (CO17-1A, EGP, KS1-4, KSA, and Ep-CAM). J Biol Chem 276:5804–5813
110. Pauli C, Münz M, Kieu C, Mack B, Breinl P, Wollenberg B, Lang S, Zeidler R, Gires O (2003) Tumor-specific glycosylation of the carcinoma-associated epithelial cell adhesion molecule EpCAM in head and neck carcinomas. Cancer Lett 193:25–32
111. Munz M, Fellinger K, Hofmann T, Schmitt B, Gires O (2008) Glycosylation is crucial for stability of tumour and cancer stem cell antigen EpCAM. Front Biosci 13:5195–5201
112. Zhang D, Liu X, Gao J, Sun Y, Liu T, Yan Q, Yang X (2017) The role of epithelial cell adhesion molecule N-glycosylation on apoptosis in breast cancer cells. Tumour Biol 39:1–8
113. Corfield AP (1850) Mucins: a biologically relevant glycan barrier in mucosal protection. Biochim Biophys Acta 2015:236–252
114. Curry JM, Thompson KJ, Rao SG, Besmer DM, Murphy AM, Grdzelishvili VZ, Ahrens WA, McKillop IH, Sindram D, Iannitti DA (2013) The use of a novel MUC1 antibody to identify cancer stem cells and circulating MUC1 in mice and patients with pancreatic cancer. J Surg Oncol 107:713–722
115. Das S, Batra SK (2015) Understanding the unique attributes of MUC16 (CA125): potential implications in targeted therapy. Cancer Res 75:4669–4674
116. Das S, Rachagani S, Torres-Gonzalez MP, Lakshmanan I, Majhi PD, Smith LM, Wagner KU, Batra SK (2015) Carboxyl-terminal domain of MUC16 imparts tumorigenic and metastatic functions through nuclear translocation of JAK2 to pancreatic cancer cells. Oncotarget 6:5772–5787

117. Fatrai S, Schepers H, Tadema H, Vellenga E, Daenen SM, Schuringa JJ (2008) Mucin1 expression is enriched in the human stem cell fraction of cord blood and is upregulated in majority of the AML cases. Exp Hematol 36:1254–1265
118. Gautam SK, Kumar S, Cannon A, Hall B, Bhatia R, Nasser MW, Mahapatra S, Batra SK, Jain M (2017) MUC4 mucin—a therapeutic target for pancreatic ductal adenocarcinoma. Expert Opin Ther Targets 21:657–669
119. Hanson RL, Hollingsworth MA (2016) Functional consequences of differential O-glycosylation of MUC1, MUC4, and MUC16 (downstream effects on signaling). Biomol Ther 6:E34
120. Hikita ST, Kosik KS, Clegg DO, Bamdad C (2008) MUC1* mediates the growth of human pluripotent stem cells. PLoS One 3:e3312
121. Joshi S, Kumar S, Choudhury A, Ponnusamy MP, Batra SK (2014) Altered Mucins (MUC) trafficking in benign and malignant conditions. Oncotarget 5:7272–7284
122. Kaur S, Kumar S, Momi N, Sasson AR, Batra SK (2013) Mucins in pancreatic cancer and its microenvironment. Nat Rev Gastroenterol Hepatol 10:607–620
123. Krishn SR, Kaur S, Smith LM, Johansson SL, Jain M, Patel A, Gautam SK, Hollingsworth MA, Mandel U, Clausen H (2016) Mucins and associated glycan signatures in colon adenoma-carcinoma sequence: prospective pathological implication(s) for early diagnosis of colon cancer. Cancer Lett 374:304–314
124. Mimeault M, Johansson SL, Senapati S, Momi N, Chakraborty S, Batra SK (2010) MUC4 down-regulation reverses chemoresistance of pancreatic cancer stem/progenitor cells and their progenies. Cancer Lett 295:69–84
125. Stroopinsky D, Rosenblatt J, Ito K, Mills H, Yin L, Rajabi H, Vasir B, Kufe T, Luptakova K, Arnason J (2013) MUC1 is a potential target for the treatment of acute myeloid leukemia stem cells. Cancer Res 73:5569–5579
126. Zhang H, Yang Y, Wang Y, Gao X, Wang W, Liu H, He H, Liang Y, Pan K, Wu H (2015) Relationship of tumor marker CA125 and ovarian tumor stem cells: preliminary identification. J Ovarian Res 8:19
127. Zhou N, Wang R, Zhang Y, Lei Z, Zhang X, Hu R, Li H, Mao Y, Wang X, Irwin DM (2015) Staurosporine induced apoptosis may activate cancer stem-like cells (CD44(+)/CD24(−)) in MCF-7 by upregulating Mucin1 and EpCAM. J Cancer 6:1049–1057
128. Zhou N, Zhang Y, Zhang X, Lei Z, Hu R, Li H, Mao Y, Wang X, Irwin DM, Niu G (2015) Exposure of tumor-associated macrophages to apoptotic MCF-7 cells promotes breast cancer growth and metastasis. Int J Mol Sci 16:11966–11982
129. Guo H, Nagy T, Pierce M (2014) Post-translational glycoprotein modifications regulate colon cancer stem cells and colon adenoma progression in Apc(min/+) mice through altered Wnt receptor signaling. J Biol Chem 289:31534–31549
130. Schultz MJ, Holdbrooks AT, Chakraborty A, Grizzle WE, Landen CN, Buchsbaum DJ, Conner MG, Arend RC, Yoon KJ, Klug CA (2016) The tumor-associated glycosyltransferase ST6Gal-I regulates stem cell transcription factors and confers a cancer stem cell phenotype. Cancer Res 76:3978–3988
131. Terao N, Takamatsu S, Minehira T, Sobajima T, Nakayama K, Kamada Y, Miyoshi E (2015) Fucosylation is a common glycosylation type in pancreatic cancer stem cell-like phenotypes. World J Gastroenterol 21:3876–3887
132. Gupta V, Bhinge KN, Hosain SB, Xiong K, Gu X, Shi R, Ho MY, Khoo KH, Li SC, Li YT (2012) Ceramide glycosylation by glucosylceramide synthase selectively maintains the properties of breast cancer stem cells. J Biol Chem 287:37195–37205

Sialylation of Stem Cells

5

Contents

Abstract

Sialylation is the post-translational modification involving addition of the sialic acid to terminal end of glycoproteins and lipid by covalent bonding. It is a biologically significant modification observed in embryonic development, reprogramming, neurodevelopment, oncogenesis, cell–cell adhesion, cell–cell communication, and immune responses. Studies in the last decade have revealed that sialic acid and the enzymes involved in sialylation has been reported to be associated and is vital in dictating and regulating stem cell fate, maintenance of pluripotency, embryonic stem cell survival, and cellular differentiation process. In this chapter, we highlight studies on sialylation and its role in deciphering the understanding of the complex stem cell biology.

S. Chowdhury, S. Ghosh, *Stem Cells*, https://doi.org/10.1007/978-981-16-1638-9_5

Keywords

GNE · CMAS · Sialic acid · Sialyltransferase · NCAM · Polysialylation

Abbreviations

BM	Bone marrow
BM-MSC	BM derived MSC
CAMs	Cell adhesion molecules
CSCs	Cancer stem cells
ECM	Extracellular matrix
EGF	Epidermal growth factor
ESCs	Embryonic stem cells
Gal	Galactose
GalNAc	*N*-acetylgalactosamine
GPI	Glycosylphosphatidylinositol
hiPSCs	Human induced pluripotent stem cells
HPCs	Haematopoietic progenitor cells
hPSC	Human pluripotent stem cells
HSCs	Haematopoietic stem cells
IgSF	Ig superfamily
iPSC-CM	iPSC-derived cardiomyocytes
KDn	2-keto-3-deoxy-nononic acid
KO	Knock out
LECAMs	leukocyte endothelial cell adhesion molecules
MSCs	Mesenchymal stem cells
Neu	Neuraminic acid
Neu5,9Ac2	*N*-acetyl-9-*O*-acetylneuraminic acid
Neu5Ac	*N*-acetylneuraminic acid
Neu5Ac	Neuraminic acid
Neu5Ac9Lt	*N*-acetyl-9-*O*-lactylneuraminic acid
Neu5Ac9P	*N*-acetylneuraminic acid-9-phosphate
Neu5Gc	*N*-glycolylneuraminic acid
p	placenta
PC	Podocalyxin
pMSC	placenta derived MSC
PSA	Polysialic acid
PSL	*Polyporus squamosus* lectin
Sia	Sialic acids
SIA8D	CMP-N-acetylneuraminate-poly-alpha-2,8-sialyltransferase
Sia-Le^x	α2,3-sialylated $Lewis^x$ structures
sLe(x)	sialyl Lewis X
ST6Gal-1	α2,6-sialyltransferase

5.1 Introduction

Sialic acids (Sia) are *N*- or *O*-substituted derivatives of neuraminic acid (Neu) [9-Carbon (C), backbone α-keto monosaccharide] expressed widely on terminal carbohydrates on cell surface glycoproteins and glycosphingolipids (gangliosides), glycolipids of eukaryotic cells also sometimes found to be capping side chains of glycosylphosphatidylinositol (GPI) anchors. Due to its discovery from salivary mucins, they were termed as Sia. Sias are known to contain about 50 members with enormous diversity in structure, including four core Sia molecules including *N*-acetylneuraminic acid (Neu5Ac), *N*-glycolylneuraminic acid (Neu5Gc), 2-keto-3-deoxy-nononic acid (KDn), and Neu discussed in subsequent sections and the structural diversity of Sias have been known to give uniqueness to the cells in structure and function like recognition by antibodies and by different Sia-binding lectins.

The electro-negatively charged Sias have been reported to be expressed on the external surface of cell membranes, interior of lysosomal membranes and in secreted glycoproteins in blood and mucins. They play pivotal role in the stabilization of enzymes, proteins, molecules membranes, and modulating interactions with the environment. They also participate in binding and the transportation of ions and drugs, enhance viscosity of the mucins, conferring protection to molecules and cells from proteases and glycosidases digestion, regulating the receptor affinity, and modulate processes involved in transmembrane signalling, fertilization, growth, and differentiation and play role in free-radical scavenging as an antioxidative effect, on endothelia of blood vessels. Sia *O*-acetylation has been reported to inhibit apoptosis. They can cover antigenic sites, receptors and the last but one galactose residues or recognition sites. When the Sia is lost, cells and molecules can be bound, by macrophages and hepatocytes, by Gal-recognizing receptors, either taken up and degraded as in serum glycoproteins and blood cells. Sias can act as ligands for many microbial and animal lectins. Sias modification can affect ligand functions like 9-*O*-acetylation or *N*-acetyl-hydroxylation of Neu5Ac creating new receptor functions or decrease the affinity of binding.

Sialylation is the post-translational modification involving addition of the sialic acid to terminal end of glycoproteins and lipid by covalent bonding. It is a biologically significant modification observed in embryonic development, reprogramming, neurodevelopment, oncogenesis, cell–cell adhesion, cell–cell communication, and immune responses [1, 2]. Studies in the last decade have revealed that sialic acid and the enzymes involved in sialylation has been reported to be associated and is vital in dictating and regulating stem cell fate, maintenance of pluripotency, embryonic stem cell survival, and cellular differentiation process. In this chapter, we highlight studies on sialylation and its role in deciphering the understanding of the complex stem cell biology.

5.2 Sialic Acids and Their Derivatives in Humans

Sias with their electro-negatively charged, acidic monosaccharides, occupying the terminal position on macromolecules and cell membranes, being involved in many biological and pathological phenomena, contribute to the enormous structural diversity of complex carbohydrates. They are the major constituents of the lipids and proteins of cell membranes and other secreted biomolecules. The variety of glycan chains is increased due to biosynthesis of numerous types of Sias (comprising a family of >50 neuraminic acid derivatives), thus forming foundation of the "third language of life" (after nucleic acids and proteins as first and second languages, respectively) used by the glycoconjugates [2–4]. In contrast to other mammals, the glycocalyx of human cells lacks the sialic acid N-glycolylneuraminic acid (Neu5Gc, 5) and has amplified richness of its precursor Neu5Ac [5].

5.3 Distribution in Nature and Chemical Diversity

Sias comprise of a family of monosaccharides with ~50 members. All of them are derivatives of neuraminic acid (5-amino-3,5-dideoxy-D-*glycero*-D-*galacto*non-2-ulopyranosonic acid, Neu). They possess various substituents at the amino or hydroxyl groups (Fig. 5.1). The amino group of Neu in most cases is either acetylated or glycolylated, while one to many acetyl groups may be found in all non-glycosidic hydroxyl residues. Usually, Sias have only one *O*-acetyl group, frequently at C-9, but di- and tri-*O*-acetylated Sias are also known to exist, particularly in mucins from human colon and bovine submandibular gland. Phosphoryl or lactyl residues may occur at C-9 and methyl or sulphate groups at C-8. These different substituents may be combined (e.g. 8-*O*-methyl with 9-*O*-acetyl and *N*-glycolyl) producing many sorts of Sias distributed all over the animal kingdom. Other than these, unsaturated Sias as well as anhydro, lactone, and lactam forms have also been recognized from numerous biological sources. 5-Desamino-5-hydroxy-neuraminic acid or KDn is found in micro-organisms and animals. Neu itself, the de-*N*-acetylated product of Neu5Ac, are additional components of mammalian glycoconjugates [2–4].

The natural Sias are seen in the glycosidic linkages of oligosaccharides, polysaccharides (polysialic acids), gangliosides, glycoproteins, and lipopolysaccharides. In most cases they are bound to galactose (Gal) or *N*-acetylgalactosamine (GalNAc) groups via α2,3 or α2,6 bonds or to Sia itself by α2,8 or, rarely, α2,9 linkages. *N*-acetylneuraminic acid-9-phosphate (Neu5Ac9P) is an intermediate in Sia biosynthesis. Neu5Ac, *N*-acetyl-9-*O*-acetylneuraminic acid (Neu5,9Ac2), and Neu5Gc are the three most often occurring members of the Sia family, probably followed by *N*-acetyl-9-*O*-lactylneuraminic acid (Neu5Ac9Lt). Only Neu5Ac is omnipresent, while the remaining are absent in many species [1–4, 7].

Neu5Gc occurs regularly in the animal kingdom and in some human tumours, in minute quantities. In contrast, *O*-acetylated Sias are more abundant, occurring in humans, birds, reptiles, and bacteria. No studies have found all kinds of Sias in a

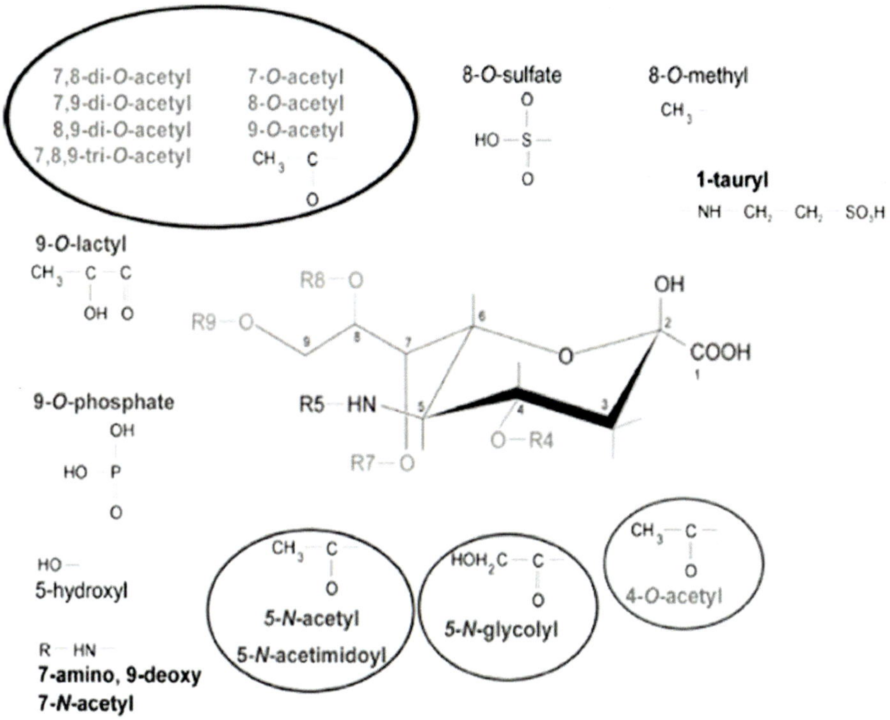

Fig. 5.1 The sialic acid comprising of over 50 naturally-occurring forms and many naturally-occurring sialic acids, many non-natural forms. Image and legend adapted with permission from Badr HA, AlSadek DM, El-Houseini ME, Saeui CT, Mathew MP, Yarema KJ, Ahmed H. Harnessing cancer cell metabolism for theranostic applications using metabolic glycoengineering of sialic acid in breast cancer as a pioneering example. Biomaterials. 2017 Feb;116:158–173 [6]

single cell or an organism. The dispersal depends on the animal and cell species as well as on the function of a cell and is tightly regulated at the genetic level. In humans, the number of Sias is reduced, with prevailing Neu5Ac, followed by derivatives which are *O*-acetylated and *O*-lactylated at the Sia side-chain [1–5, 7].

5.4 Biosynthesis and Metabolism

Neu5Ac is the maternal molecule of all other Sia derivatives, synthesized in cytosol by the condensation of phosphoenol-pyruvate with *N*-acetylmannosamine-6-phosphate for Neu5Ac9P followed by dephosphorylation. The schematic presentation of the biosynthesis and the metabolism of Sias have been shown in Figs. 5.2 and 5.3, respectively [1–5, 7, 9].

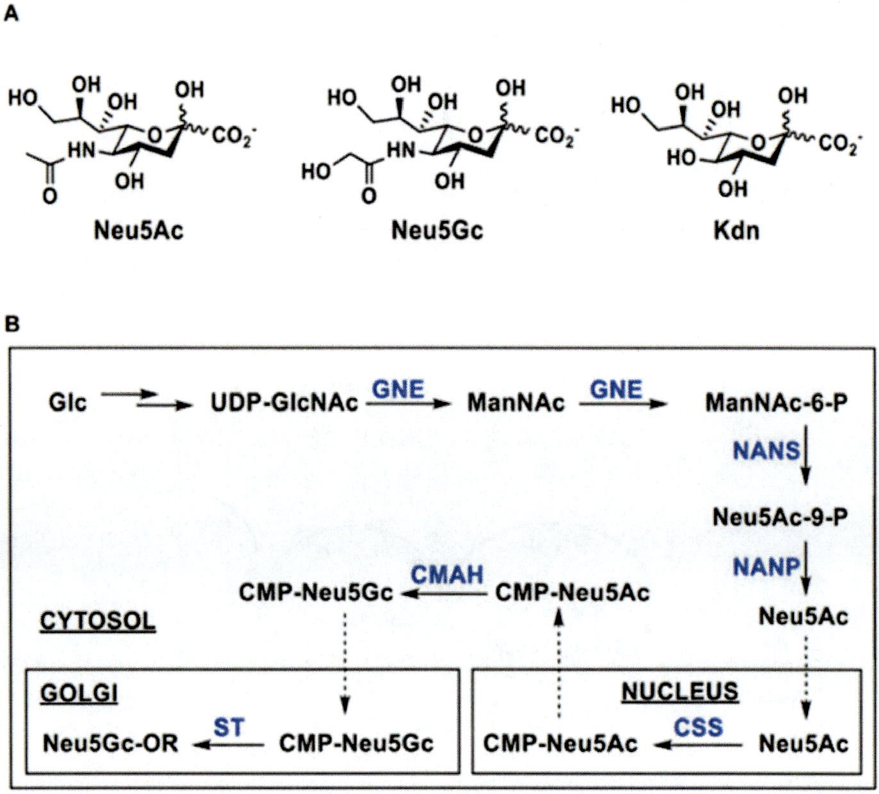

Fig. 5.2 (**a**) Schematic presentation of the pathway of the biosynthesis of Neu5Ac, Neu5Gc, 3-deoxy-D-glycero-D-galacto-2-nonulosonic acid (Kdn); (**b**) Biosynthesis of Neu5Gc and its sialosides in eukaryotic cells. Enzymes and abbreviations: GNE, UDP-GlcNAc 2-epimerase/ManNAc-6-kinase; NANS, Neu5Ac-9-P synthetase; NANP, Neu5Ac-9-P phosphatase; CSS, CMP-sialic acid synthetase; CMAH, cytidine 5′-monophosphate-N-acetylneuraminic acid hydroxylase. Image and legend adapted from open access article Kooner AS, Yu H and Chen X (2019) Synthesis of N-Glycolylneuraminic Acid (Neu5Gc) and Its Glycosides. Front. Immunol. 10:2004. doi: 10.3389/fimmu.2019.02004 [8]

5.5 Biological Role

The external location of Sias on glycoproteins and gangliosides, either alone or in oligo- or polymeric form, and congruently on the outer cell membranes implies a powerful impact in cell biology. These acidic monosaccharides easily form network with the components of other cell surfaces, extracellular substances, and effector molecules. They are involved in a multiplicity of cell signalling proceedings. The biology of Sias can be viewed from their dual role, which is that they either cover the recognition sites or represent a biological target, allowing recognition by a receptor protein, a lectin, thus representing a ligand or counter-receptor. The latter role may

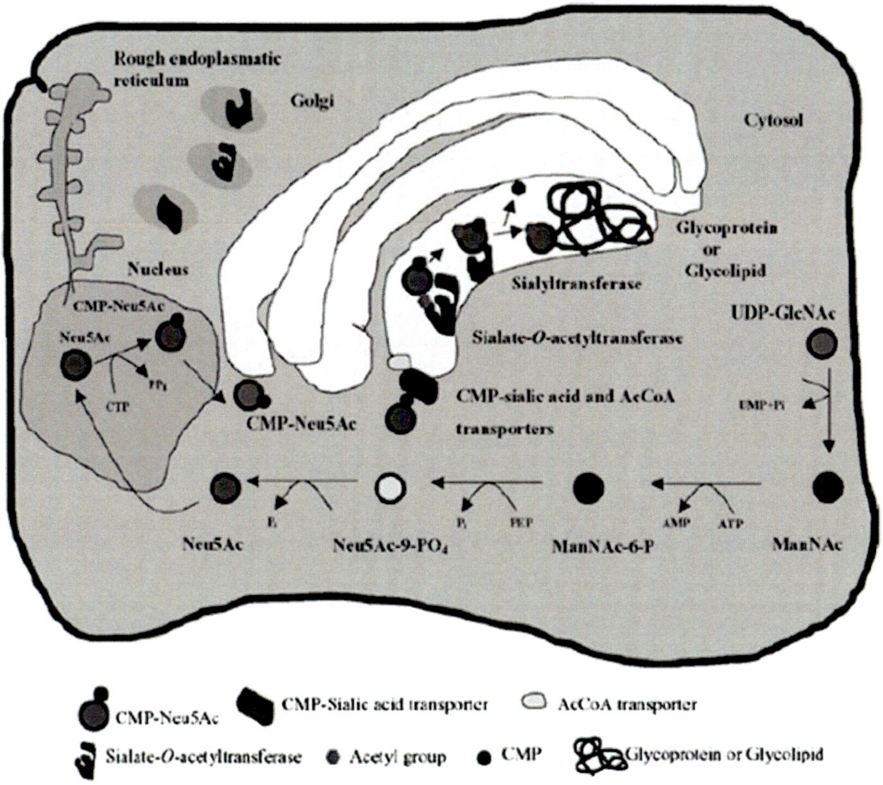

Fig. 5.3 Metabolism of sialic acids, biosynthesis of Sia and their *O*-acetylation in eukaryotes. The enzymatic reactions involved in the biosynthesis of sialic acid, activation, transfer, modification, and catabolism are shown with their intracellular localization. After Neu5Ac synthesis in the cytosol and activation in nucleus, CMP- Neu5Ac is transported into the Golgi by a transporter and is *O*-acetylated by the sialate-7-*O*-acetyltransferase and is then transported to glycoproteins or glycolipids by sialyltransferase. These transporters and transferases comprise together a protein complex. The model is based on studies from bovine submandibular gland and may differ in human lymphocytes. Image and legend adapted with permission from Mandal C, Schwartz-Albiez R, Vlasak R. Functions and Biosynthesis of *O*-Acetylated Sialic Acids. Top Curr Chem. 2015;366:1–30. doi: 10.1007/128_2011_310

be moderated or even eliminated by Sia substituents, most effectively by *O*-acetyl groups. Thus, esterification of Sias can mask their ligand function. Sias are perfect mediators of fine-tuning of the cell behaviour. They are also ideal ligands for binding of and infection of mammalian cells by micro-organisms (Fig. 5.4).

The anti-recognition effects of Sias are explained by the negative charge in amalgamation with a large, hydrophilic molecule. For example, the repulsion of cells and molecules, such as erythrocytes, by the negatively charged Sia residues, the masking of the penultimate sugars that has been designed to be recognized by receptors, like Gal, the protecting of antigenic sites in macromolecules either soluble or bound to cell membranes, the anti-proteolytic effect in glycoproteins, and the

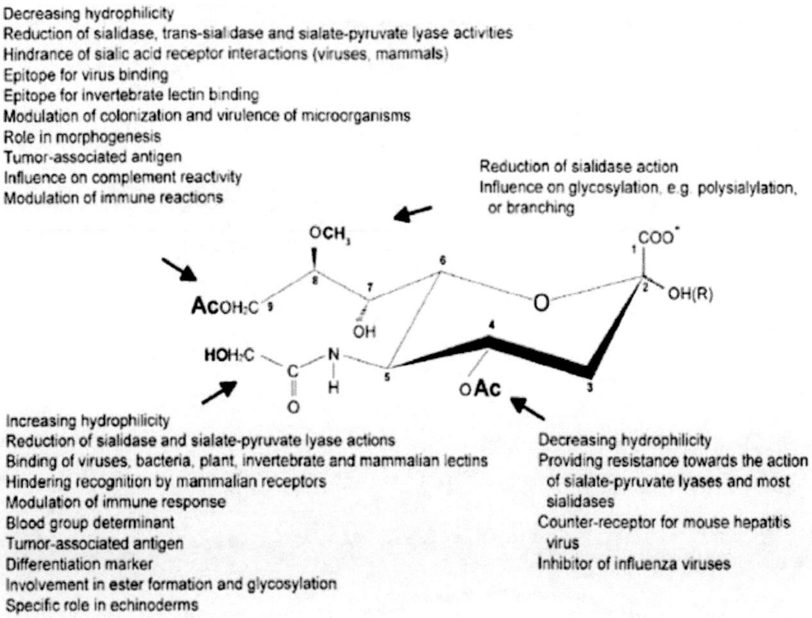

Fig. 5.4 Biological significance of the sialic acid substituents. Image and legend reproduced with permission from Schauer R. Achievements and challenges of sialic acid research. Glycoconj J. 2000, 17(7–9):485–499 [3]

hindrance of the action of some endoglycosidases. Desialylation may lead to binding to Gal-specific lectins or to identification of macromolecules and cells by the complement and immune system. Demasking of Gal can target molecules and cells to specific sites, frequently to help degradation, necessary for physiological or pathological purpose. Eradication of malignant cells by macrophages can also be in this way, and over-sialylation, often as Sialyl-Lewis^x, frequently detected in these cells, protect them from humoral and cellular defence systems, thereby increasing their malignancy and metastatic potential. Polysialic acids expressed on some bacteria and mammalian cells are powerfully anti-adhesive, thus regulating cellular contacts and movements.

A second set of examples confirming the masking effect of Sia is the generation of auto-antibodies after desialylation of cell membrane by bacterial or viral sialidases. This exposure of cellular antigens often leads to chronic diseases like neuronal disorders or glomerulonephritis. Sias and other carbohydrates may be antigenic determinants, especially in gangliosides, but they shield antigenic sites more frequently and thus deteriorate the immunoreactivity. Sias render cells as "self", not allowing recognition by one's own immune system or by macrophage lectins, e.g., the damage of these monosaccharides makes these cells "non-self" and therefore susceptible. Therefore, Sias are considered to be members of the innate immune system.

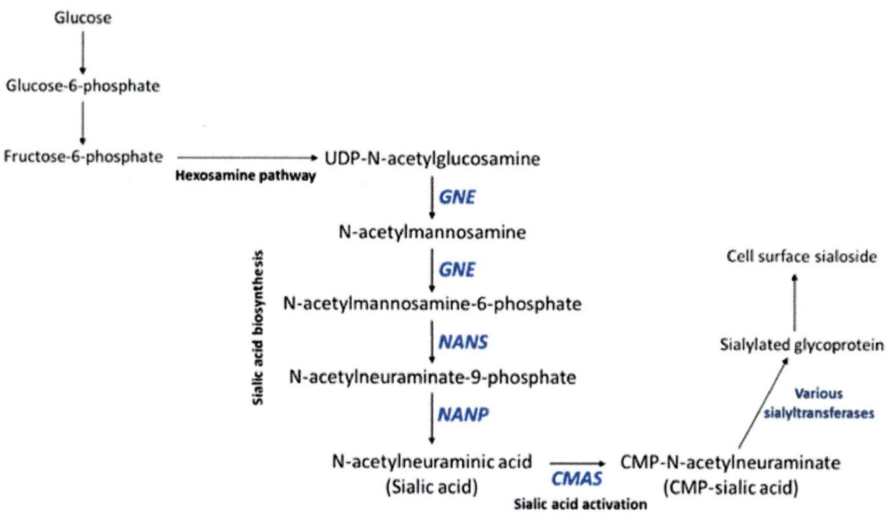

Fig. 5.5 Metabolism of sialic acid which initiates from uridine diphosphate (UDP)-N-acetylglucosamine, produced by the hexosamine pathway. GNE catalyses sialic acid biosynthesis which on coupling to cytidine monophosphate (CMP) by CMAS, produces CMP-sialic acid which is used by sialyltransferases for sialylation of glycoproteins (Image produced from Teoh ST, Ogrodzinski MP, Ross C, Hunter KW and Lunt SY (2018) Sialic Acid Metabolism: A Key Player in Breast Cancer Metastasis Revealed by Metabolomics. Front. Oncol. 8:174 [10], under Creative Commons Attribution License (CC BY)

In addition to masking, Sias participate in numerous recognition processes. It was primarily observed in micro-organisms, who use cell surface carbohydrates for attaching to their host cells. Sias are the most frequent ligands for pathogenic and non-pathogenic viruses, bacteria, and protozoa. Human influenza C virus, some corona virus, torovirus, and rotavirus species recognize Neu4, 5Ac2 or di-*O*-acetylated Sias [1–8] (Fig. 5.5).

5.6 Detection of Sialic Acids and Their Derivatives

Several Sia-recognizing lectins have been isolated from plants, mushrooms, and animals. Some of these are specialized for *O*-acetylated Sias or Neu5Gc. The second category and probably the most important Sia-recognizing receptors in higher animals are the "siglecs". The selectins form the third cluster of lectins recognizing mammalian Sias.

5.7 Lectins

The naturally occurring proteins and glycoproteins, which non-covalently bind to carbohydrate residues selectively are termed lectins. Lectin is defined as carbohydrate binding protein of non-immune origin which agglutinates cells and/or

Table 5.1 Lectins with their preferred binding glycotopes (adapted with modification from [5])

Binding glycotope	Lectin occurrence
N-Glycolylneuraminic acid	Murine and ape CD22 (siglec-2); *Crab Scylla serrata* and *Carcinoscorpius rotunda;* clam Anadara granosa; snail *Pila globosa*; mulberry *Morus alba*; mushrooms *Hericium erinaceum* and *Chlorophyllum molybdites*; malaria parasite *Plasmodium reichenowi*; *Escherichia coli* strain K99; transmissible gastroenteritis virus; rotavirus strains
N-Acetyl-9-*O*-acetylneuraminic acid	Crab *Cancer antennarius*; crustacean *Liocarcinus depurator*; snails *Achatina fulica* ("Achatinin-H") and *Cepaeahortensis*; influenza C virus Strains of coronavirus, isavirus, rotavirus, and torovirus
Neu5Ac	*Limulus polyphemus* agglutinin (LPA)
Neu5Acα2–3Gal	*Maackia amurensis* agglutinin (MAA)
Neu5Acα2–6Gal/GalNAc	*Sambucus nigra* agglutinin (SNA)
Neu5,9Ac2α2–6β-D-GalNAc	Achatinin-H

precipitates polysaccharides or glycoconjugates [11]. They can be used as probes to locate distinct monosaccharides and oligosaccharides from the astoundingly varied mixture of carbohydrate residues on or in the cells and surrounding extracellular matrix (ECM). Table 5.1 shows some lectins isolated from plant, fungus, and mushroom species [1–5, 7, 9].

5.8 Siglecs

Siglecs belong to the Ig superfamily (IgSF) with repeating extracellular domains of flexible length. Thirteen members of this family have been discovered. The first well-characterized type was macrophage sialoadhesin, associated with binding and nursing of maturing blood cells, CD22 on B lymphocytes, accountable for "cross-talk" of the immune-competent cells with the T-lymphocytes, and myelin-associated glycoprotein on oligodendrocytes and Schwann cells, contributing to the growth and myelination of neurites. Additional siglecs are involved in regulation of the functions of WBC and macrophages. CD22 mediates signalling in B-cell regulation. Mammalian siglecs differ in their identification of Sias and can discriminate between the glycosidic linkages [1–5, 7, 9].

5.9 Selectins

The selectins (like LECAMs—leukocyte endothelial cell adhesion molecules) form an additional group of mammalian Sia-recognizing lectins harbouring repeating domains. They are a family of divalent cation dependent glycoproteins comprising of leukocyte (L)-selectin, endothelial (E)-selectin, and platelet (P)-selectin. The extracellular domain of each comprises of a carbohydrate recognition motif, an epidermal growth factor (EGF)-like motif, and varying numbers of a short-repeated domain related to complement-regulatory proteins. Subsequent to the transmembrane region, each has a short cytoplasmic domain. Expressed primarily on endothelial cells, WBC, and platelets, they contribute in the primary stages of leukocyte adhesion. Cytokines play a pivotal part in selectin expression. Selectins preferably identify α2,3-sialylated Lewis ^{x} structures (Sia-Le^{x}), a fucosylated tetrasaccharide attached to glycolipids or glycoproteins. Since these oligosaccharide moieties are expressed on tumour cells (e.g. on variant isoforms of CD44), selectins may be involved in the formation of metastases [1–5, 7, 9].

5.10 Stem Cells and Sialylation

Mesenchymal stem cells (MSCs) are the multipotent progenitor cells which can be easily isolated and cultured. They mature into several tissue types, including fat, migrate to diseased organs, secrete regenerative factors, and have immunosuppressive properties. This makes them ideal for regenerative medicine [12]. An all-inclusive glycan profiling of adipose-derived MSCs using high-density lectin microarray confirmed that *α*2–6-sialylation is a differentiation marker of these cells [12–14]. Sialyl-lactotetra is an indicator of undifferentiated human pluripotent stem cells (hPSC) [15] while GD1a ganglioside is a novel marker of neural differentiation [10]. Podocalyxin (PC) is a deeply sialylated transmembrane protein of the glomerular podocytes, acting as a universal glyco-carrier and is associated with several functions of normal tissue, human embryonic stem cells (ESCs), human induced pluripotent stem cells (hiPSCs), and cancer cells. PC is a chief pluripotent marker of the ESCs and hiPSCs [16, 17]. Assemblage of the sialyllactosaminyl-decorated CD44s glycovariant is a preserved feature of the MSCs isolated from adipose tissue and marrow, thus identifying a CD44 glyco-signature of the MSCs which is supportive of the role of cell surface α(1,3)-fucosylation in programming migration of systemically administered MSCs to sites of tissue inflammation or damage [18]. *N*-glycans H6N5F1 and H7N6F1 are suggestively overexpressed in undifferentiated MSCs, while H3N4F1 and H5N4F3 are upregulated in adipogenically differentiated MSCs. These glycan structures are auspicious candidate markers for identification of the MSCs and their adipogenic progeny [12].

CD133 is extensively used as a marker for the identification, isolation, and characterization of normal and cancer stem cells. The dynamic alternation of the glycosylation of CD133 helps in the former process and is related with cell differentiation [19].Glycosylation pattern linked to MSCs include linear poly-N-acetyllactosamine chains, high-mannose type *N*-glycans and α2–3-sialylation. MSCs express sialyl-Lewis^{x} and SSEA-4 epitopes. Typical glycosylation features seen in differentiated osteoblasts include abundant sulphate ester modifications [20]. *N*-glycans isolated from hiPSCs, iPSC-derived cardiomyocytes (iPSC-CM), and original C57BL/6 mouse myocardium showed high-mannose type (immature) *N*-glycans on the hiPSCs, it decreased with cardiomyogenic differentiation, but did not attain the low levels noted in heart. A comparable decrease in neutral *N*-glycans and an upsurge in fucosylated or sialyl *N*-glycans was observed. Few physical differences have been perceived between iPSC-CM and the heart. NeuGc structures were completely lacking in iPSC-CM, while its frequent was observed in the heart, consistent with the expression of the cytidine monophosphate-N-acetylneuraminic acid hydroxylase. Additionally, numerous glycans containing Galα1–6 Gal, seldom identified in the other cells, can be detected in the iPSC-CM [21].

Potential trafficking mechanisms used by MSCs isolated from bone marrow (BM) or the placenta (p) showed that expression of an extensive range of cell adhesion molecules (CAMs) like CD44 and β1-integrins (CD29). Array data showed that MSCs derived from the above sources expressed mRNA for the CAMs like CD166 (ALCAM), CD54 (ICAM-1), E-cadherin, CD56 (NCAM), CD49a, b, c, e, and f (integrins α1, 2, 3, 4, and 6), CD106 (VCAM-1), integrin alpha11, CD51 (integrin αV), and CD29 (integrins β1). Functional binding of human pMSC, but not BM-MSC to VCAM-1 was confirmed using recombinant chimeric constructs. Neither BM nor placenta derived MSCs expressed ligands to endothelial selectins, namely PSGL-1 or sialyl-Lewis X (sLe(x)) carbohydrates and neither could bind functionally to chimeric constructs of the endothelial selectins CD62E (E-selectin) and CD62P (P-selectin). Human MSCs also expressed a limited range of transferases essential for the expression of sLe(x), lacked expression of fucosyltransferases IV or VII. Placental MSC, express mRNA for the chemokine receptors CCR1 and CCR3, and both BM-MSC and pMSC express mRNA for CCR7, CCR8, CCR10, CCR11, CXCR4, and CXCR6. Intracellular chemokine receptor protein expression of CCR1, CCR3, CXCR3, CXCR4, and CXCR6 was detected in both human BM-MSC and pMSC. Cell surface expression of chemokine receptors are restricted with only the CXCR6 exhibiting a strong signal on human pMSC and BM-MSC. Although cell surface expression of CXCR4 was not noticed, MSCs migrated to its ligand, in response CXCL12 (SDF-1). Thus, human BM-MSC and pMSC exhibit an identical profile for chemokine receptor molecules and CAMs at the mRNA and protein levels. However, at functional level, human pMSC use VLA-4-mediated binding more than BM-MSC and therefore, has better engraftment properties compared to the latter in vivo [22].

An enhanced cell surface expression of P- and E-selectins, α2,3-linked sialic acid and intercellular adhesion molecule along with reduced expression of L-selectin distinguished CD34+ haematopoietic stem cells (HSCs) of umbilical cord blood (UCB) from the leukaemic samples. High expression of *O*-acetylated sialoglycoproteins, a trademark of the lymphoblasts, is radically reduced in the CD34+ HSCs of UCB and is substantiated by the low activity of sialylate-*O*-acetyltransferase and high sialidase activity [23].

Cell recognition molecule L1 (CD171) performs a significant role in neuronal differentiation, survival, migration, neurite outgrowth, synaptic plasticity, myelination, and regeneration after damage. Overexpression of L1 augments cell survival and proliferation of mouse ESCs by endorsing the expression of FUT9 and ST3Gal4, which increase cell surface sialylation and fucosylation. Neuronal differentiation from ESCs is promoted by L1 through similar process [24].

E-selectin plays a pivotal role in facilitating the homing of tissue-specific T cells into skin, and of primitive haematopoietic progenitor cells (HPCs) into BM. Although the E-selectin ligand(s) of human HPCs are unknown, a glycoform of PSGL-1 (CLA) serves as the major ligand of E-selectin on human T lymphocytes. Later it was observed that the HPCs express a formerly unrecognized E-selectin ligand, CD44. The E-selectin ligand activity of CD44 is due to the elaboration of sialylated and fucosylated binding determinants on *N*-glycans. This glycoform of CD44 is expressed on primitive CD34+ HPCs, but not on mature hematopoietic cells. At physiologic condition, CD44 mediates E-selectin-dependent rolling interactions over a wider shear range than that of PSGL-1, promoting human HPC rolling interactions on E-selectin expressed on human BM endothelial cells. These discoveries propose novel insights into the physiology and structural biology of CD44, and also into the molecular basis of E-selectin-dependent adhesive interactions that direct migration of human HPC to BM [25].

Sialylation is indispensable for the early embryonic development and deletion of UDP-GlcNAc 2-epimerase, a rate-limiting enzyme in sialic acid biosynthesis, is embryonically deadly. α2,6-sialyltransferase (ST6GAL1) activity is crucial for reprogramming of somatic cell; its downregulation is related with decreased reprogramming efficiency. In addition, sialylation patterns and levels are transformed during cancer progression, indicating the potential of these sialylated molecules as cancer biomarkers. Thus, sialylation is associated with crucial cell fate decision [1].

Glycans occupy cell surface boundary between the hematopoietic cells and their marrow niches and glycosyltransferases are found within the intracellular secretory apparatus generating glycans. An alternative pathway to generate these cell surface glycans by remotely occurring glycosyltransferases has been proposed wherein extracellular ST6Gal-, from liver, enables the synthesis of the cell surface α2,6-linked sialic acids on HSCs and HPCs in the BM. Lectin binding studies involving *Sambucus nigra* agglutinin (SNA) and *Polyporus squamosus* lectin (PSL) together with mass spectrometric studies have confirmed such findings. HSCs and HPCs, phenotypically showing Lin-c-Kit+ and Lin-Sca-1 + c-Kit+ populations, express insignificant endogenous ST6Gal-1 and animals with lesser circulatory ST6Gal-1 have marrow Lin-Sca-1 + c-Kit+ cells with reduced SNA reactivity. BM chimeras have revealed that α2,6-sialylation of HSCs and HPCs is reliant on circulatory ST6Gal-1 and independent of the ability of these cells to express endogenous ST6Gal-1. Biologically, the abundance of HSCs and HPCs in the BM is inversely related to the status of circulatory ST6Gal-1, and this connection is repeated in the BM chimeras. Thus distantly produced ST6Gal-1 is assumed to be the modifier of the glycans of HSCs and HPCs, for α2,6-sialic acids and liver-produced ST6Gal-1 may be a potent systemic regulator of haematopoiesis [26].

5.11 Role of Sialylation Related Enzymes in Stem Cells

Numerous studies show the implication of glycosyltransferase-mediated macromolecule glycosylation in the regulation of pluripotent states in hPSCs. ST6Gal-I, known for adding α2–6-linked sialic acids to the substrate glycoproteins, is preferentially expressed in undifferentiated hPSCs compared to non-pluripotent cells. Moreover, downregulation of ST6GAL1 in undifferentiated hPSCs leads to reduction in POU5F1 (also known as OCT4) protein and significantly alters many gene expression which arrange cell morphogenesis during differentiation, indicating that ST6GAL1 is imperative in the regulation of pluripotency and differentiation in hPSCs, and the pluripotent state in human cells can be modulated using pharmacological tools to target sialyltransferase activity. Knockdown of the St6Gal1 gene along with the incidence of a sialyltransferase inhibitor diminished the efficiency of somatic cell reprograming [27].

5.12 Cancer Stem Cells and Sialylation

Most of the cancers comprise of heterogeneous populations of cells with extensive difference in tumorigenicity. Cells with the properties of self-renewal and tumour-initiation are termed as the cancer stem cells (CSCs). Since the CSCs trigger tumour reappearance and metastasis and are resilient to the existing anti-cancer therapies, novel therapeutic strategies are desirable to competently target and destroy this subset of cells. Anomalous glycosylation is a hallmark of cancer. Several cancer-associated glycans are associated with tumour progression and metastasis, and are used as tumour markers. Many such glycans have been identified on the CSCs utilizing current advances in glycoproteomic technologies [22].

ST6Gal-I has also been implicated in carcinogenesis. It is over-expressed in ovarian and pancreatic carcinomas, enriched in metastatic tumours, and associated with reduced patient survival. Remarkably, ST6Gal-I upregulation in cancer cells deliberated trademark CSC like characteristics. ST6Gal-I encourages expression of the key tumour-promoting transcription factors like Sox9 and Slug [28, 29].

5.13 Role of Sialylation in Reprogramming of Somatic Cells and Maintenance of Stem Cell Pluripotency

The role of sialylation in regulating early embryonic development in mice was established for the first time in 2002. Since then, its role in human development has progressed considerably with the application of advanced technologies [1, 30]. Studies in knock out (KO) mice devoid completely of UDP-GlcNAc 2-epimerase encoding gene *GNE, probably due to aberrant sialylation thereby affected functions of* cell–cell adhesion, cell migration, and affected cellular signalling pathways of survival, migration, and differentiation, indicating important role of *GNE* in dictating life of the ESCs [31]. *Cmas* coding for CMP-Sia synthase converting Neu5Ac to cytidine-monophosphate were reported to be vital in early murine embryonic development in vitro [32] as *Cmas* KO murine ESCs revealed lack of cell surface sialylation.

The sialylation of cells not only plays a vital role in regulating ESC survival, but has also been known regulate stem cell maintenance, since early embryonic impaired sialylation has been reported to lead to aberrant embryonic development. CAMs including sialylated glycoproteins like E-cadherin, integrin, and catenin have been reported from early ESCs [33], revealing abrupt interaction between CAMs and their receptors, delaying signal transduction affecting the process of development.

Sias can form polymers like disialic acid (diSia), oligosialic acid (oligoSia), and PSA like the synaptic cell adhesion molecule (SynCAM-1), neural cell adhesion molecule (NCAM), neuropilin-2 (NRP-2), the C–C chemokine receptor type 7 (CCR7), E-selectin ligand-1 of which NCAM expression is predominant in the neuronal cell membranes over other tissues and due to its poly-negative charge affects cell adhesion and signalling. PSA is known to bind to neurotrophins, growth factors, and neurotransmitters and play role in synaptic plasticity and neurogenesis, memory formation. Lack or reduced expression of PSA has been associated with schizophrenia. N-acetylneuraminate-9-phosphate synthase, is important in brain and skeletal development and bi-allelic mutations in *NANS*, has been reported to be associated with the infantile-onset of severe developmental delay and skeletal dysplasia [1].

Sialylation maintains the stem cell pluripotency. Overexpression of ST6GAL1 has been reported from undifferentiated human PSCs compared to the non-pluripotent cells and *St6Gal1*KO revealed reduced somatic cell reprogramming [1]. ST6GAL1 inhibition in human PSCs have been revealed to downregulate OCT4 altering morphogenesis during differentiation [1]. Altered cell surface sialylation has been related to mediating epithelial-mesenchymal transition (EMT) affecting embryonic development, organ formation and regulation of pluripotency [1]. Sialonome of differentiated hiPSCs and differentiated human dermal fibroblasts reveal alteration from α-2,3 to α-2,6 linked Sias on *N*-linked glycans at the time of differentiation, correlating with the expression of sialyltransferase genes [34].

5.14 Discussion

The study of sialylation of stem cells is an emergent area of research and finds importance in regulating the destiny of stem cells, maintenance of their differentiation and potency. Its importance in stem cell biology is gradually unfolding. Recently with the development of technology, α2–6-sialylated N-glycans, has been implicated with role as markers of the differentiation of human MSCs [1]. An overexpression of CMP-N-acetylneuraminate-poly-alpha-2,8-sialyltransferase (SIA8D) playing role in regulating the sialonome of stem cells were reported from human ESCs and NSCs indicative of their role in regulating stem cell fate [1]. Our understanding of the role of sialylation, enzymes, genes, pathways, signalling, regulation involved in controlling the stem cell fate and cellular programming and functions involved is just at the tip of the iceberg and remains a major area of focus of research in the years to come.

References

1. Li F, Ding J (2019) Sialylation is involved in cell fate decision during development, reprogramming and cancer progression. Protein Cell 10(8):550–565
2. Ghosh S (2020) Sialic acids and sialoglycoconjugates in the biology of life, health and disease, 1st edn. Elsevier, London
3. Schauer R (2000) Achievements and challenges of sialic acid research. Glycoconj J 17 (7–9):485–499
4. Schauer R (2004) Sialic acids: fascinating sugars in higher animals and man. Zoology (Jena) 107(1):49–64
5. Schauer R (2006) The diversity of sialic acids and their interplay with lectins. Glycobiology 08 (13):136–149
6. Badr HA, AlSadek DM, El-Houseini ME, Saeui CT, Mathew MP, Yarema KJ, Ahmed H (2017) Harnessing cancer cell metabolism for theranostic applications using metabolic glycoengineering of sialic acid in breast cancer as a pioneering example. Biomaterials 116:158–173
7. Altman MO, Gagneux P (2019) Absence ofNeu5Gc and presence of anti-Neu5Gc antibodies in humans-an evolutionary perspective. Front Immunol 10:789. https://doi.org/10.3389/fimmu.2019.00789
8. Kooner AS, Yu H, Chen X (2019) Synthesis of N-glycolylneuraminic acid (Neu5Gc) and its glycosides. Front Immunol 10:2004. https://doi.org/10.3389/fimmu.2019.02004
9. Mandal C, Schwartz-Albiez R, Vlasak R (2015) Functions and biosynthesis of O-acetylated sialic acids. Top Curr Chem 366:1–30
10. Teoh ST, Ogrodzinski MP, Ross C, Hunter KW, Lunt SY (2018) Sialic acid metabolism: a key player in breast cancer metastasis revealed by metabolomics. Front Oncol 8:174
11. Goldstein IJ, Hughes RC, Monsigny M, Ozawa T, Sharon N (1980) What should be called a lectin? Nature 285:60
12. Hamouda H, Ullah M, Berger M, Sittinger M, Tauber R, Ringe J, Blanchard V (2013) N-glycosylation profile of undifferentiated and adipogenically differentiated human bone marrow mesenchymal stem cells: towards a next generation of stem cell markers. Stem Cells Dev 22(23):3100–3113
13. Hasehira K, Hirabayashi J, Tateno H (2017) Structural and quantitative evidence of α2-6-sialylated N-glycans as markers of the differentiation potential of human mesenchymal stem cells. Glycoconj J 34(6):797–806
14. Tateno H, Saito S, Hiemori K, Kiyoi K, Hasehira K, Toyoda M, Onuma Y, Ito Y, Akutsu H, Hirabayashi J (2016) α2-6 sialylation is a marker of the differentiation potential of human mesenchymal stem cells. Glycobiology 26(12):1328–1337
15. Barone A, Säljö K, Benktander J, Blomqvist M, Månsson JE, Johansson BR, Mölne J, Aspegren A, Björquist P, Breimer ME, Teneberg S (2014) Sialyl-lactotetra, a novel cell surface marker of undifferentiated human pluripotent stem cells. J Biol Chem 289(27):18846–18859
16. Säljö K, Barone A, Vizlin-Hodzic D, Johansson BR, Breimer ME, Funa K, Teneberg S (2017) Comparison of the glycosphingolipids of human-induced pluripotent stem cells and human embryonic stem cells. Glycobiology 27(4):291–305
17. Toyoda H, Nagai Y, Kojima A, Kinoshita-Toyoda A (2017) Podocalyxin as a major pluripotent marker and novel keratan sulfate proteoglycan in human embryonic and induced pluripotent stem cells. Glycoconj J 34(6):817–823
18. Pachón-Peña G, Donnelly C, Ruiz-Cañada C, Katz A, Fernández-Veledo S, Vendrell J, Sackstein R (2017) A Glycovariant of human CD44 is characteristically expressed on human mesenchymal stem cells. Stem Cells 35(4):1080–1092. https://doi.org/10.1002/stem.2549
19. Zhou F, Cui C, Ge Y, Chen H, Li Q, Yang Z, Wu G, Sun S, Chen K, Gu J, Jiang J, Wei Y (2010) Alpha2,3-sialylation regulates the stability of stem cell marker CD133. J Biochem 148 (3):273–280

20. Heiskanen A, Hirvonen T, Salo H, Impola U, Olonen A, Laitinen A, Tiitinen S, Natunen S, Aitio O, Miller-Podraza H, Wuhrer M, Deelder AM, Natunen J, Laine J, Lehenkari P, Saarinen J, Satomaa T, Valmu L (2009) Glycomics of bone marrow-derived mesenchymal stem cells can be used to evaluate their cellular differentiation stage. Glycoconj J 26(3):367–384
21. Kawamura T, Miyagawa S, Fukushima S, Yoshida A, Kashiyama N, Kawamura A, Ito E, Saito A, Maeda A, Eguchi H, Toda K, Lee JK, Miyagawa S, Sawa Y (2014) N-glycans: phenotypic homology and structural differences between myocardial cells and induced pluripotent stem cell-derived cardiomyocytes. PLoS One 9(10):e111064
22. Brooke G, Tong H, Levesque JP, Atkinson K (2008) Molecular trafficking mechanisms of multipotent mesenchymal stem cells derived from human bone marrow and placenta. Stem Cells Dev 17(5):929–940
23. Chowdhury S, Bhattacharya K, Mandal C, Mondal S, Sarkar S, Chandra S, Banerjee S, Mandal C (2018) AGlycomic approach towards identification of signature molecules in CD34+ haematopoietic stem cells from umbilical cord blood. Adv Exp Med Biol 1112:309–318
24. Li Y, Huang X, An Y, Ren F, Yang ZZ, Zhu H, Zhou L, He X, Schachner M, Xiao Z, Ma K, Li Y (2013) Cell recognition molecule L1 promotes embryonic stem cell differentiation through the regulation of cell surface glycosylation. Biochem Biophys Res Commun 440(3):405–412
25. Dimitroff CJ, Lee JY, Rafii S, Fuhlbrigge RC, Sackstein R (2001) CD44 is a major E-selectin ligand on human hematopoietic progenitor cells. J Cell Biol 153(6):1277–1286
26. Nasirikenari M, Veillon L, Collins CC, Azadi P, Lau JT (2014) Remodeling of marrow hematopoietic stem and progenitor cells by non-self ST6Gal-1 sialyl-transferase. J Biol Chem 289(10):7178–7189
27. Wang YC, Stein JW, Lynch CL, Tran HT, Lee CY, Coleman R, Hatch A, Antontsev VG, Chy HS, O'Brien CM, Murthy SK, Laslett AL, Peterson SE, Loring JF (2015) Glycosyltransferase ST6GAL1 contributes to the regulation of pluripotency in human pluripotent stem cells. Sci Rep 5:13317
28. Sawanobori A, Moriwaki K, Takamatsu S, Kamada Y, Miyoshi E (2016) A glycoproteomic approach to identify novel glycomarkers for cancer stem cells. Proteomics 16(24):3073–3080
29. Schultz MJ, Holdbrooks AT, Chakraborty A, Grizzle WE, Landen CN, Buchsbaum DJ, Conner MG, Arend RC, Yoon KJ, Klug CA, Bullard DC, Kesterson RA, Oliver PG, O'Connor AK, Yoder BK, Bellis SL (2016) The tumor-associated glycosyltransferase ST6Gal-I regulates stem cell transcription factors and confers a cancer stem cell phenotype. Cancer Res 76 (13):3978–3988
30. Melo-Braga MN, Schulz M, Liu Q et al (2014) Comprehensive quantitative comparison of the membrane proteome, phosphoproteome, and sialiome of human embryonic and neural stem cells. Mol Cell Proteomics 13(1):311–328
31. Hinderlich S, Weidemann W, Yardeni T, Horstkorte R, Huizing M (2015) UDP-GlcNAc 2-epimerase/ManNAc kinase (GNE): a master regulator of sialic acid synthesis. Top Curr Chem 366:97–137
32. Abeln M, Borst KM, Cajic S et al (2017) Sialylation is dispensable for early murine embryonic development in vitro. Chembiochem 18(13):1305–1316
33. Li L, Bennett SA, Wang L (2012) Role of E-cadherin and other cell adhesion molecules in survival and differentiation of human pluripotent stem cells. Cell Adhes Migr 6(1):59–70
34. Berger RP, Dookwah M, Steet R, Dalton S (2016) Glycosylation and stem cells: regulatory roles and application of iPSCs in the study of glycosylation-related disorders. BioEssays 38 (12):1255–1265

Markers of Stem Cells

6

Contents

Abstract

The role of stem cells in regenerative medicine is imperative. The population of these stem cells is infinitesimally small and therefore, suitable markers are needed for identifying them. In this chapter, we have made an attempt to list majority of the markers of the different types of stem cells which will help readers in isolating them.

S. Chowdhury, S. Ghosh, *Stem Cells*, https://doi.org/10.1007/978-981-16-1638-9_6

Keywords

TRA-1-60 and TRA-1-81 · Stem cell factor (SCF or c-Kit) · Ligand transcription factors

Abbreviations

ABCG2	ATP-binding cassette transporters
ALDH	Aldehyde dehydrogenase
ALP	Alkaline phosphatase
ATF-2	Activation transcription factor-2
BM	Bone marrow
CD	Cluster of differentiation
CESCs	Corneal epithelial stem cells
DPPA	Developmental pluripotency-associated genes
DSCs	Dental stem cells
ECAT	ESC Associated transcript genes
EGFR	Epidermal growth factor receptor
EpCAM	Epithelial cell adhesion molecule
ESC	Embryonic stem cell
FAK	Focal adhesion kinase
FBXO15	F-box 15
FCM	Flow cytometry
FoxD3	Forkhead Box D3
Fzd	Frizzled
GCNF	Germ cell nuclear factor
GPCR	G-protein-coupled receptor
HAP	Hair follicle-associated-pluripotent stem cells
HMGA2	High mobility group A2
HNF3-beta	Hepatocyte nuclear factor 3-beta
HPCs	Haematopoietic progenitor cells
HSCs	Haematopoietic stem cells
ICM	Inner cell mass
iPSCs	Induced pluripotent stem cells
Klf	Krupple-like Factor
KSC	Keratinocyte stem cells
LEF1	Lymphoid enhancer factor 1
McSC	Melanocyte stem cell
MHC-I	Class 1 major histocompatibility antigens
MSC	Mesenchymal stem cell
NAC1	Nucleus Accumbens-1
NCAM	Neural cell adhesion molecule
NSC	Neural stem cell
POU	Pit-Oct-Unc family
Rex1 or Zfp-42	Reduced expression 1 gene

SCF	Stem cell factor
SGSCs	Sweat gland stroma-derived stem cells
SKP	Skin derived precursors
Sox	Sry-related high-mobility group (HMG) box-containing family
SSEA	Stage specific embryonic antigens
SVZ	Subventricular zone
Tbn	Taube nuss
TCF	T-cell factor 1
TDGF-1	Teratocarcinoma-derived growth factor-1
TGF-β	Epidermal growth factor
TRA	Tumour recognition antigen
UCB	Umbilical cord blood stem cells
UTF1	Undifferentiated embryonic cell transcription factor
VEGF	Vascular endothelial growth factor
ZFX	X-linked zinc finger protein

6.1 Introduction

Stem cells with inherent properties of self-renewal and differentiation have been discovered. Human stem cell technology offers immense potential in regenerative medicine and treatment of various human diseases [1]. However, these stem cells comprise of an infinitesimally small population in the human body and remain in dormant condition in most cases. Identifying such a cell is similar to finding a needle in a haystack. Thus, identification of stem cell using appropriate markers expressed by them finds importance.

6.2 Markers for Stem Cell Identification

Specific genes and their proteins, expressed by the stem cells are known as stem cells markers and have been used for identification and isolation and stem cells. Stem cell markers can be of several types. The cell surface markers are specialized proteins coating the cell surface and selectively bind or adhere to other signal molecules, some of which are uniquely present, expressed or secreted in specific cell types, thereby, making them cell type specific. At present, the marker-based flow cytometry (FCM) technique, and magnetic cell sorting are the current methods of cell isolation and a thorough panel of markers enable to initially identify and isolate the stem cells easily [2].

These cell surface proteins of different classes, like adhesion molecules, integrins, glycoproteins, and receptors which help to differentiate one cell from another are designated by cluster of differentiation (CD) antigens. Different cells express different and specific CD antigens. Antibodies recognizing CD antigens are frequently used as an efficient tool in cell sorting, identifying, characterizing, and isolating different cell populations. Apart from the aforesaid markers, there are transcription factors which also help to distinguish one cell type from another [2].

6.2.1 Carbohydrate Based Cell Surface Markers

Cell surface carbohydrate antigens participate in many biomedical phenomena like cell differentiation and immune reactivity, which finds importance in clinical applications of stem cells [1]. These include tumour recognition antigens TRA-1-60 and TRA-1-81, and the stage-specific embryonic antigens (SSEA)-3 and SSEA-4 [3–5].

6.2.1.1 Stage Specific Embryonic Antigens (SSEA)

Stage Specific Embryonic Antigens (SSEAs) are developmentally regulated, carbohydrate-associated cell surface molecules [3]. SSEA-1, SSEA-3, and SSEA-4 contain carbohydrate epitopes associated with the lacto- and globo-series glycolipids [4]. They usually play a role in cellular adhesion and differentiation [6]. SSEA-1/CD15 is expressed on neural stem cells (NSCs, 7). SSEA-3 and SSEA-4 are expressed on human ESCs [7]. The dental stem cells (DSCs) are SSEA-4^+ [8].

6.2.1.2 TRA-1-60 and TRA-1-81 Antigens

TRA-1-60 and TRA-1-81 are high molecular weight, proteoglycan-based antigens on the human ESCs [5]. While TRA-1-60 antibody reacts with a neuraminidase-sensitive epitope of a proteoglycan, TRA-1-81 reacts with a neuraminidase insensitive epitope of the same molecule. This proteoglycan molecule has been proposed to be a form of the podocalyxin protein [2].

6.2.2 Protein Based Cell Surface Markers

6.2.2.1 Alkaline Phosphatase (ALP)

Alkaline phosphatase (ALP) is an enzyme that play role in different physiological processes and finds importance in functions of the liver function and bone development processes. ALP is high in human embryonic stem cells (ESCs) [2].

6.2.2.2 Thy-1 Antigen or CDw90

The antigen Thy-1 discovered by Reif and Allen in 1964 as the first T cell marker from thymus [9]. It is a 25–35 kDa phosphatidylinositol-anchored cell surface molecule and is a marker of different stem cells including ESCs [10, 11]. It is known to play major role in cell–cell adhesion and cell–cell communication in stem cells and in cancer progression in some malignancies by increasing proliferation, migration, and adhesion [2].

6.2.2.3 CD9

CD9 is a cell surface glycoprotein, member of the tetraspanin family which is involved in cell–cell adhesion, motility, signal transduction, differentiation tumour cell invasion, apoptosis, and resistance to chemotherapy [12]. In haematopoietic stem cells (HSCs) and progenitor cell, this antigen is involved in megakaryocyte differentiation and adhesion between myeloid and stromal cells [13] (Fig. 6.1).

6.2.2.4 CD24

CD24 is a sialoglycoprotein expressed on human ESCs, mature granulocytes and B cells functioning as cell adhesion molecule, modulating growth, and differentiation signals to these cells [14] (Fig. 6.2).

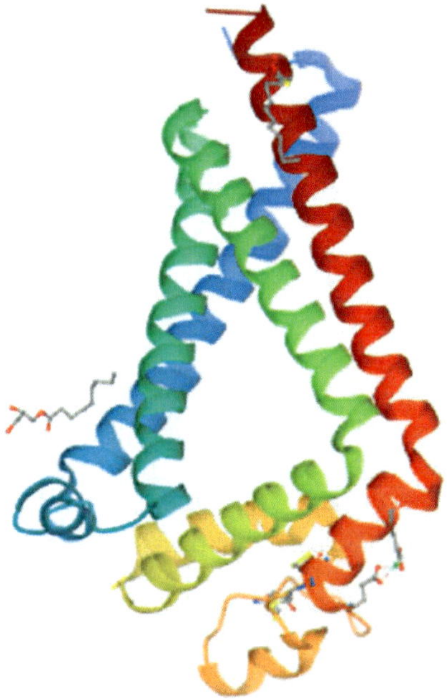

Fig. 6.1 6K4J: Crystal Structure of the CD9, Structural insights into tetraspanin CD9 function. Umeda, R., Satouh, Y., Takemoto, M., Nakada-Nakura, Y., Liu, K., Yokoyama, T., Shirouzu, M., Iwata, S., Nomura, N., Sato, K., Ikawa, M., Nishizawa, T., Nureki, O. (2020) Nat Commun 11: 1606–1606

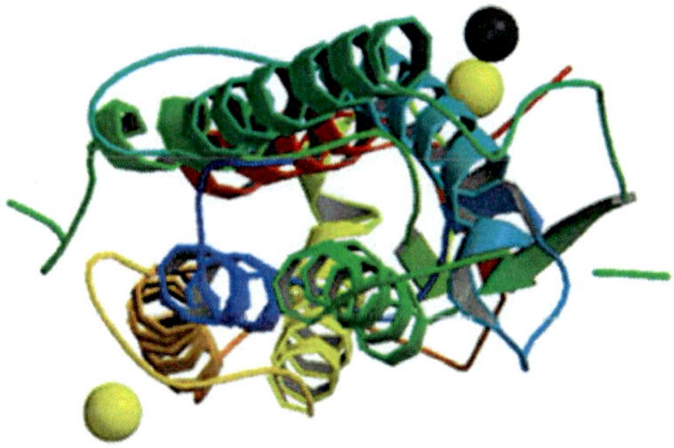

Fig. 6.2 1EXZ, Structure of Stem cell factor, Crystal structure of human stem cell factor: implication for stem cell factor receptor dimerization and activation. Zhang, Z., Zhang, R., Joachimiak, A., Schlessinger, J., Kong, X.P. (2000) Proc Natl Acad Sci USA 97: 7732–7737

6.2.2.5 Stem Cell Factor (SCF or c-Kit Ligand)

The stem cell factor (or SCF, kit-ligand, KL or steel factor) is a cytokine that binds with its receptor- the c-Kit receptor (SCFR, CD117), being a 145 kDa transmembrane glycoprotein, encoded stem cell factor by the c-kit proto-oncogene [15]. SCF is known to exist both as a transmembrane and a soluble form and known to transduce signals by ligand mediated dimerization of its receptor, Kit, leading to receptor dimerization and thereby followed by activation of specific protein kinases. Various fibroblast-type cells, and sites where haematopoiesis occurs, including foetal liver and BM, all are reported to express SCF. This cytokine plays an important role in haematopoiesis, spermatogenesis, and melanogenesis [2].

SCF and its ligand is reported to play an important role in haematopoiesis. Receptor-ligand binding induces the formation of receptor homodimers, activation of tyrosine kinase, and phosphorylation of several intracellular targets including auto-phosphorylation followed by activation of a signal transduction cascade. Up to 60–70% of $CD34^+$ haematopoietic progenitor cells (HPCs) co-express CD117, and the cell expressing the markers $CD34^+CD117^+$ fraction includes the clonogenic cells predominantly. Expression of CD117 during haematopoietic development is reported to be the highest at the early and initial stages and then reported to continuously diminish with maturation [16].

6.2.2.6 CD133

Human AC133 (CD133), a stem cell glycoprotein antigen with a molecular weight of 120 kDa, is known to be selectively expressed on $CD34^+$ cells, marker of human HSCs. cDNA coding for AC133 antigen has been isolated, coding for a 97 kDa polypeptide with 865 amino acids. This antigen is modelled as a 5-transmembrane molecule, a structure that is novel among known cell surface structures. $CD133^+$ cells are capable of forming colonies in vitro and also demonstrate in vivo haematopoietic repopulation ability. It plays a role in maintaining the cell membrane topology [17, 18].

6.2.2.7 Integrins

Integrins are α/β heterodimeric cell surface receptors facilitating the attachment of cell to its surrounding tissues. They play pivotal role in cell adhesion, signalling, and migration, and in cell growth and survival [19]. Different cell surfaces express numerous types of integrins which play an important role in constructing the stem cell niches [20]. The integrin family of proteins contains at least 18 α- and eight β-subunits that form 24 known integrins with distinct tissue distributions and overlapping ligand specificities [21].

Integrin β1 (CD29)

The human Integrin β1 (ITGB1), also known as CD29, is a cell surface receptor that is encoded by the *ITGB1*gene and participates in cell adhesion and signalling [22].

Integrin α6 (CD49f)

Integrin $\alpha 6$, or CD49f, is reported to be the only biomarker found to be expressed in more than 30 different stem cells populations, including some cancer stem cells. It is known to play role in maintaining self-renewal of pluripotent stem cells and breast and glioblastoma cancer stem cells [23].

6.2.2.8 Frizzled (Fzd)

Frizzled (Fzd) belongs to the G-protein-coupled receptor (GPCR) superfamily of proteins and is known to play role in transducing Wnt mediated signals [24]. The mammalian Fzd subfamily of proteins has ten members (Fzd1 to Fzd10) and mediates signalling through different pathways [2].

6.2.2.9 Cripto (TDGF-1)

The Cripto gene, or teratocarcinoma-derived growth factor-1 (TDGF-1), encodes a novel human growth factor structurally related to epidermal growth factor (EGF). Transforming Growth Factor-β, (TGF-β) plays critical role during embryogenesis. Oncogene Cripto is reported for its high expression in tumours and to play role in promotion of tumorigenesis by mechanisms of activation of mitogenic signalling pathways and antagonism of activin signalling [25].

6.2.2.10 CD34

CD34 is a 110–115 kDa transmembrane phosphoglycoprotein, also known as a marker of HSCs and HPCs was first discovered in 1984 [26]. However, CD34 is now reported as a marker of several other cells of non-haematopoietic lineages, including vascular endothelial progenitors [27] and epithelial progenitors [28]. Studies in haematopoietic cells suggest its roles in cyto-adhesion and regulation of cell differentiation and proliferation [29, 30]. A population of the melanocyte stem cells (McSCs) are CD34+ and exhibit an ability to myelinate neurons [31].

6.2.2.11 Nestin

Nestin or the neuroepithelial stem cell protein is a cytoskeletal intermediate filament, constituting a major component of the cytoskeleton. It plays major role in essential stem cell functions including self-renewal, differentiation, proliferation and migration in the context of the cytoskeleton. It was initially characterized in the NSCs but is also found in other cells with progenitor functions like the BM [31].

6.2.2.12 Epidermal Growth Factor Receptor (EGFR)

Epidermal growth factor receptor (EGFR) is a transmembrane protein and receptor tyrosine kinase. EGFR signalling plays pivotal role in maintaining the balance between NSC and neural progenitor cells in the subventricular zone (SVZ) [34].

6.2.2.13 Notch Proteins

The Notch proteins are a group of highly conserved, type-1 transmembrane proteins forming a core component of the Notch signalling pathway. Among various functions, Notch signalling is also involved in maintaining the balance between the NSCs and progenitor cell populations in the SVZ along with EGFR [32]. Additionally, Notch signalling helps in the expansion of the HSCs compartment during bone development and regeneration [33].

6.2.2.14 CD81

Human CD81 is a 26 kDa cell surface protein belonging to the tetraspanin family of proteins, also known as Target of the Antiproliferative Antibody 1 (TAPA-1) and Tetraspanin-28 [34]. This protein mediates signal transduction events and regulates cell development, activation, growth, and motility. It is known to play role in cellular proliferation, migration, and differentiation in NSCs [35, 36].CD81 plays an essential role when HSCs enter into quiescence [37].

6.2.2.15 Vascular Endothelial Growth Factor (VEGF)

Vascular endothelial growth factor (VEGF) is a protein that promotes the growth of new blood vessels. It also promotes a wide range of neuronal functions, both in vitro and in vivo, including neurogenesis, neuronal migration, survival, and axon guidance [38].

6.2.2.16 CD71

CD71 or the transferrin receptor protein is a glycoprotein which is required to import iron from transferrin into cells by endocytosis [39]. Lower expression of CD71 is seen in keratinocytes and corneal stem cells [40, 41].

6.2.2.17 Vimentin

Vimentin is the most abundant protein in the intermediate filament of the cytoskeleton [42]. It has been reported to be associated with the adhesion, differentiation of cells, migration and invasion, signalling transmission and cytoskeletal rearrangement, and the regulation of cell morphology and is expressed in the corneal epithelial stem cells (CESCs) [45].

6.2.2.18 Flk-1

Flk-1 is a cell surface receptor for VEGF and is expressed by the retinal progenitor cells and neural progenitors [112]. It is also expressed on the DSCs [8].

6.2.2.19 CD13

CD13 or alanyl aminopeptidase is an enzyme that in humans is encoded by the ANPEP gene. This multifunctional cell surface peptidase plays pivotal role in activation of focal adhesion kinase (FAK) to promote mesenchymal stem cell (MSC) migration and activation [43].

6.2.2.20 CD105

CD105 or endoglin is a type I transmembrane protein reported to play role in activation and proliferation of endothelial cells. CD105 binds several factors of the Transforming Growth Factor (TGF)-β superfamily, a pleiotropic cytokine that regulates different cellular functions including proliferation, differentiation, and migration. It is expressed in HSCs [44] and DSCs [8].

6.2.2.21 CD44

CD44 is a cell surface adhesion receptor expressed profusely in many cancers and plays role in regulation of metastasis by recruitment of CD44 to the cell surface. CD44 has also been identified as a marker for stem cells of several types including HSCs, cancer stem cells. The universally expressed cell surface protein is primarily involved in aggregation, migration, and activation of cells, these functions are mediated through the adhesive properties of CD44 [45].

6.2.2.22 CD59

CD59 is studied as a glycosylphosphatidylinositol (GPI)-anchored receptor that interacts with the membrane attack complex (MAC) and stops it from being able to permeabilize the membrane. It is a mesenchymal marker and is expressed in DSCs [46, 47] (Fig. 6.3).

6.2.2.23 Stro-1

Stro-1, an endothelial antigen is predominantly a marker for MSCs [48].

6.2.3 Transcription Factors

Transcription factors are vital for the regulation of genes. Most of them are inactive under normal conditions and bind to their recognition sequences in response to specific signal transduction events. This results in specific gene expression. Therefore, tracking these genes expression can be used as a marker for a specific cell situation [2].

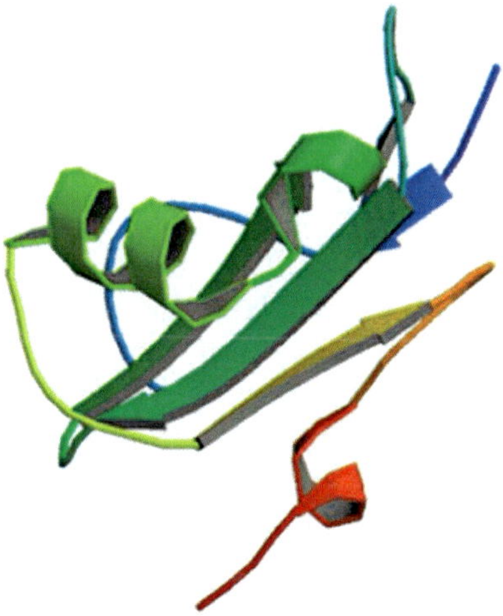

Fig. 6.3 High resolution structure of human CD59. High-Resolution Structures of Bacterially Expressed Soluble Human CD59. Leath, K.J., Johnson, S., Roversi, P., Hughes, T.R., Smith, R.A.G., Mackenzie, L., Morgan, B.P., Lea, S.M. (2007) Acta Crystallogr Sect F Struct Biol Cryst Commun 63: 648

6.2.3.1 Octamer-Binding Protein 4 (Oct4)

Oct4 (Oct3/4 or POU5F1) is a member of the Oct family of Pit-Oct-Unc (POU) family of transcription factors and is known to play a crucial role in regulating stem cell pluripotency and differentiation [49]. The expression of Oct4 is restricted to pluripotent germ line cells and the inner cell mass (ICM) of blastocysts [50].

6.2.3.2 Sry-Related High-Mobility Group (HMG) Box-Containing (Sox) Family

The Sox2 belongs to the Sox gene families, which are HMG box transcription factors which interact functionally with POU domain proteins. The Sox2 gene bears at least two regulatory regions that specifically function in pluripotent embryonic cells [51]. Similar to Oct3/4, the Sox family of genes maintains pluripotency. However, it is associated with multipotent and unipotent stem cells, while Oct3/4 is exclusively expressed in pluripotent stem cells. It is one of the first genes to be used for induction of induced pluripotent stem cells (iPSCs) [52].

6.2.3.3 Krupple-Like Factor (Klf) Family

The Krupple-like factor (Klf) family is characterized by three Cys2 His2 zinc fingers located at the C-terminus, separated by a highly conserved H/C link. KLFs have similar affinities for different GC-rich sites, and can compete with each other for the occupation of such sites. It regulates many biological processes, including cell proliferation, differentiation, development, and apoptosis [53].

6.2.3.4 Nanog

Homeobox protein Nanog is a transcription factor which plays a crucial role in the maintenance of pluripotency and self-renewal in human ESCs. In 2003, Chambers et al. established a central role for Nanog in the transcription factor hierarchy that defines ESC identity [54].

6.2.3.5 Reduced Expression 1 Gene (Rex1 or Zfp-42)

The Rex1 gene encodes for a zinc finger family transcription factor and contains four nucleic acid zinc finger motifs (ZFM) and an acidic domain. It is profusely expressed in human ESCs and is a marker of pluripotency in various stem cells, such as multipotent adult progenitor cells and amniotic fluid cells [55].

6.2.3.6 Undifferentiated Embryonic Cell Transcription Factor (UTF1)

The undifferentiated embryonic cell transcription factor (UTF1) is a transcriptional co-activator that interacts with the metal-binding motif of activation transcription factor-2 (ATF-2). UTF1 is a strongly chromatin-associated protein involved in the initiation of ESC differentiation and its absence delays or blocks differentiation [56]. The co-expression of UTF1 with the reprogramming factors c-Myc, Oct4, Sox2, and KLF4, along with the siRNA knockdown of p53, is known to increase the efficiency of iPSC generation [57].

6.2.3.7 X-Linked Zinc Finger Protein (ZFX)

The X-linked Zinc Finger Protein (ZFX) gene is expressed from the inactive X chromosome having similarity with its homologue on the Y chromosome, ZFY. Transcripts of ZFX and ZFY encode for proteins with a highly acidic amino-terminal domain and a carboxy-terminal zinc-finger motif commonly associated with nucleic acid-binding [58]. Among other functions, ZFX is also required in self-renewal of ESCs and HSCs [59].

6.2.3.8 Taube Nuss (Tbn)

Taube Nuss (Tbn) is a highly conserved class of proteins with an important function in development. Tbn is restricted to the ESCs of the ICM and is necessary for their survival and its absence leads to apoptosis [55].

6.2.3.9 Forkhead Box D3 (FoxD3)

FoxD3 is a member of the Forkhead box family and is characterized by a winged-helix DNA-binding structure. This transcriptional regulator plays an important role in the maintenance of pluripotency [60].

6.2.3.10 High Mobility Group A2 (HMGA2)

Human high mobility group A2 (HMGA2) protein is a non-histone transcription factor which binds to AT-rich sequences of DNA and regulates gene expression. HMGA2 is abundantly and universally expressed and plays a crucial role during embryonic development, promoting stem cell self-renewal. Reduced expression of HMGA2 is associated with stem cell aging [61].

6.2.3.11 Nucleus Accumbens-1 (NAC1)

Nucleus Accumbens-1(NAC1) or BTBD14 is a nuclear factor belonging to the Pox virus and zinc finger/bric-a-brac tramtrack broad complex (POZ/BTB) domain family. It was originally identified in a unique neuronal forebrain structure responsible for reward motivation and addictive behaviour [62]. NAC1 is a part of the extended transcriptional network in pluripotent cells that involves Oct4, Sox2, Nanog, Sall1, KLF4, and Sall4 [63].

6.2.3.12 Germ Cell Nuclear Factor (GCNF)

The Germ Cell Nuclear Factor (GCNF), also known as the nuclear receptor subfamily six group A member (NR6A1), is a lone member of the nuclear receptor gene superfamily. It is expressed in the nervous system during development and is a repressor of Oct4 and the protamine genes, playing a critical role in the control of gene expression during embryogenesis [64].

6.2.3.13 STAT3

The STAT3 transcription factor is an important signalling molecule for many cytokines and growth factor receptors. It sustains self-renewal and pluripotency of ESCs and its suppression results in ESC differentiation [65].

6.2.3.14 Lymphoid Enhancer Factor 1(LEF1) and T-Cell Factor 1(TCF)

Lymphoid enhancer factor (LEF1) and T-cell factor (TCF) are members of the HMG DNA-binding protein family of transcription factors, which regulate early lymphoid development and act downstream in Wnt signalling. They play important role in integrating Wnt signalling with the regulation of stem cell differentiation [66].

6.2.3.15 SALL Family

The SALL gene family comprises of SALL1, SALL2, SALL3, and SALL4 and plays significant role in regulating the developmental processes of many organisms [67].

6.2.3.16 ESC Associated Transcript (ECAT) Genes

The ESC associated transcript genes (ECAT) play an important role in regulating the biology of stem cell. ECAT4 was identified as Nanog, a master regulator for the maintenance of mouse and human ESC [68]. ECAT11 is abundantly expressed in undifferentiated human ESC and is a downstream target of Nanog, forming a useful marker in identifying undifferentiated human ESC [69].

6.2.3.17 Developmental Pluripotency-Associated (DPPA) Genes

The developmental pluripotency-associated (DPPA) genes translate into a group of five proteins, described as a set of Oct4-related genes. They are known as markers for early embryonic and germline pluripotent cells [70].

6.2.3.18 Hepatocyte Nuclear Factor 3-Beta (HNF3-Beta)

Hepatocyte nuclear factor 3-β (HNF3-β) binds to fibrinogen beta promoter and is involved in IL6-induced fibrinogen β transcriptional activation. This transcription factor is involved in embryonic development [71]. It is expressed in the DSCs [8].

6.2.3.19 C-Myc

Myc is a proto-oncogene whose protein is encoded by the Myc gene. It is a member of the Myc family of transcription factors. The Myc regulatory module/network revolves around c-Myc and includes partner co-activators and chromatin modifiers. The Myc module primarily controls self-renewal of ESCs through cell cycle, metabolism [72]. It is also reported from DSCs [8].

6.2.4 Cell Cycle Regulators

6.2.4.1 ΔNp63

A member of the p53 family, p63 plays an important role in cell cycle regulation including cell proliferation and apoptosis. The p63 gene encodes two groups of protein isoforms, namely TAp63 and ΔNp63. Its expression is seen in the CESCs [73].

6.2.4.2 ATP-Binding Cassette Transporters (ABCG2)

ATP-Binding Cassette Transporters (ABCG2), a member of the ATP-binding cassette transporter family, serves as a specific biomarker for BM stem cells [74]. De Paiva et al. found that ABCG2 was specifically expressed in the CESCs [75].

6.2.5 Signal Pathway-Related Intracellular Markers

Although cell surface markers are mostly characterized, several intracellular signal pathways play important roles in maintaining self-renewal and pluripotency of stem cells. Therefore, many signal pathway-related proteins, which are critical for these roles, have found importance as markers for identification of stem cells. The signal pathway related proteins which have important functions in determining the fate of these stem cells are discussed under each specific stem cell type in the following section [2].

6.2.5.1 Lin28

Lin28 is a RNA-binding protein playing vital role in promoting pluripotency by regulation of the micro-RNA (mi-RNA), let-7. It is a gatekeeper molecule that regulates the transition between pluripotency and committed cell lineages [76]. The DSCs are positive for this bio-marker [8].

6.2.6 Aldehyde Dehydrogenase (ALDH)

Aldehyde dehydrogenases (ALDHs) are intracellular enzymes that play role in oxidizing cellular aldehydes. Substrates for these enzymes include acetaldehyde, an intermediate in ethanol metabolism, and biogenic amines produced during catecholamine catabolism. ALDH is also known to play a crucial role in the conversion of vitamin A to its active metabolite, retinoic acid. Cellular resistance to cyclophosphamide has also been shown to result from high levels of ALDH. Of the four ALDH isozymes identified, liver cytosolic ALDH isozyme is reported to be involved in vitamin A and cyclophosphamide resistance. HPCs are known to express high levels of cytosolic ALDH activity and are accordingly relatively resistant to cyclophosphamide. The HSCs express the highest level of ALDH, while lymphocytes express the least. Immature erythroid cells express ALDH at a level intermediate between progenitor cells and lymphocytes [77–79].

6.3 Markers of Embryonic Stem Cells (ESCs)

Embryonic stem cell (ESC) markers are those macromolecules specifically expressed in ESCs. The knowledge of the functions of these markers is essential for immunophenotypic and functional characterization of these cells. ESC markers are of four types namely, the lectins and peptide markers, cell surface markers, pathway related markers, and transcription factors [2].

6.3.1 Carbohydrate Cell Surface Markers

Cell surface carbohydrate antigens expressed on the human ESCs are the SSEAs. While, SSEA-1 (CD15/Lewis x) is not expressed by human ESCs [80], SSEA-3 and SSEA-4 are synthesized during oogenesis and are expressed in human embryonic germ cells and ESCs [2, 7]. They participate in cellular differentiation and have no critical functional role in maintaining the pluripotency of the human ESCs [6]. TRA-1-60 and TRA-1-81 antigens expressed on ESCs participate in cellular differentiation [81].

6.3.2 Protein Based Cell Surface Markers

Human ESCs express high levels of class 1 major histocompatibility antigens (MHC-I) which is associated with rejection on transplantation [11, 82]. ALP is used as a marker predominantly expressed in human ESCs and can be detected by two antibodies, TRA-2-49 and TRA-2-54 [83, 84]. The presence of these enzymes decreases with differentiation [81, 85].

Other cell surface antigens include Thy1, CD9, CD30. The Thy-1 antigen is expressed by human ESCs and is downregulated upon differentiation [11]. CD9 is classically expressed on human ESCs and is downregulated on their differentiation. Therefore, it is used in identifying early stages of stem cell differentiation and finds importance in identifying cell subset within the stem cell compartment [86].

CD30 is a member of the tumour necrosis factor receptor (TNFR) superfamily of proteins and is usually expressed by human ESCs that have acquired cytogenetic abnormalities. Expression of CD30 protects these cells against apoptosis [86]. Human ESCs also express CD147 or TRA-1-85 [87] and this marker helps to identify human ESCs in mouse xenografts [5]. Other CD antigens expressed on ESCs are CD24 [14], CD117, and CD133 [2]. The c-kit ligand and its receptor (CD117) binding affect the survival of differentiating ESCs [88].

Integrin α6 (CD49f/CD29), a 120-kDa protein with two splice variants, integrins α6A and α6B, acts as a receptor for laminins and mediates cellular adhesion events on the basal membrane of ESCs [23, 89]. Some Fzds can also bind to other secreted proteins, such as Norrin and R-Spondin. Fzd 1–10 are expressed in human ESCs [2]. These markers contribute to ESC proliferation and pluripotency [90]. Cripto expression in human ESCs regulate neural differentiation and permit differentiation of ESCs to cardiac fate [91, 92].

Molecular biology studies have revealed three key functionally distinct regulatory modules/networks in ESCs which maintain the ESC state. These are CORE (active), Myc (active), and Polycomb/PRC (repressive) [93, 94]. The CORE module is composed of TFs—OCT4, SOX2, and NANOG (OSN)—that cooperate with other

ESC Transcription factors (TFs), co-activators, and chromatin regulators to activate ESC-specific genes and repress lineage-specific genes [94], whereas the PRC module consists of components PRC1 and PRC2 and suppresses lineage-specific genes to maintain the ESC like state [72, 95].

The transcription factor Oct4, expressed on ESCs is responsible for the preservation of the totipotent phenotype and are expressed on undifferentiated ESCs [49]. The Sox2 is expressed like Oct4 in human ESCs and is involved in induction process. The transcription factors Sox2, Oct4, and Nanog play pivotal role in initial phase of development and propagation of undifferentiated ESCs in culture [51]. Klf5 regulates the transcription of Oct3/4 and Nanog, which maintains the pluripotency and renewal of ESCs [53]. Nanog mRNA found in pluripotent ESCs exhibits downregulated expression during early ESC differentiation, which supports its close association with pluripotent stem cell identity [54].

The Rex1 gene is highly expressed in human ESCs and is a marker of pluripotency lacking any functional implication [55]. UTF1 is primarily expressed in pluripotent ESC, firmly associated with human ESC chromatin, and is involved in maintaining an epigenetic milieu required for differentiation [96]. Overexpression of ZFX facilitates self-renewal of ESC by opposing differentiation [59]. The Tbn protein plays chief role in upholding homeostasis between survival and apoptosis of ESCs [55]. A regulatory network is formed by FoxD3 along with Oct4 and Nanog to maintain the pluripotency of ESCs [97]. HMGA2 is a critical element of the regulatory network which governs the differentiation of ESCs [98]. In the prolonged transcriptional network in pluripotent cells involving Oct4, Sox2, Nanog, Sall1, KLF4, and Sall4, another gene NAC1 is an important participant [63]. The GCNF represses the expression of Oct4 and thus modulates the overall gene expression in ESCs [99]. The transcription factor Stat3 sustains pluripotency and self-renewal of the ESCs [65]. TCF and LEF1 play important role in regulating the differentiation of stem cells [66]. SALL4 is an important controller of Oct4 and is mandatory for the maintenance of pluripotency of the ESCs [100]. ECAT4/Nanog is a master regulator of human ESC [68]. A marker of pluripotency DPPA3 plays role in cell division, transcriptional repression, and maintenance of cell pluripotentiality [70]. BMP-SMAD, LIF-STAT3, TGFβ/Activin/Nodal, IGF-IR, FGFR, and Wnt-β-catenin are also involved in the core signal pathways in regulating ESC self-renewal and pluripotency [101, 102]. These are used as the signal pathway related intracellular markers [2].

6.4 Markers of Umbilical Cord Blood Stem Cells (UCB)

Among many other markers, two major markers of UCB stem cells are CD34 and ALDH, where the former is a cell surface marker while the latter is an enzyme whose presence attests to the functional normal nature of the cells [79, 103].

6.5 Markers of Neural Stem Cells (NSCs)

Neural stem cells (NSCs) are identified using multiple cell surface markers like CD15 [104], CD29, CD24 [105], and CD133 [106]. SSEA1/CD15 is vital in cell-to-cell recognition in the NSCs. The distinct pattern of CD29 expression advocates the role of β1-integrin signalling in neural development. A heightened expression of CD24 surface expression is an effective marker for maturing neuronal cell types. Thus, three distinct populations, using a combination of CD15/CD24/CD29 marker profiles have been identified which explain the neural lineage development of NSC, neural crest, and neuronal populations from human stem cells. NSCs express CD15 +/CD29hi/CD24lo surface antigen, neural crest-like and mesenchymal phenotypes expressed CD15−/CD29hi/CD24lo while the CD15−/CD29lo/CD24hi cells formed the neuroblasts and neurons [105]. Other molecular markers include Sox1 [107], Nestin [31], EGFR [32, 108], Notch [33], VEGF [38], CD49f [23], Flk-1 [109], CD81 [35, 36], CD184, and CD24. They lack expression of CD133, CD271, and CD44 [110]. In addition, there are several molecular markers like Ki67 and PH3 [111]. Quiescent NSCs (qNSCs) are GFAP+CD133 + Nestin- and can be separated from the GFAP+CD133 + EGFR+Nestin+ activated NSCs (aNSCs) which are [112].

6.6 Markers of Corneal Epithelial (CESCs) and Stromal Stem Cells

A trivial subpopulation of the limbal epithelial cells is integrin $\alpha6^{bri}$CD71^{dim} or CD49^{bri}CD71^{dim} showing greater clonogenic capacity and expression of stem cell markers, and less expression of differentiation markers, in comparison to other cell populations. They are the corneal epithelial stem cells (CESCs), localized in the limbal basal layer and are maintained in a state of low cell activity with low iron demand [40]. Apart from the above-mentioned biomarkers, other putative CESC biomarkers include cytoskeletal proteins, cell cycle regulators, enzymes, cell adhesion molecules, growth factors and their receptors, ATP-binding cassette transporters, and differentiation associated markers [113]. The co-expression of CESC biomarkers is p63+ [73], ABCG2+ [75], and vimentin+ [114]. A subpopulation of the limbal stromal cells express SSEA-4 [115].

6.7 Markers of Dental Stem Cells (DSCs)

The dental stem cells (DSCs) isolated from the dental pulp are Oct4+, SSEA-4+, Nanog+, FLK-1+, Nestin+, HNF3beta+, Sox2+, c-Myc+, Lin28+, CD13+, CD105 +, CD29+. These DSCs exhibit low expression of CD90, CD73, and STRO-1 [8]. DPSCs also express mesenchymal markers like CD59, CD44, CD146, CD73 [46, 47] and Stro-1, CD146, CD106, and CD73 [116].

6.8 Markers of Stem Cells from the Epidermis (Skin)

The hair follicle-associated-pluripotent (HAP) stem cells comprise of oval/round-shaped bodies (diameter of ~7μm) and 2-three long elongated processes containing club-like bodies. The nestin-expressing HAP stem cells co-expressed a Schwann cell marker (S100) and p75, a neural crest and immature Schwann-cell marker, as well as nestin. The club bodies of the nestin-expressing hair follicle cells co-expressed p75, S100, and b-III tubulin, a neuron marker. HAP stem cells also express the stem-cell marker CD34, are K15-negative, suggesting their relatively undifferentiated state [117].

Human sweat gland stroma comprises nestin positive stem cells called sweat gland stroma-derived stem cells (SGSCs). SGSCs express Nestin, Oct4, and Nanog and stem cell markers, namely, Klf4, Sox2, and cMyc. The proliferation marker Ki67 is also expressed. The skin derived precursors (SKP) express nestin and Sox2 and also exhibit long-term proliferation potential when cultured in suspension [118].

The keratinocyte stem cells (KSC) are demarcated based on two cell-surface markers, namely, α6-integrin (α6) and CD71 [119, 120]. Thus, these KSCs are $\alpha6^{bri}CD71^{dim}$ or $CD49^{bri}CD71^{dim}$ expression [40, 41]. They are also positive for p63+ [73]. One of the markers of McSCs is CD34 [121] while the merkel cell stem cells can be identified as *Atoh1*-positive skin cells [81].

6.9 Markers of Thymus Epithelial Stem Cells (TESC)

The two supposed adult bipotent TEC stem/progenitor populations are clearly discrete and includes markers like FoxN1 and Epithelial Cell Adhesion Molecule (EpCAM): one is $FoxN1^{+}EpCAM^{+}$ and the other $FoxN1^{-}$ $EpCAM^{-}$ [122].

6.10 Markers of Oral Epithelial Stem Cells (OESC)

The oral epithelial stem cells (OESCs) are $Ki\text{-}67^{+}CD29^{+}CD49f^{hi}keratin\ 13^{low}$. Additionally, expression of other stem cell markers like p63 and K19 have been observed. The OESCs can be isolated from $\alpha6\beta4^{+}$ $CD71^{-}$ gingival keratinocytes which have high colony forming efficiency and can form a fully stratified epithelium, the oral epithelial equivalent (OEE), derived from isolated oral epithelial cells that is grown in vitro [123–125].

6.11 Markers of Oesophageal Epithelial Stem Cells (OeSC)

The oesophageal epithelial stem cells (OeSCs) are characterized by lesser expression of Integrin ß1 and higher levels of the ß2 laminin chain. OeSCs are less proliferative (Ki67+), and divide asymmetrically to generate one stem cell and one differentiated daughter cell in response to signals from the underlying basement membrane

[126]. Interestingly, according to Croagh et al. the oesophageal basal cells have three subpopulations based on their expression levels of Integrin α6 and transferrin receptor CD71. These groups include quiescent stem cells (Itgα6^{high}CD71^{low}), transit-amplifying cells (Itgα6^{high}CD71^{high}), and cells destined to differentiate (Itgα6^{low} CD71^{high}) [127]. These cells are also positive for CD49f [128].

6.12 Markers of Salivary Gland Stem Cells (SGSCs)

Salivary gland stem cells (SGSCs) express EpCAM + cells and nuclear β-catenin. Additionally, they also express cell-surface markers like CD49f, c-Kit, CD133, CD24, and CD29 [129–131].

6.13 Markers of Epithelial Intestine Stem Cells (ISC)

Epithelial intestine stem cells (ISC) are identified by the presence of pleckstrin homology-like domain family A member 1 (PHLDA1+), CD44 along with CD49f, CD24, CD166 and lack of GRP78 and c-kit [132–134].

6.14 Markers of Haematopoietic Stem Cells (HSCs)

Zfx is a shared transcriptional regulator of HSCs [59]. ABCG2, which is a member of the ATP-binding cassette transporter family, serves as a specific biomarker for BM stem cells [74]. Integrin α6 (CD49f/CD29) plays significant role in HSCs and HPCs homing to the BM and human prostate carcinoma cells [2].CD105 is expressed on 30–80% of CD34+ cells and is a marker for human long-term repopulating HSCs [44]. The common cell surface markers of HSCs are CD34 [26, 135], CDw90 [10, 136], CD117 [137], CD133 [18, 138], CD44 [45], and ALDH [77, 78] to name a few.

6.15 Markers of Gallbladder Stem Cells

Gallbladder stem cells are EpCAM+. CD44 + CD13 + cells represent the cells most enriched for clonal self-renewal from primary gallbladder [139].

6.16 Markers of Hepatic Stem Cells (HpSCs)

The hepatic stem cells (HpSCs) express the combination of EpCAM, neural cell adhesion molecule (NCAM), cytokeratin (CK) 19, albumin±, and are negative for α-fetoprotein (AFP). They are precursors of hepatoblasts, which differ from HpSCs in morphology, size, and in expressing the combination of EpCAM, intercellular cell adhesion molecule (ICAM-1), albumin^{++}, CK19, and AFP^{++} [140].

6.17 Markers of Lung Epithelial Stem Cells

The basal epithelial stem cell population from human lung tissue is KRT5 + TP63 + [141].

6.18 Markers of Mammary Epithelial Stem Cells (MaSCs)

A novel Wnt target in the mammary gland, protein C receptor (Procr), is the main marker of MaScs. Procr-positive cells are present in the basal layer, exhibit epithelial-to-mesenchymal transition characteristics, and also show low levels of basal keratins. Procr-expressing cells have a greater capacity of regeneration in transplantation assays and differentiate into all lineages of the mammary epithelium by lineage tracing [142].

6.19 Markers of Spermatogonial Stem Cells

The expression of inhibitor of differentiation 4 (ID4) [143], paired box 7(PAX7) [144], and B cell-specific Moloney murine leukaemia virus integration site1 (BMI1) are markers for the spermatogonial stem cells [145].

6.20 Markers of Mesenchymal Stem Cells

Cell surface markers of the mesenchymal stem cells (MSCs) include positive expression of CD90, CD73, and CD105, and negative expression of CD34, CD14, CD45, and human leukocyte antigen-DR (HLA-DR) [146]. Stro-1, an endothelial antigen is the best-known MSC marker [48].

6.21 Markers of Cardiac Stem Cell

The cardiac progenitors/stem cells are recognized by the presence of a set of cell surface markers like SCA-1 (Ly6A), c-kit, and ABC transporter cells [126, 147].

6.22 Markers of Muscle Stem Cells (MuSC)

Satellite cells can be identified by the explicit expression of intracellular markers like the nuclear membrane proteins lamin A/C (LMNA), transcription factors PAX7 and emerin (EMD). Cell membrane surface expression include syndecans 3 and 4 (SDC3 and SDC4), muscle (M)-cadherin, calcitonin receptor (CALCR), C-X-C chemokine receptor type 4 (CXCR4), calveolin 1 (CAV1), a7- and b1-integrins, NCAM1, vascular cell adhesion molecule 1 (VCAM1), and CD34 [148–150].

6.23 Markers of Adipose Stem Cells

Human adipose-derived stem cells (ADSCs) are the adult stem cells isolated from fat tissues during procedures such as liposuction, abdominoplasty, or breast reduction. They are found in the stromal-vascular junction. These cells also express genes that constitute the core circuitry of self-renewal such as Sox2, Oct4, Nanog, and neurogenic lineage genes such as Neurod1, Pax6, and Sox3. They find importance in diverse therapeutic procedures aimed at treating different forms of tissue damage, the autologous transplant of adipose tissue being an added advantage [151].

References

1. Breimer ME, Säljö K, Barone A, Teneberg S (2017) Glycosphingolipids of human embryonic stem cells. Glycoconj J 34(6):713–723. https://doi.org/10.1007/s10719-016-9706-y
2. Zhao W, Ji X, Zhang F, Li L, Ma L (2012) Embryonic stem cell markers. Molecules 17 (6):6196–6236. https://doi.org/10.3390/molecules17066196
3. Henderson JK, Draper JS, Baillie HS, Fishel S, Thomson JA, Moore H, Andrews PW (2002) Preimplantation human embryos and embryonic stem cells show comparable expression of stage-specific embryonic antigens. Stem Cells 20(4):329–337
4. Shamblott MJ, Axelman J, Wang S, Bugg EM, Littlefield JW, Donovan PJ, Blumenthal PD, Huggins GR, Gearhart JD (1998) Derivation of pluripotent stem cells from cultured human primordial germ cells. Proc Natl Acad Sci U S A 95(23):13726–13731. Erratum in: Proc Natl Acad Sci USA 1999 Feb 2;96(3):1162
5. Wright AJ, Andrews PW (2009) Surface marker antigens in the characterization of human embryonic stem cells. Stem Cell Res 3(1):3–11. https://doi.org/10.1016/j.scr.2009.04.001
6. Brimble SN, Sherrer ES, Uhl EW, Wang E, Kelly S, Merrill AH Jr, Robins AJ, Schulz TC (2007) The cell surface glycosphingolipids SSEA-3 and SSEA-4 are not essential for human ESC pluripotency. Stem Cells 25(1):54–62
7. Kannagi R, Cochran NA, Ishigami F, Hakomori S, Andrews PW, Knowles BB, Solter D (1983) Stage-specific embryonic antigens (SSEA-3 and -4) are epitopes of a unique globo-series ganglioside isolated from human teratocarcinoma cells. EMBO J 2(12):2355–2361
8. Atari M, Barajas M, Hernández-Alfaro F, Gil C, Fabregat M, FerrésPadró E, Giner L, Casals N (2011) Isolation of pluripotent stem cells from human third molar dental pulp. Histol Histopathol 26(8):1057–1070. https://doi.org/10.14670/HH-26.1057
9. Reif AE, Allen JM (1964) The AKR thymic antigen and its distribution in leukemias and nervous tissues. J Exp Med 120:413–433

10. Craig W, Kay R, Cutler RL, Lansdorp PM (1993) Expression of Thy-1 on human hematopoietic progenitor cells. J Exp Med 177(5):1331–1342
11. Draper JS, Pigott C, Thomson JA, Andrews PW (2002) Surface antigens of human embryonic stem cells: changes upon differentiation in culture. J Anat 200(Pt 3):249–258
12. Ullah M, Akbar A, Thakor AS (2019) An emerging role of CD9 in stemness and chemoresistance. Oncotarget 10(40):4000–4001. https://doi.org/10.18632/oncotarget.27021
13. Brosseau C, Colas L, Magnan A, Brouard S (2018) CD9 tetraspanin: a new pathway for the regulation of inflammation? Front Immunol 9:2316. https://doi.org/10.3389/fimmu.2018.02316
14. Lian Q, Lye E, Suan Yeo K, Khia E, Salto-Tellez M, Liu TM, Palanisamy N, El Oakley RM, Lee EH, Lim B, Lim SK (2007) Derivation of clinically compliant MSCs from CD105+, CD24- differentiated human ESCs. Stem Cells 25(2):425–436
15. Ullrich A, Schlessinger J (1990) Signal transduction by receptors with tyrosine kinase activity. Cell 61(2):203–212
16. Sperling C, Schwartz S, Büchner T, Thiel E, Ludwig WD (1997) Expression of the stem cell factor receptor C-KIT (CD117) in acute leukemias. Haematologica 82(5):617–621
17. Yin T, Li L (2006) The stem cell niches in bone. J Clin Invest 116(5):1195–1201
18. Bonanno G, Perillo A, Rutella S, De Ritis DG, Mariotti A, Marone M, Meoni F, Scambia G, Leone G, Mancuso S, Pierelli L (2004) Clinical isolation and functional characterization of cord blood CD133+ hematopoietic progenitor cells. Transfusion 44(7):1087–1097
19. Ruoslahti E, Pierschbacher MD (1987) New perspectives in cell adhesion: RGD and integrins. Science 238(4826):491–497
20. Watt FM, Hogan BL (2000) Out of Eden: stem cells and their niches. Science 287 (5457):1427–1430
21. Fässler R, Pfaff M, Murphy J, Noegel AA, Johansson S, Timpl R, Albrecht R (1995) Lack of beta 1 integrin gene in embryonic stem cells affects morphology, adhesion, and migration but not integration into the inner cell mass of blastocysts. J Cell Biol 128(5):979–988
22. Goodfellow PJ, Nevanlinna HA, Gorman P, Sheer D, Lam G, Goodfellow PN (1989) Assignment of the gene encoding the beta-subunit of the human fibronectin receptor (beta-FNR) to chromosome 10p11.2. Ann Hum Genet 53(1):15–22. https://doi.org/10.1111/j.1469-1809.1989.tb01118.x
23. Krebsbach PH, Villa-Diaz LG (2017) The role of Integrin α6 (CD49f) in stem cells: more than a conserved biomarker. Stem Cells Dev 26(15):1090–1099. https://doi.org/10.1089/scd.2016.0319
24. Katoh Y, Katoh M (2007) Conserved POU-binding site linked to SP1-binding site within FZD5 promoter: transcriptional mechanisms of FZD5 in undifferentiated human ES cells, fetal liver/spleen, adult colon, pancreatic islet, and diffuse-type gastric cancer. Int J Oncol 30 (3):751–755
25. Gray PC, Shani G, Aung K, Kelber J, Vale W (2006) Cripto binds transforming growth factor beta (TGF-beta) and inhibits TGF-beta signaling. Mol Cell Biol 26(24):9268–9278. Erratum in: Mol Cell Biol. 2008 Dec;28(23):7260
26. Civin CI, Strauss LC, Brovall C, Fackler MJ, Schwartz JF, Shaper JH (1984) Antigenic analysis of hematopoiesis. III. A hematopoietic progenitor cell surface antigen defined by a monoclonal antibody raised against KG-1a cells. J Immunol 133(1):157–165
27. Fina L, Molgaard HV, Robertson D, Bradley NJ, Monaghan P, Delia D, Sutherland DR, Baker MA, Greaves MF (1990) Expression of the CD34 gene in vascular endothelial cells. Blood 75 (12):2417–2426
28. Blanpain C, Lowry WE, Geoghegan A, Polak L, Fuchs E (2004) Self-renewal, multipotency, and the existence of two cell populations within an epithelial stem cell niche. Cell 118 (5):635–648. https://doi.org/10.1016/j.cell.2004.08.012

29. Healy L, May G, Gale K, Grosveld F, Greaves M, Enver T (1995) The stem cell antigen CD34 functions as a regulator of hemopoietic cell adhesion. Proc Natl Acad Sci U S A 92 (26):12240–12244. https://doi.org/10.1073/pnas.92.26.12240
30. Sandeep SJ, Tandukar B, Pan L, Huang JM, Livak F, Smith BJ, Hodges T, Mahurkar AA, Hornyak TJ (2019) CD34 defines melanocyte stem cell subpopulations with distinct regenerative properties. PLoS Genet 15(4):e1008034. https://doi.org/10.1371/journal.pgen.1008034
31. Bernal A, Arranz L (2018) Nestin-expressing progenitor cells: function, identity and therapeutic implications. Cell Mol Life Sci 75(12):2177–2195. https://doi.org/10.1007/s00018-018-2794-z
32. Aguirre A, Rubio ME, Gallo V (2010) Notch and EGFR pathway interaction regulates neural stem cell number and self-renewal. Nature 467(7313):323–327. https://doi.org/10.1038/nature09347
33. Cao J, Wei Y, Lian J, Yang L, Zhang X, Xie J, Liu Q, Luo J, He B, Tang M (2017) Notch signaling pathway promotes osteogenic differentiation of mesenchymal stem cells by enhancing BMP9/Smadsignaling. Int J Mol Med 40(2):378–388. https://doi.org/10.3892/ijmm.2017.3037
34. Andria ML, Hsieh CL, Oren R, Francke U, Levy S (1991) Genomic organization and chromosomal localization of the TAPA-1 gene. J Immunol 147(3):1030–1036
35. Tiwari-Woodruff SK, Buznikov AG, Vu TQ, Micevych PE, Chen K, Kornblum HI, Bronstein JM (2001) OSP/claudin-11 forms a complex with a novel member of the tetraspanin super family and beta1 integrin and regulates proliferation and migration of oligodendrocytes. J Cell Biol 153(2):295–305. https://doi.org/10.1083/jcb.153.2.295
36. Klassen H, Schwartz MR, Bailey AH, Young MJ (2001) Surface markers expressed by multipotent human and mouse neural progenitor cells include tetraspanins and non-protein epitopes. Neurosci Lett 312(3):180–182. https://doi.org/10.1016/s0304-3940(01)02215-7
37. Lin KK, Rossi L, Boles NC, Hall BE, George TC, Goodell MA (2011) CD81 is essential for the re-entry of hematopoietic stem cells to quiescence following stress-induced proliferation via deactivation of the Akt pathway. PLoS Biol 9(9):e1001148. https://doi.org/10.1371/journal.pbio.1001148
38. Wada T, Haigh JJ, Ema M, Hitoshi S, Chaddah R, Rossant J, Nagy A, van der Kooy D (2006) Vascular endothelial growth factor directly inhibits primitive neural stem cell survival but promotes definitive neural stem cell survival. J Neurosci 26(25):6803–6812. https://doi.org/10.1523/JNEUROSCI.0526-06.2006
39. Aisen P (2004) Transferrin receptor 1. Int J Biochem Cell Biol 36(11):2137–2143. https://doi.org/10.1016/j.biocel.2004.02.007
40. Hayashi R, Yamato M, Saito T, Oshima T, Okano T, Tano Y, Nishida K (2008) Enrichment of corneal epithelial stem/progenitor cells using cell surface markers, integrin alpha6 and CD71. Biochem Biophys Res Commun 367(2):256–263
41. Metral E, Bechetoille N, Demarne F, Rachidi W, Damour O (2017) $\alpha 6$ integrin ($\alpha 6^{high}$)/transferrin receptor $(CD71)^{low}$ keratinocyte stem cells are more potent for generating reconstructed skin epidermis than rapid adherent cells. Int J Mol Sci 18(2):282. https://doi.org/10.3390/ijms18020282
42. Fletcher DA, Mullins RD (2010) Cell mechanics and the cytoskeleton. Nature 463 (7280):485–492. https://doi.org/10.1038/nature08908
43. Rahman MM, Subramani J, Ghosh M, Denninger JK, Takeda K, Fong GH, Carlson ME, Shapiro LH (2014) CD13 promotes mesenchymal stem cell-mediated regeneration of ischemic muscle. Front Physiol 4:402. https://doi.org/10.3389/fphys.2013.00402
44. Kays SK, Kaufmann KB, Abel T, Brendel C, Bonig H, Grez M, Buchholz CJ, Kneissl S (2015) CD105 is a surface marker for receptor-targeted gene transfer into human long-term repopulating hematopoietic stem cells. Stem Cells Dev 24(6):714–723. https://doi.org/10.1089/scd.2014.0455

45. Senbanjo LT, Chellaiah MA (2017) CD44: a multifunctional cell surface adhesion receptor is a regulator of progression and metastasis of cancer cells. Front Cell Dev Biol 5:18. https://doi.org/10.3389/fcell.2017.00018
46. Nuti N, Corallo C, Chan BM, Ferrari M, Gerami-Naini B (2016) Multipotent differentiation of human dental pulp stem cells: a literature review. Stem Cell Rev Rep 12(5):511–523
47. Kanafi MM, Ramesh A, Gupta PK, Bhonde RR (2014) Dental pulp stem cells immobilized in alginate microspheres for applications in bone tissue engineering. Int Endod J 47(7):687–697. https://doi.org/10.1111/iej.12205
48. Ning H, Lin G, Lue TF, Lin CS (2011) Mesenchymal stem cell marker Stro-1 is a 75 kd endothelial antigen. Biochem Biophys Res Commun 413(2):353–357. https://doi.org/10.1016/j.bbrc.2011.08.104
49. Pesce M, Schöler HR (2000) Oct-4: control of totipotency and germline determination. Mol Reprod Dev 55(4):452–457
50. Schoorlemmer J, Kruijer W (1991) Octamer-dependent regulation of the kFGF gene in embryonal carcinoma and embryonic stem cells. Mech Dev 36(1–2):75–86
51. Boyer LA, Lee TI, Cole MF, Johnstone SE, Levine SS, Zucker JP, Guenther MG, Kumar RM, Murray HL, Jenner RG, Gifford DK, Melton DA, Jaenisch R, Young RA (2005) Core transcriptional regulatory circuitry in human embryonic stem cells. Cell 122(6):947–956
52. Takahashi K, Yamanaka S (2006) Induction of pluripotent stem cells from mouse embryonic and adult fibroblast cultures by defined factors. Cell 126(4):663–676
53. Ghaleb AM, Nandan MO, Chanchevalap S, Dalton WB, Hisamuddin IM, Yang VW (2005) Krüppel-like factors 4 and 5: the yin and yang regulators of cellular proliferation. Cell Res 15 (2):92–96. Review
54. Chambers I, Colby D, Robertson M, Nichols J, Lee S, Tweedie S, Smith A (2003) Functional expression cloning of Nanog, a pluripotency sustaining factor in embryonic stem cells. Cell 113(5):643–655
55. Koestenbauer S, Zech NH, Juch H, Vanderzwalmen P, Schoonjans L, Dohr G (2006) Embryonic stem cells: similarities and differences between human and murine embryonic stem cells. Am J Reprod Immunol 55(3):169–180
56. Okita K, Ichisaka T, Yamanaka S (2007) Generation of germline-competent induced pluripotent stem cells. Nature 448(7151):313–317
57. Zhao Y, Yin X, Qin H, Zhu F, Liu H, Yang W, Zhang Q, Xiang C, Hou P, Song Z, Liu Y, Yong J, Zhang P, Cai J, Liu M, Li H, Li Y, Qu X, Cui K, Zhang W, Xiang T, Wu Y, Zhao Y, Liu C, Yu C, Yuan K, Lou J, Ding M, Deng H (2008) Two supporting factors greatly improve the efficiency of human iPSC generation. Cell Stem Cell 3(5):475–479. https://doi.org/10.1016/j.stem.2008.10.002
58. Schneider-Gädicke A, Beer-Romero P, Brown LG, Mardon G, Luoh SW, Page DC (1989) Putative transcription activator with alternative isoforms encoded by human ZFX gene. Nature 342(6250):708–711
59. Galan-Caridad JM, Harel S, Arenzana TL, Hou ZE, Doetsch FK, Mirny LA, Reizis B (2007) Zfx controls the self-renewal of embryonic and hematopoietic stem cells. Cell 129(2):345–357
60. Plank-Bazinet JL, Mundell NA (2016) The paradox of Foxd3: how does it function in pluripotency and differentiation of embryonic stem cells? Stem Cell Investig 3:73
61. Cleynen I, Van de Ven WJ (2008) The HMGA proteins: a myriad of functions (review). Int J Oncol 32(2):289–305
62. Kalivas PW, Duffy P, Mackler SA (1999) Interrupted expression of NAC-1 augments the behavioral responses to cocaine. Synapse 33(2):153–159
63. Kim J, Chu J, Shen X, Wang J, Orkin SH (2008) An extended transcriptional network for pluripotency of embryonic stem cells. Cell 132(6):1049–1061. https://doi.org/10.1016/j.cell.2008.02.039. Erratum in: Cell. 2008 Jun 27;133(7):1290

64. Lan ZJ, Chung AC, Xu X, DeMayo FJ, Cooney AJ (2002) The embryonic function of germ cell nuclear factor is dependent on the DNA binding domain. J Biol Chem 277 (52):50660–50667
65. Burdon T, Smith A, Savatier P (2002) Signalling, cell cycle and pluripotency in embryonic stem cells. Trends Cell Biol 12(9):432–438
66. Cole MF, Johnstone SE, Newman JJ, Kagey MH, Young RA (2008) Tcf3 is an integral component of the core regulatory circuitry of embryonic stem cells. Genes Dev 22 (6):746–755. https://doi.org/10.1101/gad.1642408
67. Kohlhase J, Wischermann A, Reichenbach H, Froster U, Engel W (1998) Mutations in the SALL1 putative transcription factor gene cause Townes-brocks syndrome. Nat Genet 18 (1):81–83. https://doi.org/10.1038/ng0198-81
68. Mitsui K, Tokuzawa Y, Itoh H et al (2003) The homeoprotein Nanog is required for maintenance of pluripotency in mouse epiblast and ES cells. Cell 113(5):631–642. https://doi.org/10.1016/s0092-8674(03)00393-3
69. Wong RC, Ibrahim A, Fong H, Thompson N, Lock LF, Donovan PJ (2011) L1TD1 is a marker for undifferentiated human embryonic stem cells. PLoS One 6(4):e19355. https://doi.org/10.1371/journal.pone.0019355
70. Maldonado-Saldivia J, van den Bergen J, Krouskos M et al (2007) Dppa2 and Dppa4 are closely linked SAP motif genes restricted to pluripotent cells and the germ line. Stem Cells 25 (1):19–28. https://doi.org/10.1634/stemcells.2006-0269
71. Verschuur M, de Jong M, Felida L, de Maat MP, Vos HL (2005) A hepatocyte nuclear factor-3 site in the fibrinogen beta promoter is important for interleukin 6-induced expression, and its activity is influenced by the adjacent -148C/T polymorphism. J Biol Chem 280 (17):16763–16771. https://doi.org/10.1074/jbc.M501973200
72. Seruggia D, Oti M, Tripathi P, Canver MC, LeBlanc L, Di Giammartino DC, Bullen MJ, Nefzger CM, Sun YBY, Farouni R, Polo JM, Pinello L, Apostolou E, Kim J, Orkin SH, Das PP (2019) TAF5L and TAF6L maintain self-renewal of embryonic stem cells via the MYC regulatory network. Mol Cell 74(6):1148–1163. https://doi.org/10.1016/j.molcel.2019.03.025
73. Pellegrini G, Dellambra E, Golisano O, Martinelli E, Fantozzi I, Bondanza S, Ponzin D, McKeon F, De Luca M (2001) p63 identifies keratinocyte stem cells. Proc Natl Acad Sci U S A 98(6):3156–3161. https://doi.org/10.1073/pnas.061032098
74. Goodell MA (2005) Stem cell identification and sorting using the Hoechst 33342 side population (SP). Curr Protoc Cytom 34:9. https://doi.org/10.1002/0471142956.cy0918s34
75. Ebrahimi M, Taghi-Abadi E, Baharvand H (2009) Limbal stem cells in review. J Ophthalmic Vis Res 4(1):40–58
76. Tsialikas J, Romer-Seibert J (2015) LIN28: roles and regulation in development and beyond. Development 142(14):2397–2404. https://doi.org/10.1242/dev.117580
77. Kastan MB, Schlaffer E, Russo JE, Colvin OM, Civin CI, Hilton J (1990) Direct demonstration of elevated aldehyde dehydrogenase in human hematopoietic progenitor cells. Blood 75 (10):1947–1950
78. Jones RJ, Barber JP, Vala MS, Collector MI, Kaufmann SH, Ludeman SM, Colvin OM, Hilton J (1995) Assessment of aldehyde dehydrogenase in viable cells. Blood 85(10):2742–2746
79. Storms RW, Green PD, Safford KM, Niedzwiecki D, Cogle CR, Colvin OM, Chao NJ, Rice HE, Smith CA (2005) Distinct hematopoietic progenitor compartments are delineated by the expression of aldehyde dehydrogenase and CD34. Blood 106(1):95–102
80. Solter D, Knowles BB, Sundberg M, Jansson L, Ketolainen J, Pihlajamäki H, Suuronen R, Skottman H, Inzunza J, Hovatta O, Narkilahti S (2009) CD marker expression profiles of human embryonic stem cells and their neural derivatives, determined using flow-cytometric analysis, reveal a novel CD marker for exclusion of pluripotent stem cells. Stem Cell Res 2 (2):113–124. https://doi.org/10.1016/j.scr.2008.08.001
81. Wright MC, Reed-Geaghan EG, Bolock AM, Fujiyama T, Hoshino M, Maricich SM (2015) Unipotent, Atoh1+ progenitors maintain the Merkel cell population in embryonic and adult mice. J Cell Biol 208(3):367–379. https://doi.org/10.1083/jcb.201407101

82. Drukker M, Katz G, Urbach A, Schuldiner M, Markel G, Itskovitz-Eldor J, Reubinoff B, Mandelboim O, Benvenisty N (2002) Characterization of the expression of MHC proteins in human embryonic stem cells. Proc Natl Acad Sci U S A 99(15):9864–9869
83. Andrews PW, Meyer LJ, Bednarz KL, Harris H (1984) Two monoclonal antibodies recognizing determinants on human embryonal carcinoma cells react specifically with the liver isozyme of human alkaline phosphatase. Hybridoma 3(1):33–39
84. Benham FJ, Andrews PW, Knowles BB, Bronson DL, Harris H (1981) Alkaline phosphatase isozymes as possible markers of differentiation in human testicular teratocarcinoma cell lines. Dev Biol 88(2):279–287
85. Štefková K, Procházková J, Pacherník J (2015) Alkaline phosphatase in stem cells. Stem Cells Int 2015:628368. https://doi.org/10.1155/2015/628368
86. Herszfeld D, Wolvetang E, Langton-Bunker E, Chung TL, Filipczyk AA, Houssami S, Jamshidi P, Koh K, Laslett AL, Michalska A, Nguyen L, Reubinoff BE, Tellis I, Auerbach JM, Ording CJ, Looijenga LH, Pera MF (2006) CD30 is a survival factor and a biomarker for transformed human pluripotent stem cells. Nat Biotechnol 24(3):351–357
87. Williams BP, Daniels GL, Pym B, Sheer D, Povey S, Okubo Y, Andrews PW, Goodfellow PN (1988) Biochemical and genetic analysis of the OKa blood group antigen. Immunogenetics 27 (5):322–329
88. Bashamboo A, Taylor AH, Samuel K, Panthier JJ, Whetton AD, Forrester LM (2006) The survival of differentiating embryonic stem cells is dependent on the SCF-KIT pathway. J Cell Sci 119(Pt 15):3039–3046
89. Xu C, Inokuma MS, Denham J, Golds K, Kundu P, Gold JD, Carpenter MK (2001) Feeder-free growth of undifferentiated human embryonic stem cells. Nat Biotechnol 19(10):971–974
90. Layden BT, Newman M, Chen F, Fisher A, Lowe WL (2010) G protein coupled receptors in embryonic stem cells: a role for Gs-alpha signaling. PLoS One 5:e9105
91. Bhattacharya B, Miura T, Brandenberger R, Mejido J, Luo Y, Yang AX, Joshi BH, Ginis I, Thies RS, Amit M, Lyons I, Condie BG, Itskovitz-Eldor J, Rao MS, Puri RK (2004) Gene expression in human embryonic stem cell lines: unique molecular signature. Blood 103 (8):2956–2964
92. Minchiotti G, Parisi S, Persico MG (2006) Criptosignaling in differentiating embryonic stem cells. Methods Mol Biol 329:151–169
93. Chen X, Xu H, Yuan P et al (2008) Integration of external signaling pathways with the core transcriptional network in embryonic stem cells. Cell 133(6):1106–1117. https://doi.org/10.1016/j.cell.2008.04.043
94. Kim J, Woo AJ, Chu J, Snow JW, Fujiwara Y, Kim CG, Cantor AB, Orkin SH (2010) A Myc network accounts for similarities between embryonic stem and cancer cell transcription programs. Cell 143(2):313–324. https://doi.org/10.1016/j.cell.2010.09.010
95. Laugesen A, Helin K (2014) Chromatin repressive complexes in stem cells, development, and cancer. Cell Stem Cell 14(6):735–751. https://doi.org/10.1016/j.stem.2014.05.006
96. van den Boom V, Kooistra SM, Boesjes M, Geverts B, Houtsmuller AB, Monzen K, Komuro I, Essers J, Drenth-Diephuis LJ, Eggen BJ (2007) UTF1 is a chromatin-associated protein involved in ES cell differentiation. J Cell Biol 178(6):913–924
97. Liu Y, Labosky PA (2008 Oct) Regulation of embryonic stem cell self-renewal and pluripotency by Foxd3. Stem Cells 26(10):2475–2484. https://doi.org/10.1634/stemcells.2008-0269
98. Navarra A, Musto A, Gargiulo A, Petrosino G, Pierantoni GM, Fusco A, Russo T, Parisi S (2016) Hmga2 is necessary for Otx2-dependent exit of embryonic stem cells from the pluripotent ground state. BMC Biol 14:24. https://doi.org/10.1186/s12915-016-0246-5
99. Wang H, Wang X, Xu X, Kyba M, Cooney AJ (2016) Germ cell nuclear factor (GCNF) represses Oct4 expression and globally modulates gene expression in human embryonic stem (hES) cells. J Biol Chem 291(16):8644–8652. https://doi.org/10.1074/jbc.M115.694208

100. Zhang J, Tam WL, Tong GQ, Wu Q, Chan HY, Soh BS, Lou Y, Yang J, Ma Y, Chai L et al (2006) Sall4 modulates embryonic stem cell pluripotency and early embryonic development by the transcriptional regulation of Pou5f1. Nat Cell Biol 8(10):1114–1123. https://doi.org/10.1038/ncb1481
101. Boiani M, Schöler HR (2005) Regulatory networks in embryo-derived pluripotent stem cells. Nat Rev Mol Cell Biol 6(11):872–884. https://doi.org/10.1038/nrm1744
102. Niwa H, Burdon T, Chambers I, Smith A (1998) Self-renewal of pluripotent embryonic stem cells is mediated via activation of STAT3. Genes Dev 12(13):2048–2060. https://doi.org/10.1101/gad.12.13.2048
103. Chowdhury S, Bhattacharya K, Mandal C, Mondal S, Sarkar S, Chandra S, Banerjee S, Mandal C (2018) A glycomic approach towards identification of signature molecules in $CD34^+$ haematopoietic stem cells from umbilical cord blood. Adv Exp Med Biol 1112:309–318. https://doi.org/10.1007/978-981-13-3065-0_21
104. Capela A, Temple S (2002) LeX/ssea-1 is expressed by adult mouse CNS stem cells, identifying them as nonependymal. Neuron 35(5):865–875. https://doi.org/10.1016/s0896-6273(02)00835-8
105. Pruszak J, Ludwig W, Blak A, Alavian K, Isacson O (2009) CD15, CD24, and CD29 define a surface biomarker code for neural lineage differentiation of stem cells. Stem Cells 27 (12):2928–2940. https://doi.org/10.1002/stem.211
106. Wang J, O'Bara MA, Pol SU, Sim FJ (2013) CD133/CD140a-based isolation of distinct human multipotent neural progenitor cells and oligodendrocyte progenitor cells. Stem Cells Dev 22(15):2121–2131. https://doi.org/10.1089/scd.2013.0003
107. Barraud P, Thompson L, Kirik D, Björklund A, Parmar M (2005) Isolation and characterization of neural precursor cells from the Sox1-GFP reporter mouse. Eur J Neurosci 22 (7):1555–1569. https://doi.org/10.1111/j.1460-9568.2005.04352.x
108. Doetsch F, Petreanu L, Caille I, Garcia-Verdugo JM, Alvarez-Buylla A (2002) EGF converts transit-amplifying neurogenic precursors in the adult brain into multipotent stem cells. Neuron 36(6):1021–1034. https://doi.org/10.1016/s0896-6273(02)01133-9
109. Yang X, Cepko CL (1996) Flk-1, a receptor for vascular endothelial growth factor (VEGF), is expressed by retinal progenitor cells. J Neurosci 16(19):6089–6099. https://doi.org/10.1523/JNEUROSCI.16-19-06089.1996
110. Yuan SH, Martin J, Elia J, Flippin J, Paramban RI, Hefferan MP, Vidal JG, Mu Y, Killian RL, Israel MA, Emre N, Marsala S, Marsala M, Gage FH, Goldstein LS, Carson CT (2011) Cell-surface marker signatures for the isolation of neural stem cells, glia and neurons derived from human pluripotent stem cells. PLoS One 6(3):e17540. https://doi.org/10.1371/journal.pone.0017540
111. Zhang J, Jiao J (2015) Molecular biomarkers for embryonic and adult neural stem cell and neurogenesis. Biomed Res Int 2015:727542. https://doi.org/10.1155/2015/727542
112. Codega P, Silva-Vargas V, Paul A, Maldonado-Soto AR, Deleo AM, Pastrana E, Doetsch F (2014) Prospective identification and purification of quiescent adult neural stem cells from their in vivo niche. Neuron 82(3):545–559. https://doi.org/10.1016/j.neuron.2014.02.039
113. Guo ZH, Zhang W, Jia YYS, Liu QX, Li ZF, Lin JS (2018) An insight into the difficulties in the discovery of specific biomarkers of Limbal stem cells. Int J Mol Sci 19(7):1982. https://doi.org/10.3390/ijms19071982
114. Schlötzer-Schrehardt U, Dietrich T, Saito K et al (2007) Characterization of extracellular matrix components in the limbal epithelial stem cell compartment. Exp Eye Res 85 (6):845–860. https://doi.org/10.1016/j.exer.2007.08.020
115. Lim MN, Hussin NH, Othman A, Umapathy T, Baharuddin P, Jamal R, Zakaria Z (2012) Ex vivo expanded SSEA-4+ human limbal stromal cells are multipotent and do not express other embryonic stem cell markers. Mol Vis 18:1289–1300

116. Aydin S, Şahin F (2019) Stem cells derived from dental tissues. Adv Exp Med Biol 1144:123–132. https://doi.org/10.1007/5584_2018_333
117. Amoh Y, Hoffman RM (2017) Hair follicle-associated-pluripotent (HAP) stem cells. Cell Cycle 16(22):2169–2175. https://doi.org/10.1080/15384101.2017.1356513
118. Danner S, Kremer M, Petschnik AE, Nagel S, Zhang Z, Hopfner U, Reckhenrich AK, Weber C, Schenck TL, Becker T, Kruse C, Machens HG, Egaña JT (2012) The use of human sweat gland-derived stem cells for enhancing vascularization during dermal regeneration. J Invest Dermatol 132(6):1707–1716. https://doi.org/10.1038/jid.2012.31. Epub 2012 Mar 1
119. Metral E, Bechetoille N, Demarne F, Damour O, Rachidi W (2018) Keratinocyte stem cells are more resistant to UVA radiation than their direct progeny. PLoS One 13(9):e0203863. https://doi.org/10.1371/journal.pone.0203863
120. Li A, Simmons PJ, Kaur P (1998) Identification and isolation of candidate human keratinocyte stem cells based on cell surface phenotype. Proc Natl Acad Sci U S A 95(7):3902–3907
121. Joshi SS, Tandukar B, Pan L, Huang JM, Livak F, Smith BJ, Hodges T, Mahurkar AA, Hornyak TJ (2019) CD34 defines melanocyte stem cell subpopulations with distinct regenerative properties. PLoS Genet 15(4):e1008034. https://doi.org/10.1371/journal.pgen.1008034
122. Blackburn CC, Manley NR, Palmer DB, Boyd RL, Anderson G, Ritter MA (2002) One for all and all for one: thymic epithelial stem cells and regeneration. Trends Immunol 23(8):391–395
123. Jones KB, Klein OD (2013) Oral epithelial stem cells in tissue maintenance and disease: the first steps in a long journey. Int J Oral Sci 5(3):121–129. https://doi.org/10.1038/ijos.2013.46
124. Igarashi T, Shimmura S, Yoshida S, Tonogi M, Shinozaki N, Yamane GY (2008) Isolation of oral epithelial progenitors using collagen IV. Oral Dis 14(5):413–418. https://doi.org/10.1111/j.1601-0825.2007.01390.x
125. Calenic B, Ishkitiev N, Yaegaki K, Imai T, Costache M, Tovaru M, Tovaru S, Parlatescu I (2010) Characterization of oral keratinocyte stem cells and prospects of its differentiation to oral epithelial equivalents. Rom J Morphol Embryol 51(4):641–645
126. Pfister O, Oikonomopoulos A, Sereti KI, Sohn RL, Cullen D, Fine GC, Mouquet F, Westerman K, Liao R (2008) Role of the ATP-binding cassette transporter Abcg2 in the phenotype and function of cardiac side population cells. Circ Res 103(8):825–835. https://doi.org/10.1161/CIRCRESAHA.108.174615
127. Croagh D, Phillips WA, Redvers R, Thomas RJ, Kaur P (2007) Identification of candidate murine esophageal stem cells using a combination of cell kinetic studies and cell surface markers. Stem Cells 25(2):313–318
128. Zhao R, Quaroni L, Casson AG (2012) Identification and characterization of stemlike cells in human esophageal adenocarcinoma and normal epithelial cell lines. J Thorac Cardiovasc Surg 144:1192
129. Hisatomi Y, Okumura K, Nakamura K, Matsumoto S, Satoh A, Nagano K, Yamamoto T, Endo F (2004) Flow cytometric isolation of endodermal progenitors from mouse salivary gland differentiate into hepatic and pancreatic lineages. Hepatology 39(3):667–675
130. Lombaert IM, Brunsting JF, Wierenga PK, Faber H, Stokman MA, Kok T, Visser WH, Kampinga HH, de Haan G, Coppes RP (2008) Rescue of salivary gland function after stem cell transplantation in irradiated glands. PLoS One 3(4):e2063. https://doi.org/10.1371/journal.pone.0002063
131. Nanduri LS, Maimets M, Pringle SA, van der Zwaag M, van Os RP, Coppes RP (2011) Regeneration of irradiated salivary glands with stem cell marker expressing cells. Radiother Oncol 99(3):367–372. https://doi.org/10.1016/j.radonc.2011.05.085
132. Sakthianandeswaren A, Christie M, D'Andreti C, Tsui C, Jorissen RN, Li S, Fleming NI, Gibbs P, Lipton L, Malaterre J, Ramsay RG, Phesse TJ, Ernst M, Jeffery RE, Poulsom R, Leedham SJ, Segditsas S, Tomlinson IP, Bernhard OK, Simpson RJ, Walker F, Faux MC, Church N, Catimel B, Flanagan DJ, Vincan E, Sieber OM (2011) PHLDA1 expression marks

the putative epithelial stem cells and contributes to intestinal tumorigenesis. Cancer Res 71 (10):3709–3719. https://doi.org/10.1158/0008-5472.CAN-10-2342
133. Barker N, van Es JH, Kuipers J, Kujala P, van den Born M, Cozijnsen M, Haegebarth A, Korving J, Begthel H, Peters PJ, Clevers H (2007) Identification of stem cells in small intestine and colon by marker gene Lgr5. Nature 449(7165):1003–1007
134. Wang F, Scoville D, He XC, Mahe MM, Box A, Perry JM, Smith NR, Lei NY, Davies PS, Fuller MK, Haug JS, McClain M, Gracz AD, Ding S, Stelzner M, Dunn JC, Magness ST, Wong MH, Martin MG, Helmrath M, Li L (2013) Isolation and characterization of intestinal stem cells based on surface marker combinations and colony-formation assay. Gastroenterology 145(2):383–395. https://doi.org/10.1053/j.gastro.2013.04.050
135. Tindle RW, Nichols RA, Chan L, Campana D, Catovsky D, Birnie GD (1985) A novel monoclonal antibody BI-3C5 recognises myeloblasts and non-B non-T lymphoblasts in acute leukaemias and CGL blast crises, and reacts with immature cells in normal bone marrow. Leuk Res 9(1):1–9
136. Holden JT, Geller RB, Farhi DC, Holland HK, Stempora LL, Phillips CN, Bray RA (1995) Characterization of Thy-1 (CDw90) expression in CD34+ acute leukemia. Blood 86(1):60–65
137. Ashman LK, Cambareri AC, To LB, Levinsky RJ, Juttner CA (1991) Expression of the YB5. B8 antigen (c-kit proto-oncogene product) in normal human bone marrow. Blood 78(1):30–37
138. Yin AH, Miraglia S, Zanjani ED, Almeida-Porada G, Ogawa M, Leary AG, Olweus J, Kearney J, Buck DW (1997) AC133, a novel marker for human hematopoietic stem and progenitor cells. Blood 90(12):5002–5012
139. Manohar R, Li Y, Fohrer H, Guzik L, Stolz DB, Chandran UR, LaFramboise WA, Lagasse E (2015) Identification of a candidate stem cell in human gallbladder. Stem Cell Res 14 (3):258–269. https://doi.org/10.1016/j.scr.2014.12.003
140. Zhang L, Theise N, Chua M, Reid LM (2008) The stem cell niche of human livers: symmetry between development and regeneration. Hepatology 48(5):1598–1607. https://doi.org/10.1002/hep.22516
141. Gilpin SE, Charest JM, Ren X, Tapias LF, Wu T, Evangelista-Leite D, Mathisen DJ, Ott HC (2016) Regenerative potential of human airway stem cells in lung epithelial engineering. Biomaterials 108:111–119. https://doi.org/10.1016/j.biomaterials.2016.08.055
142. Wang D, Cai C, Dong X, Yu QC, Zhang XO, Yang L, Zeng YA (2015) Identification of multipotent mammary stem cells by protein C receptor expression. Nature 517(7532):81–84. https://doi.org/10.1038/nature13851
143. Komeya M, Ogawa T (2015) Spermatogonial stem cells: progress and prospects. Asian J Androl 17(5):771–775. https://doi.org/10.4103/1008-682X.154995
144. Aloisio GM, Nakada Y, Saatcioglu HD, Peña CG, Baker MD, Tarnawa ED, Mukherjee J, Manjunath H, Bugde A, Sengupta AL, Amatruda JF, Cuevas I, Hamra FK, Castrillon DH (2014) PAX7 expression defines germline stem cells in the adult testis. J Clin Invest 124 (9):3929–3944. https://doi.org/10.1172/JCI75943
145. Komai Y, Tanaka T, Tokuyama Y, Yanai H, Ohe S, Omachi T, Atsumi N, Yoshida N, Kumano K, Hisha H, Matsuda T, Ueno H (2014) Bmi1 expression in long-term germ stem cells. Sci Rep 4:6175. https://doi.org/10.1038/srep06175
146. Pendleton C, Li Q, Chesler DA, Yuan K, Guerrero-Cazares H, Quinones-Hinojosa A (2013) Mesenchymal stem cells derived from adipose tissue vs bone marrow: in vitro comparison of their tropism towards gliomas. PLoS One 8:e58198
147. Oh H, Bradfute SB, Gallardo TD, Nakamura T, Gaussin V, Mishina Y, Pocius J, Michael LH, Behringer RR, Garry DJ, Entman ML, Schneider MD (2003) Cardiac progenitor cells from adult myocardium: homing, differentiation, and fusion after infarction. Proc Natl Acad Sci USA 100(21):12313–12318

148. Fukada S, Yamaguchi M, Kokubo H, Ogawa R, Uezumi A, Yoneda T, Matev MM, Motohashi N, Ito T, Zolkiewska A, Johnson RL, Saga Y, Miyagoe-Suzuki Y, Tsujikawa K, Takeda S, Yamamoto H (2011) Hesr1 and Hesr3 are essential to generate undifferentiated quiescent satellite cells and to maintain satellite cell numbers. Development 138 (21):4609–4619. https://doi.org/10.1242/dev.067165
149. Gnocchi VF, White RB, Ono Y, Ellis JA, Zammit PS (2009) Further characterisation of the molecular signature of quiescent and activated mouse muscle satellite cells. PLoS One 4(4): e5205. https://doi.org/10.1371/journal.pone.0005205
150. Wang YX, Dumont NA, Rudnicki MA (2014) Muscle stem cells at a glance. J Cell Sci 127 (Pt 21):4543–4548. https://doi.org/10.1242/jcs.151209
151. Gentile P, Piccinno MS, Calabrese C (2019) Characteristics and Potentiality of human adipose-derived stem cells (hASCs) obtained from enzymatic digestion of Fat graft. Cell 8 (3):e282. https://doi.org/10.3390/cells8030282

Cancer Stem Cells

7

Contents

Abstract

Cancer stem cells (CSC) are cancerous cells expressing stem cell like properties of self-renewal and differentiation and are now reported from almost all cancers. They share some markers and developmental signalling pathways in common to normal stem cells and also reveal aberrant signalling pathways as compared to normal stem cells. Biomarkers expressed by these cells enable their identification and isolation. The stem cell niche, cell–cell interactions, and signalling pathways in the stem cell environment enable them to maintain their stemness property in cancerous tissue. They are associated with epithelial to mesenchymal transition (EMT) in cancerous state. They play role in conferring increased proliferation, growth, metastasis of cancer cells, multidrug resistance (MDR), resistance to radio and chemotherapy, relapse and minimal residual disease in cancer. They are promising therapeutic targets in cancer therapy and chemicals, inhibitors, drugs and natural products finds importance in their targeting. In this chapter, we discuss (1) epigenetics, (2) presence in different cancers, (3) biology, (4) evolution, (5) markers, (6) microRNA (miRNA), (7) mitochondria, (8) niche and signalling pathways and EMT transition, (9) metabolism, (10) redox and hypoxia, (11) therapy and challenges, (12) targeting, (13) natural products and targeting of CSC.

S. Chowdhury, S. Ghosh, *Stem Cells*, https://doi.org/10.1007/978-981-16-1638-9_7

Keywords

Cancer stem cells · EMT · Niche · Normal stem cells · Evolution of CSC · Natural products · Angiogenesis

Abbreviations

3D	Three-dimensional
ABCG2	ATP-binding cassette sub family G member 2
ADH	Aldehyde dehydrogenase
AML	Acute myeloid leukaemia
ASC	Amniotic stem cells
ATM	Ataxia telangiectasia mutated
ATX	Autotaxin
Aurora-A	Aurora kinase A
BCSCs	Breast cancer stem cells
BDNF	Brain-derived neurotrophic factor
BRCA1	Breast Cancer Type 1
CAFs	Cancer-associated fibroblasts
CBP	CREB-binding protein
CPE	*Clostridium perfringens (C perfringens)* enterotoxin
CRC	Colorectal cancer
CSC	Cancer Stem cells
CTC	Circulating cancer cells
CXCR4	C-X-C motif receptor 4
Dclk1	Doublecortin like kinase protein 1
DUB	Deubiquitylating enzymes
EC	Endothelial cells
EGCG	(−)-epigallocatechin gallate
EMT	Epithelial to mesenchymal transition
ESC	Embryonic stem cells
FACS	Fluorescence activated cell sorting
GBM	Glioblastoma multiforme
GCSCs	Gastric CSCs
GF	Growth Factors
gi	Gastrointestinal
GPCRs	G protein-coupled receptors
HSC	Haematopoietic stem cells
iPSC	Induced pluripotent stem cells
LESC	Lingual epithelial stem cells
lncRNAs	Long non-coding RNAs
LPA	Lysophosphatidic acid
LPA	Lysophosphatidic acid
MACS	Magnetic activated cell sorting
MaSCs	Mammary gland stem cells

MB	Medulloblastoma
MDR	Multidrug resistance
miRNA	microRNA
MRD	Minimal residual disease
MSC	Mesenchymal stem cells
ncRNAs	Non-coding RNAs
NF-κB	Nuclear factor kappa-light-chain-enhancer of activated B cells
NGF	Nerve growth factor
NP	Natural products
NSC	Neural stem cells
NSC	Normal stem cells
NSSC	Normal somatic stem cell
OXPHOS	Oxidative phosphorylation
PCa	Prostate cancer
PCSCs	Pancreatic cancer stem cells
PML	Promyelocytic leukaemia
SHH	Sonic hedgehog
SP	Side population
TAMs	Tumour-associated macrophages
TCF-LEF	T cell factor-lymphoid enhancer factor
TME	Tumour microenvironment
USP-1	Ubiquitin-specific protease 1
VEGF	Vascular endothelial growth factor
VSELSC	Very small embryonic-like stem cells
Wnt	Wingless-related integration site

7.1 Introduction

Cancer Stem Cells (CSCs) are cancerous cells located within tumours revealing the stem cell like property of self-renewal and differentiation on one hand and properties of tumorigenicity on the other so that when they are transplanted into an animal host they can divide relentlessly and form tumours.

CD24, CD44, and CD133 are markers expressed by CSCs that enable their identification. MicroRNAs (miRNAs) and signalling pathways including the Hedgehog pathway, Notch pathway, and Wnt/β-catenin pathways are known to regulate their properties.

CSCs have been known to play role in origin in cancer metastasis and are known to be chemo and radiotherapy resistant.

7.1.1 History

The occurrence and concept of CSCs came to light almost 40 years ago and were identified in 1994 [1]. They were identified as a group of tumour initiating cells

(TICs) cells that occur in particular niches of tumourous tissue driving the initiation, growth, repopulation, and maintainance of cancerous cells after injury and metastasis [2] by self-renewal properties [3, 4] and revealed pluripotency and involve common signalling pathways [5] and can give rise to distant metastasis [5] and relapse. CSCs can also give rise to differentiated progeny cells, lacking these features [2].

CSC have been reported from several solid tumours, like breast cancer, brain tumours, colon cancer, lung cancer, and melanoma [5]. Lineage-tracing and different experimental study methods has revealed properties of plasticity, quiescence, renewal of CSC [3]. The epithelial to mesenchymal transition (EMT) of cells in cancerous tissue is known to confer CSC like properties [6].

7.1.2 Chemoresistance

They reveal chemoresistance [3] and pose challenge as resistance conferred to therapies and therefore difficult to eradicate cancer completely [4]. The 3D culture of tumour-derived spheroid cultures derived from glioma, breast, colon, ovary, and prostate cancers are used to enrich the CSCs and find application in search of novel therapeutics targeting the CSC [7] (Fig. 7.1).

We discuss in this chapter (1) epigenetics, (2) study in different cancers, (3) biology, (4) evolution, (5) markers, (6) miRNA, (7) mitochondria, (8) niche, signalling pathways and EMT transition, (9) metabolism, (10) redox and hypoxia, (11) therapy and challenges, (12) targeting, (13) natural products and targeting of CSC.

7.2 Epigenetics

CSC are known to promote tumorigenesis and lead to therapeutic resistance. Studies of epigenomics have revealed that deregulation of various epigenetic pathways, altered chromatin and DNA lead to altered cellular plasticity and cellular reprogramming that contribute to CSC survival thereby promoting cancer initiation and tumorigenesis and finds application in the search of new novel therapeutic

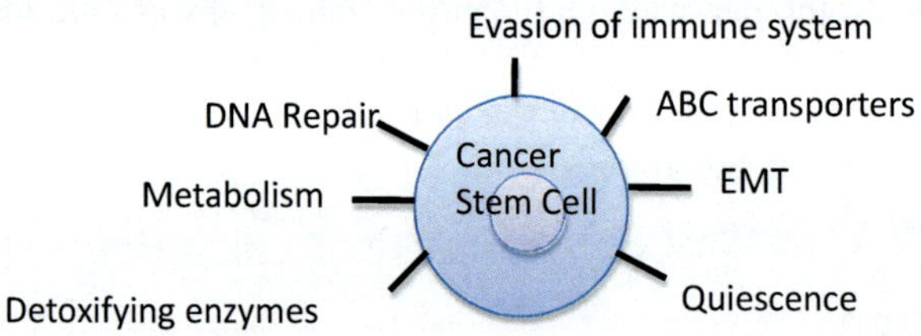

Fig. 7.1 CSC and its hallmark properties including quiescence, EMT, increased DNA repair and detoxifying enzymes, altered metabolism, immune evasion and overexpression of ABC transporters. Image adapted and modified from [8] Open access article

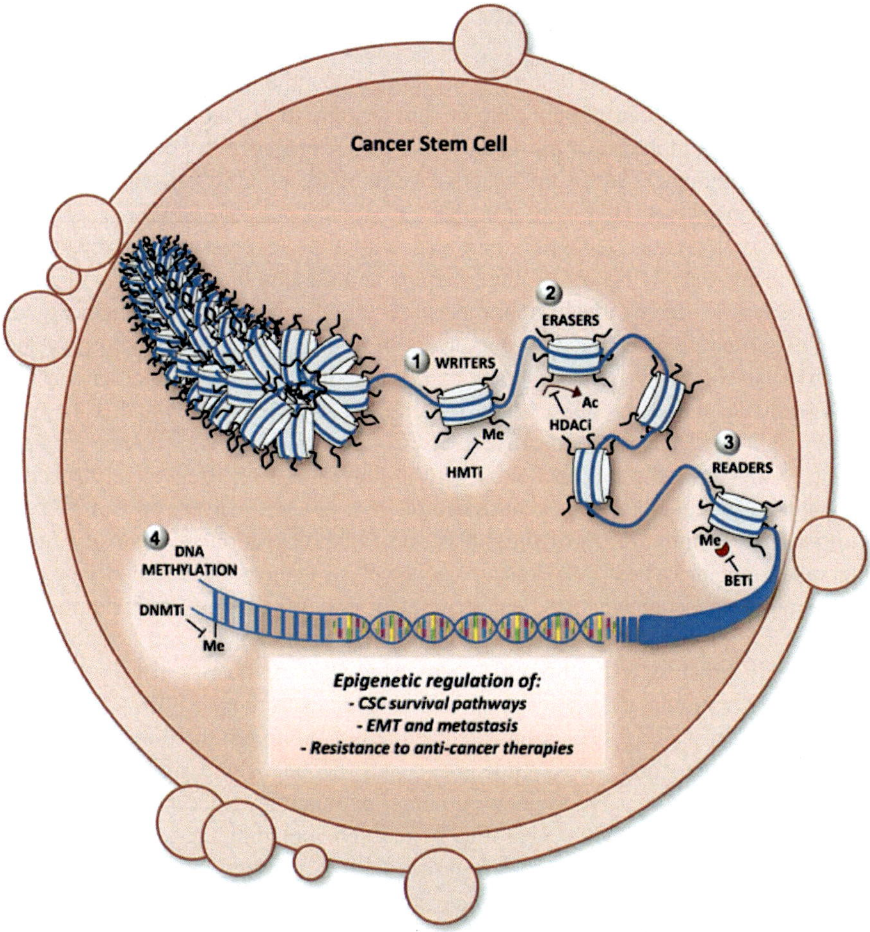

Fig. 7.2 Model of epigenetic regulation in CSCs and the potential therapeutic approaches. The chromatin fibre and the nucleosome are represented in the nucleus of a CSC. Epigenetic enzymes designated as writers, erasers, readers, and DNA methyltransferases regulate the major survival pathways in CSCs, by Notch, Wnt, and Hedgehog signalling. Epigenetic alterations guide (EMT), metastasis in CSCs, thus conferring CSC resistance to therapy. These find application in design of targets to therapy. Image and legend adapted from [8] Open access article under CC BY License

strategies to target the CSC [9]. Epigenetic changes are known to affect cells at different stages of the disease. With progressing cancer, epigenetic changes enable interaction of cancer cells with the microenvironment thereby affecting the cell to cell signalling, cancer cell characteristics, tumour characteristics and their biological function [10] (Fig. 7.2).

7.3 CSC in Different Cancers

CSC were first discovered in human myeloid leukaemia in 1960s, and later in solid tumours including breast, brain cancers, and recently to melanoma [11, 12]. Cervical carcinoma with high cancer mortality rate among females reveal occurrence of CSC in niche. Although CSC from cancers express markers, CSC related markers vary with the cancer type [13]. Breast cancer reveals the presence of CSC [14]. Triple-negative breast cancer (TNBC) enriched with CSC expressing markers $CD44^+$/$CD24^-$ and $ALDH1^+$ lacks targeted therapy and CSC in breast cancer contributing to resistance to therapy and tumour metastasis pose further more challenge to therapy [15]. The mammary gland also comprises of adult self-renewing mammary gland stem cells (MaSCs) that are regulated by the signalling pathways, microenvironment of mammary gland, and non-coding RNAs (ncRNAs). However whether these cells lead to breast cancer stem cells (BCSCs) like progeny is a matter of debate [16]. Malignant liver cancer is fatal with a high rate of recurrence and property of chemoresistance and reveals occurrence of CSC [17, 18]. Liver cancer CSCs finds importance in monitoring of prognosis and is being exploited for designing new therapeutic approaches [19]. Prostate cancer (PCa) is the most commonly occurring cancer reported in men causing cancer-related mortality among men. Although anti-PCa therapies like chemotherapy, surgical methods, radiation, and hormonal ablation are in practise, it is difficult to target the CSC or PCa stem cells leading to increased chemotherapeutic response and tendencies to relapse [20].

CSC in biliary tract cancer has revealed association with the tumour microenvironment (TME), microRNAs (miRNAs), and the oncogenic signalling pathways [21]. CSC of the digestive system are being exploited as cancer therapeutic targets [22]. Gastric CSCs (GCSCs) identified by markers including CD44 and CD133 may originate from gastric stem cells in gastric tissues, or bone marrow mesenchymal stem cells, expressing drug resistance genes including multiple drug resistance (MDR) and aldehyde dehydrogenase (ALDH, [23]). Colorectal cancer (CRC) leads to 10% of total adult malignancies worldwide and reported to contain CSCs in the colon epithelium [24, 25] that are reported to play role in colon cancer in its initiation and maintenance [26].

In lung cancer both normal pluripotent cells and lung CSCs reveal similar markers including CD44, ALDH, CD133, and ABCG2 (ATP-binding cassette subfamily G member 2). Developmental signalling pathways including hedgehog pathway, Wnt/β-catenin pathway, and Notch pathway have been reported to play role in the regulation of lung CSCs [27] and strategies encompassing inhibition of signalling pathways driving stemness, can combat lung cancer. Proteomics studies of CSC are elucidating the altered expression of biomolecules in malignant tumour and carcinoma cells and deregulation of signalling pathways [28].

Application of techniques of Fluorescence (FACS) and magnetic (MACS) activated cell sorting, laser cell capture microdissection, and three-dimensional (3D) spheroid or organoid cell culture enable enrichment and isolation of CSC using antibodies against known CSC specific cell surface molecules, like clusters of differentiation (CD) 44, 24, 133 (CD44, CD24, CD133), epithelial cell adhesion molecule (EpCAM), aldehyde dehydrogenase 1 (ALDH1), however, lack of strictly CSC specific surface markers is a bottleneck to antibody mediated staining of CSC [29, 30].

Melanoma CSCs have been identified and characterized by cell surface and intracellular markers [31] and is regulated interactions of the tumour microenvironment (TME) involving adipocytes, tumour-associated fibroblasts, endothelial cells, mesenchymal stem cells (MSC), and immune cells through cytokines and growth factors (GF) and finds application in therapy [32]. Thyroid carcinomas (TCs, [33]) and head and neck squamous cell carcinoma (HNSCC, [34, 35]) have been reported to express CSCs. Fluorescent based methods have enabled identification of lingual epithelial stem cells (LESCs, [36]). Cancer Stem cell like cells have been isolated from several solid tumors. Brain cancers, such as glioblastoma multiforme (GBM), medulloblastomas, and astrocytomas, also contain CSC [37, 38]. CSC is reported from thyroid gland [39]. Adult human skin/keratinocyte stem cells expressing and trancriptionally regulated by p63alphaDeltaN has been reported to play role in cancer development, ataxia telangiectasia mutated (ATM) kinase and key regulatory events that take place on p53 [40].

7.4 Biology of CSC

Cancer originates from normal cells that lose the ability to control cell growth and proliferate profusely and finally becomes malignant. CSCs with its self-renewal property can regenerate tumours. Studies with advanced imaging tools have revealed that CSC comprise of 0.1% of tumour cells [41]. CSC share similar features with that of normal stem cells (NSC), including self-renewal and differentiation. Adult NSC are quiescent and generally observed to be morphologically smaller than their differentiated progeny, while increased cell size is associated with differentiated cells revealing association of small cell sizes with the stemness property of the cells. CSCs, like NSC reveal smaller cell size [42]. Developmental signalling pathways including Bmi1 and Wnt (Wingless-related integration site) function similarly and regulate functioning of NSC and CSC [43]. Signalling by Wnt, Notch, and Hedgehog pathways in NSC is reported to regulate the self-renewal and survival of CSC but Notch pathway has been implicated in oncogenic functions [44]. Defective Wnt signalling has been associated in malignancies [45].

Signalling pathway has revealed dichotomy in determining stem cell fate decisions, and β-catenin interaction with either of the two homologous Kat3 coactivator proteins, CBP, and p300 and β-catenin/Kat3 coactivator has been reported to be associated with maintenance of potency and initiation of differentiation [46].

NSC function in maintanance and repair of many adult tissues have been revealed by in vitro clonogenic assays, transplantation and in vivo lineage tracing assays. These methods have enabled study of CSC in solid tumours that contribute to tumour growth and relapse after therapy [47].

Although stem cells reveal the property of self-renewal and differentiation into different tissues, studies have revealed that a differentiated cell can gain back its pluripotency and can be somatically reprogrammed to its plouripotent state by expressing the specific transcription factors.Yamanaka factors mediated somatic reprograming of oncogenes have elucidated the origin of CSC [48]. Doublecortin like kinase protein 1 (Dclk1), a microtubule-associated protein with C-terminal serine/threonine kinase domain involved in microtubule polymerization, neuronal migration, and early neurogenesis is a marker for radial glia cells, quiescent gastro-intestinal (gi) cells, pancreatic stem cells, tuft cells in the gi tract. Although Dclk1+ tuft cells find importance as niche or sensory cells in the normal gut, they are recently being implicated to give rise to malignancies, modulating several signalling pathways including Kras signalling [49].

CSCs are regulated by signalling pathways, and inflammation involved is now reported to play role in cancer progression. Cytokines released by the tumour-associated cells initiate the CSC progressing through the EMT and metastasis. These CSC can secrete factors and generate a secondary tumour in a distant site [50].

In solid tumours, 30% of primary tumour cells have revealed stem cell like characteristics and interaction between SCS and non-SCSs lead to tumour progression [51].

Normal animal cells vary in size from 10 to 20 μm in diameter and PI3K/Akt/mTOR, Myc, and Hippo pathways are known to control and coordinate normal cell growth, number, size, and division during homeostasis which gets deregulated during tumorigenesis leading to cell size variations and increased proliferation in cancer cells.

The property of stemness, survival, growth, proliferation, invasion, tumor angiogenesis are regulated by signals in the cancer cell, vascular niche [52] and microenvironment including factors like neurotrophins like and brain-derived neurotrophic factor (BDNF, [53]) and nerve growth factor (NGF, [53]) (Fig. 7.3).

The interplay between the host immune system and the CSCs can evade immune detection and elimination [54]. While a normal somatic stem cell (NSSC) gives rise to differentiated cell progeny throughout its life, with ageing the divisions may reveal aberrations and the decision for an old somatic stem cell to undergo either a symmetric division or an asymmetric one might lead to the differences of formation of NSC or CSC. The symmetric or asymmetric divisions in NSSC and factors including Kat3 coactivators, CREB-binding protein (CBP), and p300 find importance in dictating CSC fate [55].

The Hippo tumour suppressor pathway has been implicated to play role in breast cancer and other cancers [56, 57]. However, aberrations in the Hippo activity, or Hippo-independent mechanisms, has been known to activate Hippo transducers TAZ and YAP rendering the cancer cells become more aggressive, with self-renewal, drug resistance, and progression of metastasis like properties [56]. In gastric cancer signalling pathways and environmental factors known contribute to the CSC population [57].

G protein-coupled receptors (GPCRs) transduce signals into normal cells by heterotrimeric G proteins but their aberrant expression has been associated with

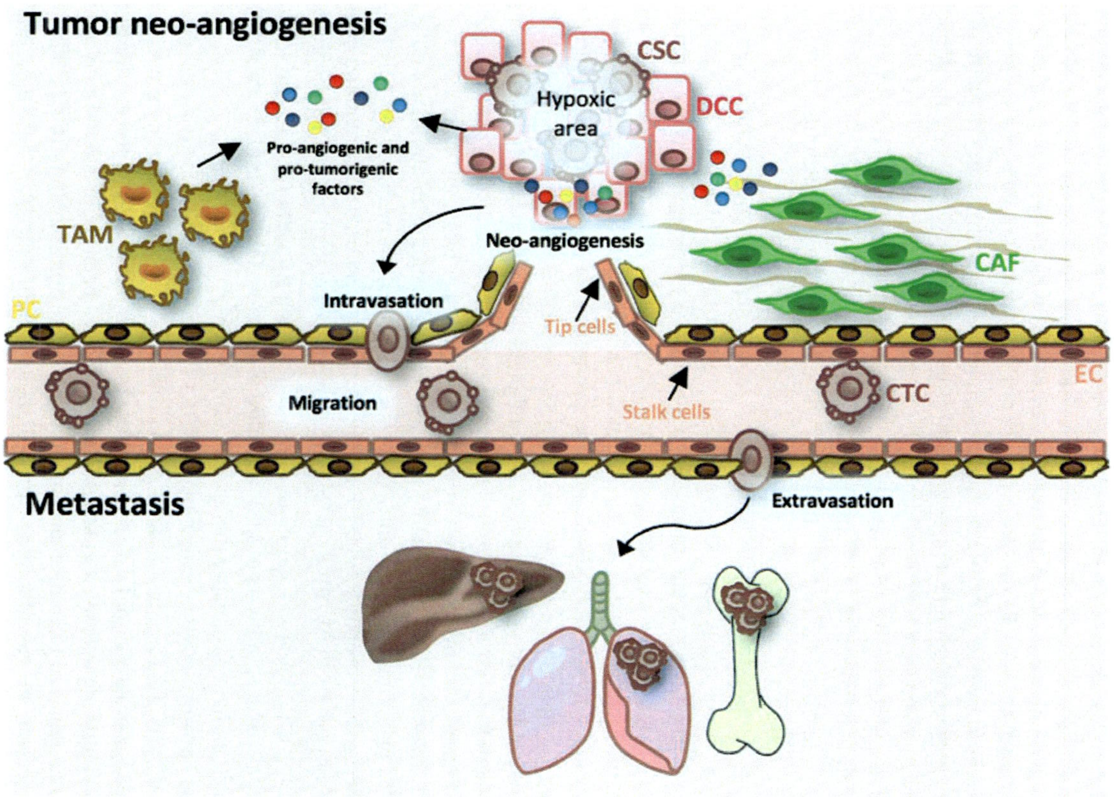

Fig. 7.3 Tumour angiogenesis and metastatic process. Cancer cells secrete pro-angiogenic and pro-tumorigenic factors and VEGF-A activates endothelial cells (ECs) in tip cells, which enables the sprouting vessels, and stalk cells, cytokines released by cancer cells, activate cancer-associated fibroblasts (CAFs) and activated tumour-associated macrophages (TAMs), favouring CSC intravasation. Circulating cancer cells (CTCs) reach the target organ, transported by blood, extravasate and start proliferation and dissemination. Image and legend adapted from [8]. Open access article under CC BY license

the CSC and reported in various cancers. G protein-coupled receptor 84 (GPCR84) and G protein subunit Gαq has been implicated in the maintenance of CSC in acute myeloid leukaemia (AML, [58]).

CSC revealing properties of being genetically instable with increased drug resistance is a challenge to be targeted. Lipid mediator regulatory system comprising of autotaxin (ATX), lysophosphatidic acid (LPA), together with their signalling pathways has been associated with cancer progression, metastasis, and resistance to therapy and is a potential therapeutic target in CSC [59].

Aurora kinase A (Aurora-A), a member of the Aurora family of serine-threonine kinases, is known to play role in mitotosis, microtubule stability in G1 phase of cell cycle, chromosome alignment, segregation, and cytokinesis. Its aberrant expression has been associated with self-renewal of embryonic stem cells (ESC) through negative regulation of p53, oncogenic transformation and is reported to induce stem cell like properties in oestrogen receptor α-positive breast cancer cells. Silencing of Aurora-A has been implicated in elimination of leukaemia stem cells in vivo [60].

Transcription factor NF-κB (nuclear factor kappa light chain enhancer of activated B cells) has been known to play role in the proliferation, inflammation, angiogenesis, invasion, and metastasis. Constitutive expression of NF-κB has been reported from CSCs of different cancers, playing role in their expansion, maintenance, proliferation, and survival. Targeted break in the NF-κB signalling pathway by specific inhibitors of NF-κB may be promising tool to remove CSC and find therapeutic importance [61].

CSC has been reported to possess the inherent ability to evade apoptosis [62] and autophagy [63, 64] thereby contributing to its resistance to therapy, thus manipulation of apoptosis and autophagy pathways in CSC may target the CSCs.

The enzyme Ubiquitin-specific protease 1 (USP1) can deubiquitinate and stabilize IDs in normal cells and thus prevents differentiation mediated by bHLH thus regulating CSC properties. USP1 bound to protein WD repeat-containing protein (WDR48) or USP1/WDR48 complex has been associated in CSC tumorigenesis, inhibiting differentiation and increasing proliferation [65] and regulates DNA damage repair. Thus inhibitors of USP including Pimozide, GW7647, C527, SJB2–043, ML323 can have promising therapeutic effects [65].

The promyelocytic leukaemia (PML) protein, as a component of the PML-retinoic acid receptor alpha fusion protein, is known to play role in oncogenesis, tumour progression, DNA repair, alternative lengthening of telomeres, transcriptional control, apoptosis, and senescence. PML has been implicated to play role in CSC maintenance from haematopoietic cancers and solid tumours [66]. CSCs with angiogenesis in vascular niche enable their growth, survival, and proliferation [67]. Targeting CSC together with targeting the microenvironment is a novel therapeutic strategy against solid tumour [67]. Mutant p53 is associated with generation of CSC [68]. Although it is hard to differentiate between NSC and CSC telomere dynamics and telomerase activity in NSC and CSC provide insights into origin of tumours [69]. CSC in breast cancer reveal expression of markers like CD44 (+)-CD24(−/low)-Lin(−) and ALDH1 positive cells and mammosphere formation assay in vitro [70]. CSCs are known to initiate tumour neovascularization and promote invasion but the mechanisms [71] involved are yet to be studied. Myc-induced epigenetic reprogramming has been reported to enhance the CSC phenotypes [72]. Mouse models have demonstrated that integrins play role in biology of CSC [73]. Overexpressed cytokines, including IL-6, -8, CCL2, and TGF-β, in breast tumours may hold promise as potential therapeutic targets to eradicate CSCs [74]. Overexpression of polycomb group proteins including repressive complexes 1 and 2 (PRC1 and PRC2) have been reported from different tumours and their epigenetic misregulation leads to block or activation of developmental pathways, thus enabling the cell to proliferate and is associated with formation of CSC [75]. CSC in AML has been associated with relapse [76].

7.4.1 CSC and Evolution

More than heterogeneity in CSCs in a single tumour, DNA mutations are thought to be the origin of CSC. Altered function of epigenetic regulators is thought to underlie the malignant populations. Tumours owe their origin to single cells in specific tissues that has undergone a series of genetic and or epigenetic transformation events that cause genomic unstable state giving rise to the idea that cancer evolves from abnormal clones of cells (Fig. 7.4).

Observations of early stage or pre-malignant conditions suggest that they can differentiate normally and clonal amplification occurs by self-renewal and proliferation and or survival of primitive cells and differentiation into cells incapable of further division.

Neoplasia is a population of clone of cells that display altered growth properties, hypothesized to comprise of two distinct population including (1) pre-malignant conditions with a greater clonal stability and no immediate cancer-generating potential but may become malignant with time and (2) overtly malignant conditions with lesser stability which retain CSC like features which are cells that produce invasive tumour or leukaemia (Fig. 7.5).

7.4.2 CSC and Markers

Understanding the CSC phenotype is a challenge. Side Population (SP) cells, detected by Hoechst dye exclusion in flow cytometry based studies, are considered enriched in CSC [78].

CD133, CD44, and ALDH are common markers to both NSC and CSC [79]. Although CSCs express heterogeneity, different CSC markers include CD 26, CD133, CD24, CD44, and ALDH1. CD26 has been correlated with in vitro sphere formation, new tumour formation, and resistance to chemotherapy [80]. CD133 has been identified in CSC of different types of cancer and cell expressing CD133^{high} population of cancer cells reveal property of self-renewal, proliferation, metastasis, and resistance to therapy [81]. CD133 is also brain CSC but

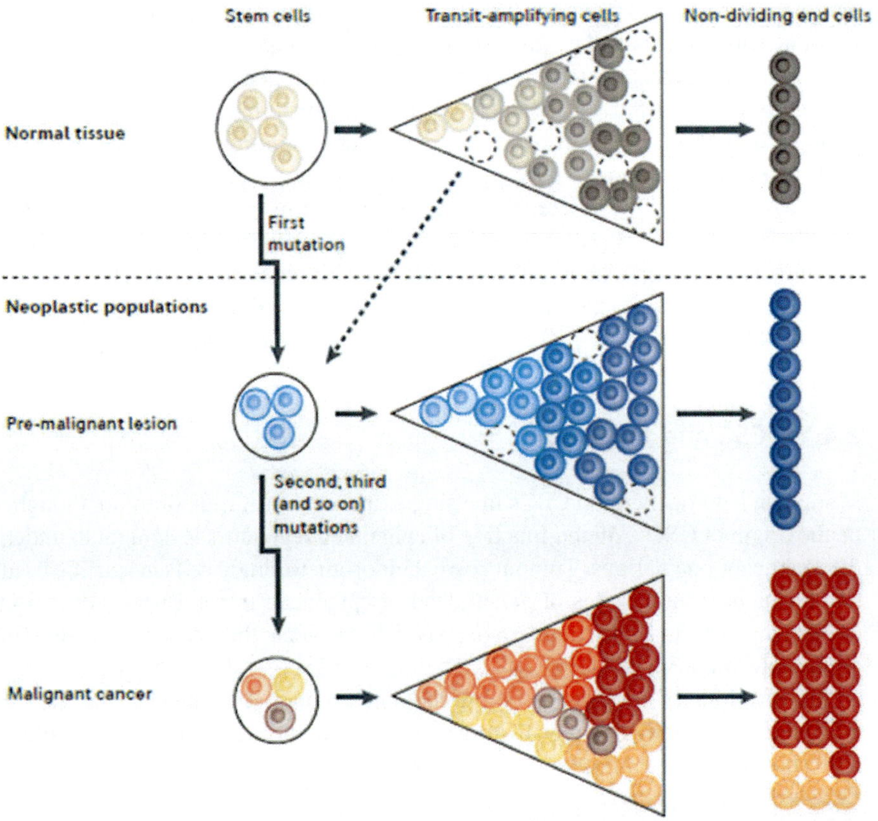

Fig. 7.4 Schematic models of cancer stem cell evolution revealing a regulated hierarchy of normal tissue populations in which stem cells give rise to amplifying cells that are differentiating to produce mature cells. Malignant cells are generated by sequential acquisition of mutations and/or epigenetic changes Image with permission from [77]

CD133(−) cells also reveal tumorogenicity [82]. CSCs reveal molecular pathways including Wnt, STAT, TGF-β, and Hippo-YAP-TAZ [79]. Oct-4 and embryonic signalling cascades, common markers to CSC and NSC reveal differential expression in CSCs [83].

ALDH activity and specifically ALDH1A1 of the 19 isoforms have been detected by the aldefluor assay in CSC from murine haematopoietic stem cells (HSC), murine progenitor pancreatic cells. Human breast CSCs reveal predominance of ALDH1A3 isoform [84]. Three main isotypes of ALDH1 including ALDH1A1, ALDH1A2, and ALDH1A3 are known to be markers of NSC and CSC in murine and human cells, involved in self-renewal, differentiation, and self-protection. ALDH1A1 expression bears correlation to cancerous phenotype. A high ALDH1 activity and overexpression of ALDH1A1 has been associated with CSC, poor cancer prognosis, in studies from breast, lung, oesophagus, colon, and stomach cancer [85]. Transcription factor in embryonic stem cells (ESCs), Nanog playing role in maintaining

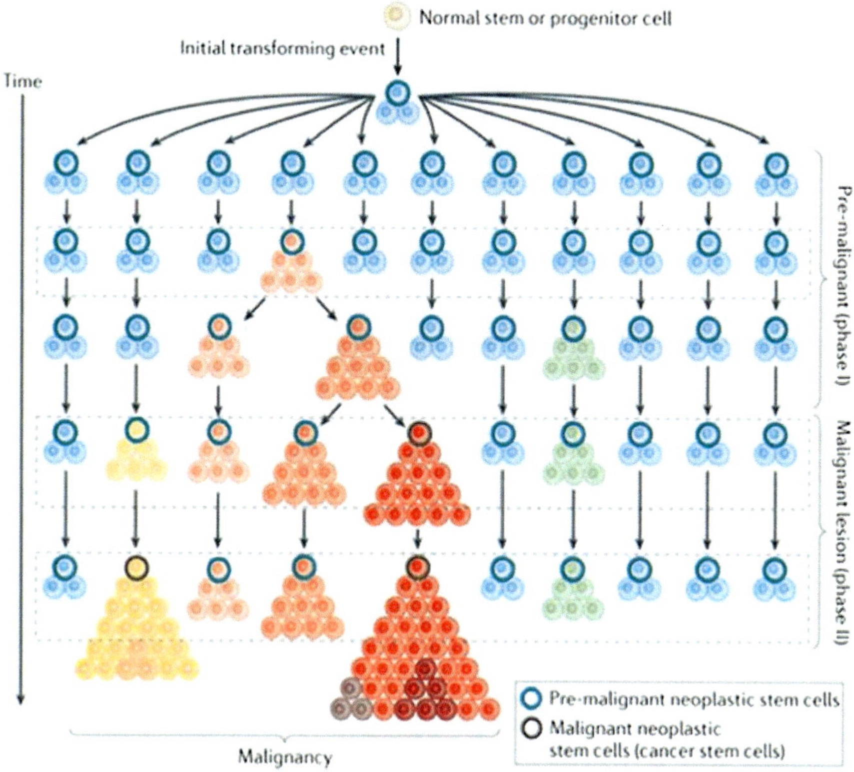

Fig. 7.5 A schematic representation of malignant transformation. Image with permission from [77]

pluripotency, embryonic development, and cellular reprogramming, has been recently implicated in promoting functions of self-renewal and long-term proliferative potential in CSC, thereby acting as biomarkers to CSC and probable therapeutic agents [86]. The chemokine C-X-C motif receptor 4 (CXCR4) is reported to be a prognostic marker in different cancers, being involved in chemotaxis, stemness, and drug resistance. Interaction of CXCR4 with chemokine C-X-C motif ligand 12 (CXCL12) has been known to modulate TME, angiogenesis, and CSC niche and finds importance in targeting of CSC [87]. CXCR4/CXCL12 axis and transcription factors including Nanog, NF-κB, and Bmi-1 have been reported to play role in maintenance of stemness and induction of metastasis in CSC [88].

Lipid metabolic enzymes stearoyl-CoA desaturase 1 (SCD1) and 3-hydroxy-3-methylglutharyl-coenzyme A reductase (HMG-CoAR) together with Hippo and Wnt signalling pathway have been known to control CSC fate decision [89]. Pancreatic CSC reveal expression of CD24(+)CD44(+)CD326 (ESA)(+) and CD133 [90, 91] and express genes including Sonic hedgehog (SHH) and BMI-1 but represent 0.5–1.0% of all pancreatic cancer cells. Pancreatic CSC reveal upregulated signalling cascades of pathways involved in tumour metastasis [91]. Renal cell carcinoma CSC reveal expression of CXCR4, CD105, and CD133 and side population by

Table 7.1 Markers of CSC

Tumor type	Cancer stem cell markers
Breast cancer	$CD133^+$, $CD44^+$, $CD24^+$, $EpCAM^+$, $ALDH^{high}$
Colon cancer	$CD133^+$, $CD44^+$, $CD24^+$, $CD166^+$, $EpCAM^+$, $ALDH^{high}$, ESA^+
Gastric cancer	$CD133^+$, $CD44^+$, $CD24^+$
Glioblastoma	$CD133^+$
Head and neck cancer	SSEA-1^+, $CD44^+$, $CD133^+$
Leukemia (AML)	$CD34^+$, $CD38^-$, $CD123^+$
Liver cancer	$CD133^+$, $CD44^+$, $CD49f^+$, $CD90^+$, $ALDH^{high}$, $ABCG2^+$, $CD24^+$, ESA^+
Lung cancer	$CD133^+$, $CD44^+$, $ABCG2^+$, $ALDH^{high}$, $CD87^+$, $CD90^+$
Melanoma	$ABCB5^+$, $CD20^+$
Ovarian cancer	$CD133^+$, $CD44^+$
Pancreatic cancer	$CD133^+$, $CD44^+$, $CD24^+$, $ABCG2^+$, $ALDH^{high}$, $EpCAM^+$, ESA^+
Prostate cancer	$CD133^+$, $CD44^+$, $\alpha 2\beta 1^+$, $ABCG2^+$, $ALDH^{high}$

Table adapted from [8] Open access article

staining with dye, Hoechst 33342, ALDH1, dye cycle violet, and rhodamine 123 [92]. Glycolipids including stage-specific embryonic antigens, SSEA-1, SSEA-3, and SSEA-4 and gangliosides including GD3, GD2, and A2B5 antigens are expressed by pluoripotent stem cells including amniotic stem cells (ASC), embryonic stem cells (ESC), induced pluripotent stem cells (iPSC), multipotent stem cells, neural stem cells (NSC), Mesenchymal stem cells (MSC), foetal liver multipotent progenitor cells, haematopoietic stem cells (HSC), very small embryonic-like stem cells (VSELSC), and cancer stem cells including brain cancer stem cells and breast cancer stem cells [72]. Endometrial somatic stem cells are known to express CD31, CD34, and CD144, 150, Oct-4, Musashi-1, and reveal features of somatic stem cells, including high proliferative potential, self-renewal, differentiation into one or more lineages [93]. BRCA1 expression is involved in normal embryonic development, mammary stem cell differentiation, and prostate cancer development. Inactivation of BRCA1 plays an important role in CSCs and also leads to the development of high-grade, basal-like tumour [94]. Chemokines like stromal cell-derived factor-1 (SDF-1) enable growth of primary tumours and play role in metastatic process. Overexpression of CXCR4 and ligand SDF-1 has been known to play role in cell metastasis, stem cell proliferation, their survival, and homing. CSC has been reported to express CXCR4 receptor and responding to specific ligand SDF-1, they can initiate metastasis. Thus SDF-1/CXCR4 axis provides insights into migratory mechanism involving cancer cells and CSC [95] (Table 7.1).

7.4.3 miRNA and CSC

microRNAs (miRNA)s are small non-coding regulatory RNA molecules that have gained importance in playing role in regulating the self-renewal and differentiation in CSC, their migration and tumour formation [96] and has been studied in breast cancer stem cells, colorectal cancer stem cells, gastric cancer stem cells, head and neck cancer stem cells, lung cancer stem cells, prostate cancer stem cells and pancreatic cancer stem cells, etc. [96] and also regulate stem cell fate and behaviour and their dysregulated behaviour [97], directly or indirectly coordinate the cell cycle progression [98]. The miRNA profile of CSCs differs from non-stem cancer cells and may be of therapeutic importance [99, 100].

Long non-coding RNAs (lncRNAs) can act in DNA damage repair, inflammation, metabolism, cell growth and differentiation, survival, signalling, in normal cells. In cRNA including ATB, CTCF7, CUDR, DILC, DYNC2H1-4, HOTAIR, H19, HOTTIP, HIF2PUT, Lnc34a, Linc00617, MALAT-1, PVT1, ROR, SOX2OT, SOX4 and ARSR Uc.283+ act as key regulators of the CSC regulating the asymmetric CSC division, self-renewal, drug resistance and EMT, contributing to metastasis and relapse in cancers [101] and are being exploited as therapeutic targets. miRNAs have been implicated in CSCs, carcinogenesis and treatment response of cancers. Regulation of miRNAs has been known to affect repair of DNA damage, redistribution, repopulation, reoxygenation, and radiosensitivity of lung cancer stem cells (LCSCs, [102]).

7.4.4 CSC Survival and Mitochondria

It is reported that mitochondria and its related signalling pathways contribute to survival and maintenance [103, 104] of CSC and targeting mitochondrial activity in CSCs are new strategies to control cancer [89]. The metabolism heterogeneity of mitochondria in CSCs like mitochondrial membrane potential ($\Delta\psi m$), reactive oxygen species [105] contribute to survival of CSC.

7.4.5 CSC Niche and Signalling Pathways and EMT Transition

The signalling pathway for transformation of epithelial cells into mesenchymal cells or EMT in CSC were identified and studied [1]. CSC niche, is report to comprise of endothelial cells, fibroblasts, mesenchymal cells and immune cells and is known to play role in signal transduction, survival and maintainance of CSC and conferring drug resistance. The extracellular matrix (EMT) and tumor microenvironment or TME, has been reported to be instrumental in shielding the CSC from the activity of drugs and therapeutics [28, 106]. EMT enables acquisition and maintenance of stem cell like properties and endows differentiated normal cells and cancer cells with stem cell properties and CSCs are also known to exhibit EMT properties leading to tumour progression [107].

EMT transmission in breast cancer cells enable CSC to maintain plasticity to transition between mesenchymal like (EMT) and epithelial like (MET) states within the TME [108]. EMT transition pancreatic cancer stem cells (PCSCs) have been implicated in tumour relapse and therapy resistance [109].

7.4.6 Metabolism

Cancer cells reveal altered cellular metabolism in order to cope up with the profuse proliferation of the cells. Thus cancer cells reveal less efficient glycolysis for the production of ATP and building blocks also called as the Warburg effect instead of primarily using oxidative phosphorylation (OXPHOS). CSCs depending on the type of cancer can be highly glycolytic or OXPHOS dependent and mitochondrial function finds importance and inhibition of mitochondrial function can be exploited as therapeutic targets in preventing cancers [110]. Cancer stemness is also associated with metabolic pathways of amino acid and fatty acid metabolism, glycolysis and oxidative, glutamine affecting resistance to therapy driven by CSC [111]. However understanding the mehanism by which drugs target cancer metabolism is challenging and complex highlighting the role of meatbolism in CSCs vs. non-CSCs [112, 113].

7.4.7 Redox, Hypoxia and CSC

Hypoxia mediated by factor called as hypoxia inducible factors (HIFs) reflecting inadequate tissue oxygenation is observed during normal development and tumorigenesis and have been reported to play a significant role in the maintenance of CSC studied in colon cancer. HIF and its signalling pathways have been associated with properties of stemness, and maintaining Wnt/β-catenin signalling pathways [114]. In solid tumors, hypoxic conditions and niches have been reported to maintain CSC, indicating that hypoxia and its signaling pathways play predominant role in cancer maintanance, metastasis and radiotherapy resistance [115]. While upregulated ROS is observed in cancer cells, CSC reveal low ROS levels, similar to the corresponding normal stem cell. Redox regulation in various CSC finds importance as effective targets in cancer therapy [116].

7.5 Therapy and Challenges

CSC cells reveal variation in from patient to patient, posing a serious challenge to their targeting [117]. Radiation therapy is used to treat malignant tumours causing DNA damage mediated directly by effects of ionization or mediated indirectly by generation of reactive oxygen species (ROS), thereby destroying cancer cells. But ionizing radiation (IR) suffers from the challenges of promoting metastasis and cancer cell invasion by inducing the EMT by ROS mediated activation of EMT transcription factors including Snail, HIF-1, ZEB1, and STAT3, that are activated by signalling pathways, like TGF-β, Notch, Hedgehog, Wnt, G-CSF, EGFR-PI3K-Akt, and MAPK pathways, thereby acquiring property of stemness and undergo metabolic changes and induce CSC like properties, like dedifferentiation and self-renewal, and activating EMT-inducing pathways, affect the TME thereby affecting invasion and metastasis [118].

Different drugs are being used to target CSCs, in epithelial cell-derived cancers [119]. Ionophore antibiotic salinomycin has been reported to that selectively target breast CSCs and proves to be a promising anticancer drug and is under clinical trials [120]. Since EMT of cells by signalling pathways reveals similar features as CSCs with increased drug efflux pumps and anti-apoptotic effects and plays role in progression of cancer, metastasis, and drug resistance, targeting of EMT could be a strategy to overcome drug resistance [121].

7.6 Targeting CSC

7.6.1 Research

CSC has been revealed to express vascular endothelial growth factor receptor, VEGFR1+ and strategies of eradication of CSCs may involve the application of antibiotic metformin, salinomycin, and enterotoxin from *Clostridium perfringens (C perfringens)* (CPE) reported from breast cancer and ovarian cancer patients [122]. It is reported that resistance to therapy in cancer is associated with high levels of apoptotic sensitivity, associated with meiosis, nuclear fusion, VDJ recombination, somatic hypermutation, and gastrulation at the initiation of malignancy. CSC can arise stochastically from tumour cell population, with resistance to DNA damaging agents [107]. CSC apoptotic inducers including antibodies, synthetic and natural compounds, recombinant proteins and oligonucleotides [123] find importance in their targeting. Cancer heterogeneity is known to arise due to CSC plasticity and their effective targeting strategies enable efficient to eliminate CSCs [124].

Targeting autophagy in CSCs could apply as strategy for anticancer therapy [125]. In lung cancer strategies are being directed towards targeting of CSC and the signalling cascades [85].

Resistance to radiation therapy is attributed by CSC studied in breast cancer and glioblast and CSCs are novel therapeutic targets for efficient radiotherapy [126]. Surgery, radiotherapy, and chemotherapy are resisted by CSCs and agents including salinomycin, apopotin, etc. targeting CSC provide promises [127]. CSCs are known to contribute to minimal residual disease (MRD) leading to cancer relapse and metastasis, slow cell cycle, ability mediate efflux of cytotoxic agents, resistance to oxidative stress, resistance to therapy and thus disrupting the quiescence or redox defense capability [128] of CSC finds importance as a therapeutic strategy in elimination of cancer [129].

Strategies like blocking one or more self-renewal signalling pathways, reducing the expression of drug efflux and ATP-binding cassette efflux transporters, modulating epigenetic aberrations, and promoting CSC differentiation [130] are under clinical trials to eradicate cancer as a whole [131]. Photodynamic therapy has been studied in colorectal cancer which revealed uptake of actively targeted photosynthetic drug specifically mediated towards colon CSC thereby eradicating cancer [132]. Anticancer dietary phytochemicals are promising candidates to target CSCs [133]. Stem cell maintainance and CSC pluripotency and differentiation, post-translational modifications including mechanisms of ubiquitylation and deubiquitylation are tightly regulated. Deubiquitylating enzymes (DUBs) play role in regulation of core stem cell transcription factors and CSC related proteins and DUB inhibitors finds importance in targeting CSCs [134]. Resistance to tyrosine kinase inhibitors (TKIs) in non-small cell lung cancer has been reported to lead to CSC like phenotypes thus acting as potential therapeutic targets [135]. CSC expressing one or more markers including ALDH 1, EpCAM, CD49f, CD44, CD133, CD24, and CXCR4 and aberrant expression of Notch, Hedgehog, Wnt, and TGF-β signalling regulate their self-renewal and differentiation. Vitamin D and its analogs revealed inhibitory effects on the CSC signaling pathways, suggesting their therapeutic potential [136]. Medulloblastoma (MB) is the most common malignant brain tumour, in children and recently reported to contain CSC and are promising targets in MB therapy [137].

Potential new strategies in CSC targeted therapy include inhibiting or targeting the melanoma differentiation associated gene-9/Syntenin (MDA-9/Syntenin) or syndecan binding protein (SDCBP) or MDA-9/Syntenin (SDCBP), patched (PTCH), epigenetic targets, non-coding RNAs, and differentiation induction [138, 139] thereby inhibiting or suppressing malignancy. Targeting of CSC by nanoparticle mediated delivery of inhibitors, targeted transcriptional regulation and signalling pathways modulating ABC transporters, miRNA mediated inhibition, [140] are finding importance as effective strategies in targeting of CSC.

Targeting of deregulated pathways including Wnt, Hedgehog, and Notch has found importance as a strategy in targeting CSCs [141]. In breast cancer cells both PI3K-Akt and NF-κB pathways act in the CSC survival. Targeting of DHMEQ as a specific inhibitor of NF-κB [142] is finding importance as a strategy in targeting of the CSC.

Metabolic reprogramming of cancer and CSC in ovarian cancer is a strategy for eradication in ovarian cancer [143]. Metformin and its inhibitory effects on CSC are gradually gaining importance [144]. In solid tumours, CSCs markers including ALDH dehydrogenases, CD44, CD133, epithelial cell adhesion molecule (ECAM), epithelial cell specific antigen, ATP-binding cassette subfamily G member 2, and receptor tyrosine kinase have been proved as the useful targets [145].

7.6.2 Natural Products and Targeting of CSC

Natural products like (−)-epigallocatechin gallate (EGCG) and green tea catechins have been reported to remove tumour on mouse skin initiated with 7,12-dimethylbenz(a)anthracene followed by okadaic acid. Both EGCG and green tea extract have been reported to prevent cancer development in rodents.

Human CSCs overexpressing Nanog, Oct4, Sox2, CD44, CD133, EMT markers, Twist, Snail, vimentin, and ALDH and reduced E-cadherin levels and cyclin D1 reveal that EGCG inhibits the self-renewal of CSCs and EMT phenotypes of human CSCs and inhibits stemness of CSCs and reveal synergistic effects with other anticancer drugs [19].

Natural products (NPs) have been reported to modulate Wnt, Hedgehog, Notch, and PI3K/AKT/mTOR pathways associated with quiescence and self-renewal, function of CSCs like differentiation, multidrug resistance (MDR), and metastasis and regulate CSC fate [146]. Curcumin and its derivatives have been reported to play predominant role in eliminating CSC [147, 148]. Forty-nine different NPs have been known to influence the biology of CSC [148, 149] and NPs are also known to effect signalling pathway studied in breast CSC [150].

Identification of specific surface antigens, mammosphere formation assays, serial colony forming unit assays, xenograft transplantation, and label retention assays coupled with activity of ALDH1 find application in phytochemical efficiency in killing CSCs. EGCG, curcumin, piperine, sulforaphane, β-carotene, genistein, and the whole extract of some plants are able to kill and thereby target and remove CSCs. Most of these phytochemicals act by interfering with the canonical Wnt (β-catenin/T cell factor-lymphoid enhancer factor, TCF-LEF) pathway [149].

NPs like anthocyanins, curcumin, epigallocatechin, ellagic acid, resveratrol, quercetin, and other compounds and extracts can directly or indirectly effect molecular pathways, decreasing the pathological activities of CSCs [150]. Secondary metabolites, including, green tea catechins, soy isoflavone genistein, flavonol quercetin, find importance in targeting CSC and EMT transitions [149–151].

7.7 Discussion

The fact that cancer stem cells (CSCs) form a subpopulation of cells within a growing tumour expressing specific markers and carrying the properties of self-renewal and differentiation and revealing properties of resistance to chemo and radio therapy and associated with cancer relapse and metastasis is a major cause of concern in cancer treatment. Thus eradication of the CSC and targeting by effective means is a major subject of research in the domain of Cancer Biology. Although several strategies are being searched towards their eradication, this area in biology calls in for more research towards successful targeting and eradication of CSC and thus eradicating cancer.

References

1. Pattabiraman DR, Weinberg RA (2014) Tackling the cancer stem cells - what challenges do they pose? Nat Rev Drug Discov 13(7):497–512
2. Vlashi E, Pajonk F (2015) Cancer stem cells, cancer cell plasticity and radiation therapy. Semin Cancer Biol 31:28–35
3. Batlle E, Clevers H (2017) Cancer stem cells revisited. Nat Med 23(10):1124–1134
4. Abbaszadegan MR, Bagheri V, Razavi MS, Momtazi AA, Sahebkar A, Gholamin M (2017) Isolation, identification, and characterization of cancer stem cells: a review. J Cell Physiol 232 (8):2008–2018
5. Dawood S, Austin L, Cristofanilli M (2014) Cancer stem cells: implications for cancer therapy. Oncology (Williston Park) 28(12):1101–1107, 1110
6. Nassar D, Blanpain C (2016) Cancer stem cells: basic concepts and therapeutic implications. Annu Rev Pathol 11:47–76

7. Ishiguro T, Ohata H, Sato A, Yamawaki K, Enomoto T, Okamoto K (2017) Tumor-derived spheroids: relevance to cancer stem cells and clinical applications. Cancer Sci 108(3):283–289
8. Turdo A, Veschi V, Gaggianesi M, Chinnici A, Bianca P, Todaro M, Stassi G (2019) Meeting the challenge of targeting cancer stem cells. Front Cell Dev Biol 7:16
9. Toh TB, Lim JJ, Chow EK (2017) Epigenetics in cancer stem cells. Mol Cancer 16(1):29
10. Wainwright EN, Scaffidi P (2017) Epigenetics and Cancer stem cells: unleashing, hijacking, and restricting cellular plasticity. Trends Cancer 3(5):372–386
11. Menaa F, Houben R, Eyrich M, Broecker EB, Becker JC, Wischhusen J (2009) Stem cells, melanoma and cancer stem cells: the good, the bad and the evil? G Ital Dermatol Venereol 144 (3):287–296
12. Regenbrecht CR, Lehrach H, Adjaye J (2008) Stemming cancer: functional genomics of cancer stem cells in solid tumors. Stem Cell Rev 4(4):319–328
13. Huang R, Rofstad EK (2017) Cancer stem cells (CSCs), cervical CSCs and targeted therapies. Oncotarget 8(21):35351–35367
14. Luo M, Clouthier SG, Deol Y, Liu S, Nagrath S, Azizi E, Wicha MS (2015) Breast cancer stem cells: current advances and clinical implications. Methods Mol Biol 1293:1–49
15. O'Conor CJ, Chen T, González I, Cao D, Peng Y (2018) Cancer stem cells in triple-negative breast cancer: a potential target and prognostic marker. Biomark Med 12(7):813–820
16. Yang X, Wang H, Jiao B (2017) Mammary gland stem cells and their application in breast cancer. Oncotarget 8(6):10675–10691
17. Nio K, Yamashita T, Kaneko S (2017) The evolving concept of liver cancer stem cells. Mol Cancer 16(1):4
18. Cheng Z, Li X, Ding J (2016) Characteristics of liver cancer stem cells and clinical correlations. Cancer Lett 379(2):230–238
19. Fujiki H, Sueoka E, Rawangkan A, Suganuma M (2017) Human cancer stem cells are a target for cancer prevention using (−)-epigallocatechin gallate. J Cancer Res Clin Oncol 143 (12):2401–2412
20. Leão R, Domingos C, Figueiredo A, Hamilton R, Tabori U, Castelo-Branco P (2017) Cancer stem cells in prostate cancer: implications for targeted therapy. Urol Int 99(2):125–136
21. Mayr C, Ocker M, Ritter M, Pichler M, Neureiter D, Kiesslich T (2017) Biliary tract cancer stem cells - translational options and challenges. World J Gastroenterol 23(14):2470–2482
22. Colvin HS, Nishida N, Koseki J, Konno M, Kawamoto K, Tsunekuni K, Doki Y, Mori M, Ishii H (2014) Cancer stem cells of the digestive system. Jpn J Clin Oncol 44(12):1141–1149
23. Li K, Dan Z, Nie YQ (2014) Gastric cancer stem cells in gastric carcinogenesis, progression, prevention and treatment. World J Gastroenterol 20(18):5420–5426
24. Szaryńska M, Olejniczak A, Kmieć Z (2016) The role of cancer stem cells in pathogenesis of colorectal cancer. Postepy Hig Med Dosw (Online) 70:1469–1482
25. Fanali C, Lucchetti D, Farina M, Corbi M, Cufino V, Cittadini A, Sgambato A (2014) Cancer stem cells in colorectal cancer from pathogenesis to therapy: controversies and perspectives. World J Gastroenterol 20(4):923–942
26. Puglisi MA, Tesori V, Lattanzi W, Gasbarrini GB, Gasbarrini A (2013) Colon cancer stem cells: controversies and perspectives. World J Gastroenterol 19(20):2997–3006
27. Wang J, Li ZH, White J, Zhang LB (2014) Lung cancer stem cells and implications for future therapeutics. Cell Biochem Biophys 69(3):389–398
28. Jung HJ (2017) Chemical proteomic approaches targeting cancer stem cells: a review of current literature. Cancer Genomics Proteomics 14(5):315–327. https://doi.org/10.21873/cgp.20042
29. Cammareri P, Lombardo Y, Francipane MG, Bonventre S, Todaro M, Stassi G (2008) Isolation and culture of colon cancer stem cells. Methods Cell Biol 86:311–324
30. Skvortsov S, Debbage P, Skvortsova I (2014) Proteomics of cancer stem cells. Int J Radiat Biol 90(8):653–658
31. Kumar D, Gorain M, Kundu G, Kundu GC (2017) Therapeutic implications of cellular and molecular biology of cancer stem cells in melanoma. Mol Cancer 16(1):7

32. Korkaya H, Liu S, Wicha MS (2011) Breast cancer stem cells, cytokine networks, and the tumor microenvironment. J Clin Invest 121(10):3804–3809
33. Antonelli A, La Motta C (2017) Novel therapeutic clues in thyroid carcinomas: the role of targeting cancer stem cells. Med Res Rev 37(6):1299–1317
34. Qian X, Ma C, Nie X, Lu J, Lenarz M, Kaufmann AM, Albers AE (2015) Biology and immunology of cancer stem(-like) cells in head and neck cancer. Crit Rev Oncol Hematol 95 (3):337–345
35. Shah A, Patel S, Pathak J, Swain N, Kumar S (2014) The evolving concepts of cancer stem cells in head and neck squamous cell carcinoma. ScientificWorldJournal 2014:842491
36. Ueno H (2016) Identification of normal and neoplastic stem cells by the multicolor lineage tracing methods. Pathol Int 66(8):423–430
37. Galderisi U, Cipollaro M, Giordano A (2006) Stem cells and brain cancer. Cell Death Differ 13 (1):5–11
38. Khan IS, Ehtesham M (2014) Targeting glioblastoma cancer stem cells: the next great hope? Neurosurg Focus 37(6):E7
39. Zhang P, Zuo H, Ozaki T, Nakagomi N, Kakudo K (2006) Cancer stem cell hypothesis in thyroid cancer. Pathol Int 56(9):485–489
40. Finlan LE, Hupp TR (2006) Epidermal stem cells and cancer stem cells: insights into cancer and potential therapeutic strategies. Eur J Cancer 42(9):1283–1292
41. Hart LS, El-Deiry WS (2008) Invincible, but not invisible: imaging approaches toward in vivo detection of cancer stem cells. J Clin Oncol 26(17):2901–2910
42. Li Q, Rycaj K, Chen X, Tang DG (2015) Cancer stem cells and cell size: a causal link? Semin Cancer Biol 35:191–199
43. Lobo NA, Shimono Y, Qian D, Clarke MF (2007) The biology of cancer stem cells. Annu Rev Cell Dev Biol 23:675–699
44. Shigdar S, Li Y, Bhattacharya S, O'Connor M, Pu C, Lin J, Wang T, Xiang D, Kong L, Wei MQ, Zhu Y, Zhou S, Duan W (2014) Inflammation and cancer stem cells. Cancer Lett 345 (2):271–278
45. Duchartre Y, Kim YM, Kahn M (2017) Pharmacologic manipulation of Wnt signaling and cancer stem cells. Methods Mol Biol 1613:463–478
46. Kahn M (2018) Wnt signaling in stem cells and cancer stem cells: a tale of two coactivators. Prog Mol Biol Transl Sci 153:209–244
47. Beck B, Blanpain C (2013) Unravelling cancer stem cell potential. Nat Rev Cancer 13 (10):727–738
48. Friedmann-Morvinski D, Verma IM (2014) Dedifferentiation and reprogramming: origins of cancer stem cells. EMBO Rep 15(3):244–253
49. Westphalen CB, Quante M, Wang TC (2017) Functional implication of Dclk1 and Dclk1-expressing cells in cancer. Small GTPases 8(3):164–171
50. Wang J, Sullenger BA, Rich JN (2012) Notch signaling in cancer stem cells. Adv Exp Med Biol 727:174–185
51. Enderling H (2015) Cancer stem cells: small subpopulation or evolving fraction? Integr Biol (Camb) 7(1):14–23
52. Ping YF, Zhang X, Bian XW (2016) Cancer stem cells and their vascular niche: do they benefit from each other? Cancer Lett 380(2):561–567
53. Chopin V, Lagadec C, Toillon RA, Le Bourhis X (2016) Neurotrophin signaling in cancer stem cells. Cell Mol Life Sci 73(9):1859–1870
54. Sultan M, Coyle KM, Vidovic D, Thomas ML, Gujar S, Marcato P (2017) Hide-and-seek: the interplay between cancer stem cells and the immune system. Carcinogenesis 38(2):107–118
55. Thomas PD, Kahn M (2016) Kat3 coactivators in somatic stem cells and cancer stem cells: biological roles, evolution, and pharmacologic manipulation. Cell Biol Toxicol 32(1):61–81
56. Maugeri-Saccà M, De Maria R (2016) Hippo pathway and breast cancer stem cells. Crit Rev Oncol Hematol 99:115–122

57. Hao J, Zhang Y, Jing D, Li Y, Li J, Zhao Z (2014) Role of hippo signaling in cancer stem cells. J Cell Physiol 229(3):266–270
58. Lynch JR, Wang JY (2016) G protein-coupled receptor signaling in stem cells and cancer. Int J Mol Sci 17(5):E707
59. Lee D, Suh DS, Lee SC, Tigyi GJ, Kim JH (2018) Role of autotaxin in cancer stem cells. Cancer Metastasis Rev 37(2–3):509–518
60. Li M, Gao K, Chu L, Zheng J, Yang J (2018) The role of Aurora-a in cancer stem cells. Int J Biochem Cell Biol 98:89–92
61. Vazquez-Santillan K, Melendez-Zajgla J, Jimenez-Hernandez L, Martinez-Ruiz G, Maldonado V (2015) NF-κB signaling in cancer stem cells: a promising therapeutic target? Cell Oncol (Dordr) 38(5):327–339
62. Fulda S, Vazquez-Santillan K, Melendez-Zajgla J, Jimenez-Hernandez L, Martínez-Ruiz G, Maldonado V (2013) Regulation of apoptosis pathways in cancer stem cells. Cancer Lett 338 (1):168–173
63. Lin YH, Huang YC, Chen LH, Chu PM (2015) Autophagy in cancer stem/progenitor cells. Cancer Chemother Pharmacol 75(5):879–886
64. Zhang L, Tong X, Li J, Huang Y, Hu X, Chen Y, Huang J, Wang J, Liu B (2015) Apoptotic and autophagic pathways with relevant small-molecule compounds, in cancer stem cells. Cell Prolif 48(4):385–397
65. Goncalves JM, Cordeiro MMR, Rivero ERC (2017) The role of the complex USP1/WDR48 in differentiation and proliferation processes in cancer stem cells. Curr Stem Cell Res Ther 12 (5):416–422
66. Zhou W, Bao S (2014) PML-mediated signaling and its role in cancer stem cells. Oncogene 33 (12):1475–1484
67. Zhao Y, Bao Q, Renner A, Camaj P, Eichhorn M, Ischenko I, Angele M, Kleespies A, Jauch KW, Bruns C (2011) Cancer stem cells and angiogenesis. Int J Dev Biol 55(4–5):477–482
68. Shetzer Y, Molchadsky A, Rotter V (2016) Oncogenic mutant p53 gain of function nourishes the vicious cycle of tumor development and cancer stem-cell formation. Cold Spring Harb Perspect Med 6(10):a026203
69. Shay JW, Wright WE (2010) Telomeres and telomerase in normal and cancer stem cells. FEBS Lett 584(17):3819–3825
70. Dave B, Chang J (2009) Treatment resistance in stem cells and breast cancer. J Mammary Gland Biol Neoplasia 14(1):79–82
71. Ping YF, Bian XW (2011) Cancer stem cells switch on tumor neovascularization. Curr Mol Med 11(1):69–75
72. Yanagisawa M (2011) Stem cell glycolipids. Neurochem Res 36(9):1623–1635
73. Pontier SM (2009) Muller WJIntegrins in mammary-stem-cell biology and breast-cancer progression--a role in cancer stem cells? J Cell Sci 122(Pt 2):207–214
74. Chin AR, Wang SE (2014) Cytokines driving breast cancer stemness. Mol Cell Endocrinol 382(1):598–602
75. Richly H, Aloia L, Di Croce L (2011) Roles of the Polycomb group proteins in stem cells and cancer. Cell Death Dis 2:e204
76. Chan WI, Huntly BJ (2008) Leukemia stem cells in acute myeloid leukemia. Semin Oncol 35 (4):326–335
77. Valent P, Bonnet D, De Maria R, Lapidot T, Copland M, Melo JV, Chomienne C, Ishikawa F, Schuringa JJ, Stassi G, Huntly B, Herrmann H, Soulier J, Roesch A, Schuurhuis GJ, Wöhrer S, Arock M, Zuber J, Cerny-Reiterer S, Johnsen HE, Andreeff M (2012) Eaves C Cancer stem cell definitions and terminology: the devil is in the details. Nat Rev Cancer 12(11):767–775
78. Richard V, Nair MG, Santhosh Kumar TR, Pillai MR (2013) Side population cells as prototype of chemoresistant, tumor-initiating cells. Biomed Res Int 2013:517237
79. Ajani JA, Song S, Hochster HS, Steinberg IB (2015) Cancer stem cells: the promise and the potential. Semin Oncol 42(Suppl 1):S3–S17

80. Davies S, Beckenkamp A, Buffon A (2015) CD26 a cancer stem cell marker and therapeutic target. Biomed Pharmacother 71:135–138
81. Liou GY (2019) CD133 as a regulator of cancer metastasis through the cancer stem cells. Int J Biochem Cell Biol 106:1–7
82. Prestegarden L, Enger PØ (2010) Cancer stem cells in the central nervous system--a critical review. Cancer Res 70(21):8255–8258
83. Oren O, Smith BD (2017) Eliminating cancer stem cells by targeting embryonic signaling pathways. Stem Cell Rev Rep 13(1):17–23
84. Marcato P, Dean CA, Giacomantonio CA, Lee PW (2011) Aldehyde dehydrogenase: its role as a cancer stem cell marker comes down to the specific isoform. Cell Cycle 10(9):1378–1384
85. Tomita H, Tanaka K, Tanaka T, Hara A (2016) Aldehyde dehydrogenase 1A1 in stem cells and cancer. Oncotarget 7(10):11018–11032
86. Jeter CR, Yang T, Wang J, Chao HP, Tang DG (2015) Concise review: NANOG in cancer stem cells and tumor development: an update and outstanding questions. Stem Cells 33 (8):2381–2390
87. Mancini R, Noto A, Pisanu ME, De Vitis C, Maugeri-Saccà M, Ciliberto G (2018) Metabolic features of cancer stem cells: the emerging role of lipid metabolism. Oncogene 37 (18):2367–2378
88. Es-Haghi M, Soltanian S, Dehghani H (2016) Perspective: cooperation of Nanog, NF-κB, and CXCR4 in a regulatory network for directed migration of cancer stem cells. Tumour Biol 37 (2):1559–1565
89. Loureiro R, Mesquita KA, Oliveira PJ, Vega-Naredo I (2013) Mitochondria in cancer stem cells: a target for therapy. Recent Pat Endocr Metab Immune Drug Discov 7(2):102–114
90. Zou GM (2008) Cancer initiating cells or cancer stem cells in the gastrointestinal tract and liver. J Cell Physiol 217(3):598–604
91. Lee CJ, Dosch J, Simeone DM (2008) Pancreatic cancer stem cells. J Clin Oncol 26 (17):2806–2812
92. Khan MI, Czarnecka AM, Helbrecht I, Bartnik E, Lian F, Szczylik C (2015) Current approaches in identification and isolation of human renal cell carcinoma cancer stem cells. Stem Cell Res Ther 6:178
93. Giannone G, Attademo L, Scotto G et al (2019) Endometrial cancer stem cells: role, characterization and therapeutic implications. Cancers (Basel) 11(11):1820
94. Somasundaram V, Srinivas P (2012) Insights into the targeted elimination of BRCA1-defective cancer stem cells. Med Res Rev 32(5):948–967
95. Gelmini S, Mangoni M, Serio M, Romagnani P, Lazzeri E (2008) The critical role of SDF-1/CXCR4 axis in cancer and cancer stem cells metastasis. J Endocrinol Investig 31(9):809–819
96. Chakraborty C, Chin KY, Das S (2016) miRNA-regulated cancer stem cells: understanding the property and the role of miRNA in carcinogenesis. Tumour Biol 37(10):13039–13048
97. Mallick B, Chakrabarti J, Ghosh Z (2011) MicroRNA reins in embryonic and cancer stem cells. RNA Biol 8(3):415–426
98. Mens MMJ, Ghanbari M (2018) Cell cycle regulation of stem cells by MicroRNAs. Stem Cell Rev Rep 14(3):309–322
99. Chhabra R, Saini N (2014) MicroRNAs in cancer stem cells: current status and future directions. Tumour Biol 35(9):8395–8405
100. Wu K, Watabe K (2014) MicroRNAs in cancer stem cells: new regulators of stemness. Xing FCurr Pharm Des 20(33):5319–5327
101. Castro-Oropeza R, Melendez-Zajgla J, Maldonado V, Vazquez-Santillan K (2018) The emerging role of lncRNAs in the regulation of cancer stem cells. Cell Oncol (Dordr) 41 (6):585–603
102. Xu YM, Liao XY, Chen XW, Li DZ, Sun JG, Liao RX (2015) Regulation of miRNAs affects radiobiological response of lung cancer stem cells. Biomed Res Int 2015:851841
103. Loureiro R, Mesquita KA, Magalhães-Novais S, Oliveira PJ, Vega-Naredo I (2017 Dec) Mitochondrial biology in cancer stem cells. Semin Cancer Biol 47:18–28

104. Relation T, Dominici M, Horwitz EM (2017 May) Concise review: an (im)penetrable shield: how the tumor microenvironment protects cancer stem cells. Stem Cells 35(5):1123–1130
105. Zhang BB, Wang DG, Guo FF, Xuan C (2015 Mar) Mitochondrial membrane potential and reactive oxygen species in cancer stem cells. Familial Cancer 14(1):19–23
106. Prieto-Vila M, Takahashi RU, Usuba W, Kohama I, Ochiya T (2017) Drug resistance driven by cancer stem cells and their niche. Int J Mol Sci 18(12):E2574
107. Liu X, Fan D (2015) The epithelial-mesenchymal transition and cancer stem cells: functional and mechanistic links. Curr Pharm Des 21(10):1279–1291
108. Luo M, Brooks M, Wicha MS (2015) Epithelial-mesenchymal plasticity of breast cancer stem cells: implications for metastasis and therapeutic resistance. Curr Pharm Des 21 (10):1301–1310
109. Zhou P, Li B, Liu F, Zhang M, Wang Q, Liu Y, Yao Y, Li D (2017) The epithelial to mesenchymal transition (EMT) and cancer stem cells: implication for treatment resistance in pancreatic cancer. Mol Cancer 16(1):52
110. Liberti MV, Locasale JW (2016) The Warburg effect: how does it benefit cancer cells? [published correction appears in Trends Biochem Sci. 2016 Mar;41(3):287]. Trends Biochem Sci 41(3):211–218
111. Peixoto J, Lima J (2018) Metabolic traits of cancer stem cells. Dis Model Mech 11(8): dmm033464
112. Wong TL, Che N, Ma S (2017) Reprogramming of central carbon metabolism in cancer stem cells. Biochim Biophys Acta Mol basis Dis 1863(7):1728–1738
113. Chae YC, Kim JH (2018) Cancer stem cell metabolism: target for cancer therapy. BMB Rep 51(7):319–326. https://doi.org/10.5483/bmbrep.2018.51.7.112
114. Vadde R, Vemula S, Jinka R, Merchant N, Bramhachari PV, Nagaraju GP (2017) Role of hypoxia-inducible factors (HIF) in the maintenance of stemness and malignancy of colorectal cancer. Crit Rev Oncol Hematol 113:22–27
115. Peitzsch C, Perrin R, Hill RP, Dubrovska A (2014) Hypoxia as a biomarker for radioresistant cancer stem cells. Kurth IInt J Radiat Biol 90(8):636–652
116. Ding S, Li C, Cheng N, Cui X, Xu X, Zhou G (2015) Redox regulation in cancer stem cells. Oxidative Med Cell Longev 2015:750798. https://doi.org/10.1155/2015/750798
117. Wang T, Shigdar S, Gantier MP, Hou Y, Wang L, Li Y, Shamaileh HA, Yin W, Zhou SF, Zhao X, Duan W (2015) Cancer stem cell targeted therapy: progress amid controversies. Oncotarget 6(42):44191–44206
118. Lee SY, Jeong EK, Ju MK, Jeon HM, Kim MY, Kim CH, Park HG, Han SI, Kang HS (2017) Induction of metastasis, cancer stem cell phenotype, and oncogenic metabolism in cancer cells by ionizing radiation. Mol Cancer 16(1):10
119. Ahmed M, Chaudhari K, Babaei-Jadidi R, Dekker LV, Shams NA (2017) Concise review: emerging drugs targeting epithelial cancer stem-like cells. Stem Cells 35(4):839–850
120. García-Rubiño ME, Lozano-López C, Campos JM (2016) Inhibitors of cancer stem cells. Anti Cancer Agents Med Chem 16(10):1230–1239
121. Du B, Shim JS (2016) Targeting epithelial-mesenchymal transition (EMT) to overcome drug resistance in cancer. Molecules 21(7):E965
122. Markowska A, Sajdak S, Huczyński A, Rehlis S, Markowska J (2018) Ovarian cancer stem cells: a target for oncological therapy. Adv Clin Exp Med 27(7):1017–1020
123. He YC, Zhou FL, Shen Y, Liao DF, Cao D (2014) Apoptotic death of cancer stem cells for cancer therapy. Int J Mol Sci 15(5):8335–8351
124. Eun K, Ham SW, Kim H (2017) Cancer stem cell heterogeneity: origin and new perspectives on CSC targeting. BMB Rep 50(3):117–125
125. Lei Y, Zhang D, Yu J, Dong H, Zhang J, Yang S (2017) Targeting autophagy in cancer stem cells as an anticancer therapy. Cancer Lett 393:33–39
126. Rycaj K, Tang DG (2014) Cancer stem cells and radioresistance. Int J Radiat Biol 90 (8):615–621

127. Shi AM, Tao ZQ, Li H, Wang YQ, Zhao J (2015) Cancer stem cells targeting agents--a review. Eur Rev Med Pharmacol Sci 19(21):4064–4067
128. Yoshida GJ, Saya H (2016) Therapeutic strategies targeting cancer stem cells. Cancer Sci 107 (1):5–11
129. Chang L, Graham P, Hao J, Ni J, Deng J, Bucci J, Malouf D, Gillatt D, Li Y (2016) Cancer stem cells and signaling pathways in radioresistance. Oncotarget 7(10):11002–11017
130. Li Y, Atkinson K, Zhang T (2017) Combination of chemotherapy and cancer stem cell targeting agents: preclinical and clinical studies. Cancer Lett 396:103–109
131. Colak S, Medema JP (2014) Cancer stem cells--important players in tumor therapy resistance. FEBS J 281(21):4779–4791
132. Hodgkinson N, Kruger CA, Abrahamse H (2017) Targeted photodynamic therapy as potential treatment modality for the eradication of colon cancer and colon cancer stem cells. Tumour Biol 39(10):1010428317734691
133. Oh J, Hlatky L, Jeong YS, Kim D (2016) Therapeutic effectiveness of anticancer phytochemicals on cancer stem cells. Toxins (Basel) 8(7):E199
134. Haq S, Suresh B, Ramakrishna S (2018) Deubiquitylating enzymes as cancer stem cell therapeutics. Biochim Biophys Acta Rev Cancer 1869(1):1–10
135. Del Re M, Arrigoni E, Restante G, Passaro A, Rofi E, Crucitta S, De Marinis F, Di Paolo A, Danesi R (2018) Concise review: resistance to tyrosine kinase inhibitors in non-small cell lung cancer: the role of Cancer stem cells. Stem Cells 36(5):633–640
136. So JY, Suh N (2015) Targeting cancer stem cells in solid tumors by vitamin D. J Steroid Biochem Mol Biol 148:79–85
137. Huang GH, Xu QF, Cui YH, Li N, Bian XW, Lv SQ (2016) Medulloblastoma stem cells: promising targets in medulloblastoma therapy. Cancer Sci 107(5):583–589
138. Talukdar S, Emdad L, Das SK, Sarkar D, Fisher PB (2016) Evolving strategies for therapeutically targeting cancer stem cells. Adv Cancer Res 131:159–191
139. Xu Q, Le AD (2016) Targeting cancer stem cells in oral cancer. J Calif Dent Assoc 44 (2):112–120
140. Moitra K (2015) Overcoming multidrug resistance in cancer stem cells. Biomed Res Int 2015:635745
141. Kaur G, Sharma P, Dogra N, Singh S (2018) Eradicating cancer stem cells: concepts, issues, and challenges. Curr Treat Options in Oncol 19(4):20
142. Ukaji T, Umezawa K (2014) Novel approaches to target NF-κB and other signaling pathways in cancer stem cells. Adv Biol Regul 56:108–115
143. Kim TH, Suh DH, Kim MK, Song YS (2014) Metformin against cancer stem cells through the modulation of energy metabolism: special considerations on ovarian cancer. Biomed Res Int 2014:132702
144. Bao B, Azmi AS, Ali S, Zaiem F, Sarkar FH (2014) Metformin may function as anti-cancer agent via targeting cancer stem cells: the potential biological significance of tumor-associated miRNAs in breast and pancreatic cancers. Ann Transl Med 2(6):59. https://doi.org/10.3978/j.issn.2305-5839.2014.06.05
145. Ranji P, Salmani Kesejini T, Saeedikhoo S, Alizadeh AM (2016) Targeting cancer stem cell-specific markers and/or associated signaling pathways for overcoming cancer drug resistance. Tumour Biol 37(10):13059–13075
146. Moselhy J, Srinivasan S, Ankem MK, Damodaran C (2015) Natural products that target cancer stem cells. Anticancer Res 35(11):5773–5788
147. Zang S, Liu T, Shi J, Qiao L (2014) Curcumin: a promising agent targeting cancer stem cells. Anti Cancer Agents Med Chem 14(6):787–792

148. Marucci C, Fumagalli G, Calogero F, Silvani A, Christodoulou MS, Martinet N, Passarella D (2015) Natural products and cancer stem cells. Curr Pharm Des 21(38):5547–5557
149. Scarpa ES, Ninfali P (2015) Phytochemicals as innovative therapeutic tools against cancer stem cells. Int J Mol Sci 16(7):15727–15742
150. Cianciosi D, Varela-Lopez A, Forbes-Hernandez TY, Gasparrini M, Afrin S, Reboredo-Rodriguez P, Zhang J, Quiles JL, Nabavi SF, Battino M, Giampieri F (2018) Targeting molecular pathways in cancer stem cells by natural bioactive compounds. Pharmacol Res 135:150–165
151. Sak K, Everaus H (2015) Role of flavonoids in future anticancer therapy by eliminating the cancer stem cells. Curr Stem Cell Res Ther 10(3):271–282

Next Generation Sequencing and Stem Cells

8

Contents

Abstract

Advent of sequencing technologies and improvement in the sequencing studies through approaches of next generation sequencing (NGS) have enabled study of stem cells, their development, niche, signalling, self-renewal and differentiation, transcriptional network and their regulation. Application of NGS to Stem cell biology has enabled better scope and application in regenerative medicine and tumorigenesis. It has also been possible to study the biology of cancer stem cells (CSC). In this chapter we highlight (1) NGS, (2) transcriptomics and stem cells, (3) single cells transcriptomics, and (4) study of CSC.

Keywords

Next generation sequencing · Illumina · Ion torrent · RNA sequencing (RNA seq)

Abbreviations

A dubia	*Amoeba dubia*
ACc	Adipose tissue stem cells
AML	Acute myeloid leukaemia
bp	Base pairs

S. Chowdhury, S. Ghosh, *Stem Cells*, https://doi.org/10.1007/978-981-16-1638-9_8

C elegans	*Caenorhabditis elegans*
CRT	Cyclic reversible termination
CSC	Cancer stem cells
D melanogaster	*Drosophila melanogaster*
DSC	Dental stem cell
ES	Embryonic stem cells
EST	Expressed sequence tags
GBC	Glioma stem cells
GBM	Glioblastoma
H influenza	*Haemophilus influenzae*
hESCs	Human embryonic stem Cells
iPSc	Induced pluripotent stem cells
MRD	minimal residual disease
MScs	Mesenchymal stromal cells
Mus musculus	*M musculus*
NGS	Next generation sequencing
NSC	Neural stem cells
PCR	Polymerase chain reaction
PSC	Pluripotent stem cells
RNA Seq	RNA sequencing
S cerevisiae	*Saccharomyces cerevisiae*
scRNA-seq	Single-cell RNA-sequencing
SCT	Single-cell transcriptomics
SMRT	Single molecule real time
snv	Single nucleotide variants
SP1	Specificity protein 1
TP53	Tumour protein p53
ZBTB16	Zinc finger and BTB domain containing 16

8.1 Introduction

As early as in the 1970s and 1980s the manual DNA sequencing methods had its origin and included Maxam–Gilbert sequencing and Sanger sequencing. Around 1990, automated sequencing methods enabled sequencing of whole genomes. The earliest sequenced organism wherein the whole genome was sequenced was *Haemophilus influenzae* (*H influenza*) completed and released in 1995. After the initial studies, genomes of other bacteria and some Archaea were sequenced. Later both unicellular organisms such as *Amoeba dubia* (*A dubia*) with genome of 700 billion nucleotide pairs and multicellular organisms like humans (*Homo sapiens, H sapiens*) were studied. In 1996 the first eukaryotic genome of model organism commonly called as yeast (*Saccharomyces cerevisiae, S cerevisiae*) was sequenced with 12 million nucleotide pairs with whole genome sequenced. In 1998, the first multicellular nematode sequenced was *Caenorhabditis elegans* (*C elegans*) which is a model organism. In 1999, the entire DNA sequence of human chromosome 22, the

Table 8.1 Important landmarks in the sequencing of plants and animals

Year	Organism	No. of chromosomes	Importance
1976	Bacteriophage MS2	1	First genome completely sequenced
	PhiX174	1	First DNA-based genome sequenced
1995	Haemophilus influenza	1	First bacteria sequenced
	Methanococcus jannaschii	1	First archaeon sequenced
1996	*Saccharomyces cerevisiae*	32	First fungi sequenced
1998	*Caenorhabditis elegans*	12 hermaphodites 11: males	First animal sequenced
2000	*Arabidopsis thaliana*	5	First plant sequenced
	Drosophila melanogaster	4	Model organism
2001	*Homo sapiens*	23	To understand human genome in health and disease
2002	*Mus musculus*	20	95% similarity with human
	Anopheles gambiae	5	Principal vector of malaria
	Takifugu rubripes	8	Study of vertebrate evolution
	Oryza sativa	12	Staple food crop
2004	*Gallus gallus*	78	First egg-laying animal
2005	*Pan troglodyte*	23	First non-human primate
2010	*Xenopus tropicalis*	10	Study early development
2013	*Danio rerio*	25	Model organism used in research

shortest human autosome, was known and published in the year of 2000. The fruit fly *Drosophila melanogaster* (*D melanogaster*) with genome of 180 Mb in size—another model organism was sequenced in the year 2000. *Arabidopsis thaliana* (*A thaliana*) was the first plant model organism sequenced in 2000 and 2002 laboratory mouse *Mus musculus* (*M musculus*) was sequenced. In 2004, the Human Genome Project (HGP) published an incomplete version of the human genome [1, 2] (Table 8.1).

8.2 Next Generation Sequencing (NGS)

Next generation sequencing (NGS) [3, 4] is especially finding importance in sequencing of genomes highlighting genetic variation and how they influence traits as it enables fast and more accurate characterization of any species when compared to traditional methods, like Sanger sequencing.

NGS platforms currently in use are (1) Illumina performing sequencing by synthesis, (2) 454 Roche enabling pyrosequencing, which is a method of DNA sequencing based on the sequencing by synthesis, in which the sequencing is detected by the nucleotide included by a DNA polymerase, (3) Ion Torrent for ion semiconductor sequencing is based on detection of hydrogen ions released during the DNA polymerization, (4) Applied Biosystems (Life Technologies) SOLiD sequencing by ligation method, (5) Oxford nanopore performing nanopore sequencing, and (6) Pacific biosciences Single molecule real time (SMRT) sequencing.

All NGS methods are characterized by (1) sample preparation for the sequencing phase includes preparation, amplification, and sequencing and (2) a computerized stage for the analysis of the data. It includes the main steps of

1. Sample preparation, involving fragmentation of DNA from 100 to 800 base pairs (bp) based on the platform and the adapters used.
2. Amplification is based on two methods: emulsion or solid polymerase Chain Reaction (PCR) amplifications where the single DNA molecules are amplified clonally. The adapters are then bound to each DNA molecule, followed by hybridization to complementary sequences coating the surface. Amplification cycles allow fragments to be amplified clonally.
3. All the NGS platforms enable sequence and parallelly amplify the DNA molecules clonally.
4. While by Sanger method, the differentially sized fragments from each sequencing reactions are separated by electrophoresis, in the NGS technologies, sequencing is through the repetition of nucleotide extension cycles or oligonucleotide ligations.
5. The principles of sequencing and image acquisition vary with the platforms of sequencing used.
6. The cyclic reversible termination (CRT) sequencing method makes use of reversible labelled nucleotides wherein each sequencing cycle includes incorporation of nucleotide, acquisition of fluorescence, and followed by nucleotide removal.
7. Excess nucleotides are removed by step of washing.
8. Image of inserted nucleotide is captured. This is followed by removal of the terminator group and the fluorophore.
9. Post-elimination, the polymerase can continue the extension reaction and include the second nucleotide. This process is used by Illumina and Helicos, and the difference lies in the templates of sequencing as Illumina uses clonally amplified fragments on a solid surface but Helicos uses single DNA molecules. While Illumina uses four colours for labelling four reversible nucleotides which are dispensed in the sequencer at the same time, the Helicos platform uses the same fluorophore labelled nucleotides dispensed into the sequencer in a determined sequential manner (Figs. 8.1 and 8.2).

De novo sequencing is done where there is no reference sequence available for alignment and it enables our understanding of genetic details of animal's functions and its interaction with the environment. Here the sequence reads are assembled as contigs, and the coverage quality of de novo sequence data depends on the size and continuity of the contigs (i.e., the number of gaps in the data).

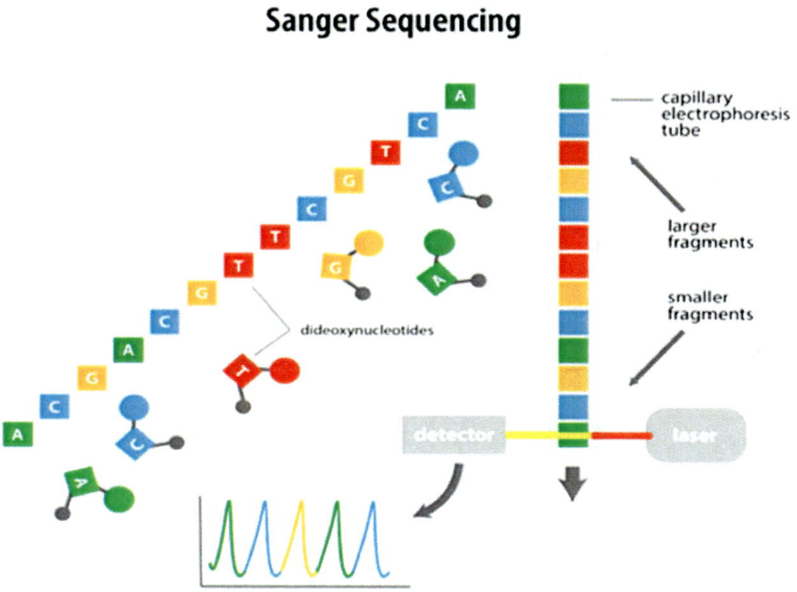

Fig. 8.1 Representative view of Sanger Sequencing. Image courtesy of Applied Biological Materials Inc. (abm) "or" Image courtesy of Applied Biological Materials Inc. (abm) at www.abmgood.com [5]

RNA-Seq enables the study of the transcriptome, highlighting the information about transcript isoforms, gene fusions, single nucleotide variants (snv) [6, 7]. This finds importance in detecting changes in disease states, response to therapeutics, under different environmental conditions and understanding epigenetic changes or study of heritable changes in gene activity caused by mechanisms other than changes in the DNA sequence DNA methylation, DNA–protein interactions, chromatin accessibility, histone modifications.

8.3 Transcriptomics and Stem Cells

Hypertranscription involved in different developmental phenomena including embryonic development, stem cell biology, regeneration could be understood using NGS based studies [8]. NGS has enabled study of molecular biology of Hematopoietic stem cells (HSCs) and their progeny and lineages, both in mouse models and human beings and finds importance in the study of homeostasis, injury and modified in hematopoietic malignancy like acute myeloid leukaemia (AML) [9]. cDNA and Expressed sequence tags (EST) sequencing and RNA-Sequencing have enabled study of transcriptome of Human Embryonic Stem Cells (hESCs) [10] and multipotent mesenchymal stromal cells (MSCs) [11]. Epigenetic modification like histone modifications, DNA methylation, chromatin remodelling, somatic cells

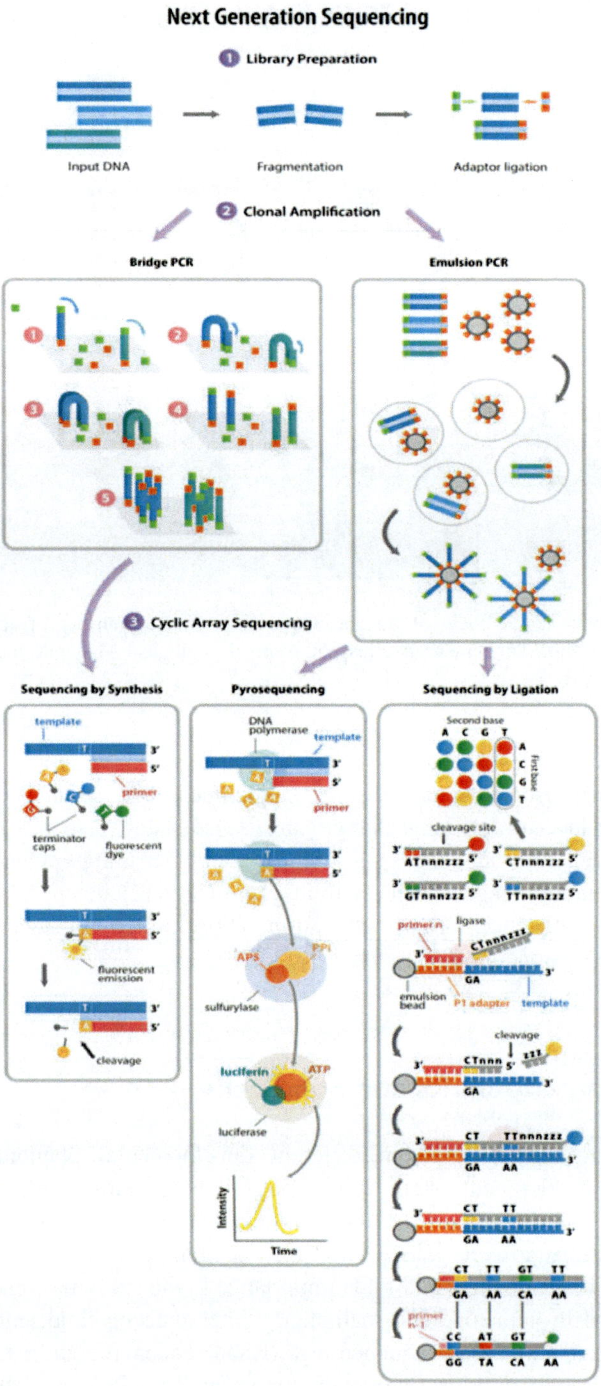

Fig. 8.2 NGS. Image adapted from Abam [5] Image courtesy of Applied Biological Materials Inc. (abm) "or" Image courtesy of Applied Biological Materials Inc. (abm) at www.abmgood.com

reprogrammed into induced pluripotent stem cells (iPSc), and biology of the embryonic stem cells (ES) cells [12] are being studied.

Role of glioma stem cells (GSCs) in the development of aggressiveness associated in glioblastoma (GBM) formation, its recurrence, and resistance to radiation and chemotherapy [13] is being studied. This advanced technique has enabled study of astrocytes as progenitors of neural stem cells (NSCs) [14]. The histone methyltransferase SETDB1 plays role in embryogenesis [15, 16] and could be detected by genome sequencing approaches [15]. The evolutionary role of neural progenitor cells and their role cortical expansion were studied by genomics and transcriptomics approaches [17], diversity in neurons from stem cells [18].

Transcriptomics has enabled indepth study of stem cell biology [19–22], its development, lineage commitment, haematopoiesis and niche [23–25], study of CSC [26], macrophage plasticity [27] and epigenetics of stem cells [28]. Transcriptomics has enabled study of transcription factors, including Zinc Finger and BTB Domain Containing 16 (ZBTB16), specificity protein 1 (SP1) and tumour protein p53, (TP53) belonging to the Sp/KLF family of transcription factors and their role in dental stem cell (DSC) differentiation [29].

8.4 Single Cell Transcriptomics (SCT) in Study of Stem Cells

Single-cell transcriptomics (SCT) of pluripotent stem cells (PSC) enabled study of cell states, heterogeneity, pluripotency, regulation of gene expression, regulatory networks, cell-fate determination, molecular mechanisms of reprogramming of somatic cells to induced pluripotent stage [30]. Transcriptome of stem cell differentiation, biology of incurable diffuse gliomas [31, 32] could be studied by single-cell RNA-sequencing (scRNA-seq) approaches which is difficult to study by conventional methods due to their complex biology, tumour heterogeneity, cancer cell lineages, signalling mechanism, tumour microenvironment, and classification [32].

It has now been possible to study the cellular nature of adipose tissue stem cells (ASCs) [33], variation and heterogeneity in stem cell population [34], mesenchymal stem cells [35], evolution and development of cells [36], naive state of human PSCs [37], oligodendrocyte lineage cells [38]. Proteomics approaches has enabled study of human embryonic stem cells (hESCs) [38], neuronal diversity [39, 40].

8.5 NGS and Cancer Stem Cells (CSC)

Renewal, differentiation, de-differentiation, signalling in CSCs could be understood by NGS approaches. CSCs epigenetic, epithelial-mesenchymal transition also called as EMT, metabolic and immunological reprogramming suitable for tumour microenvironment and survival of host defence has been studied. Signalling cascades including WNT, The Fibroblast Growth Factor (FGF), Notch, Hedgehog, and TGFβ/BMP and their role in regulation of expression of CSC markers, CD44, CD133

(PROM1), epithelial cell adhesion molecule (EPCAM), and LGR5 (GPR49) could be studied.

Aberrant canonical and non-canonical WNT signalling in breast, colorectal, gastric, lung, ovary, pancreatic, prostate and uterine cancers, leukaemia, and melanoma has been reported to play role in CSC survival, expansion and metastasis [1].

8.6 Discussion

Studies in the domain of stem cell biology and early process in developmental biology have been made possible after the application and development of NGS technologies [41–43] to the study of stem cells. The development and maintenance of the CSC and their role in pathogenesis are being understood in different Cancer. This offers an advantage over conventional forms of study, and in Cancer early detection, prediction of minimal residual disease (MRD) and in designing of therapeutics. This is also a step towards design of personal medicine.

References

1. Appels R, Nystrom-Persson J, Keeble-Gagnere G (2014) Advances in genome studies in plants and animals. Funct Integr Genomics 14(1):1–9. https://doi.org/10.1007/s10142-014-0364-5
2. Murat F, Van de Peer Y, Salse J (2012) Decoding plant and animal genome plasticity from differential paleo-evolutionary patterns and processes. Genome Biol Evol 4(9):917–928. https://doi.org/10.1093/gbe/evs066
3. Behjati S, Tarpey PS (2013) What is next generation sequencing? Arch Dis Child Educ Pract Ed 98(6):236–238. https://doi.org/10.1136/archdischild-2013-304340
4. https://www.ebi.ac.uk/training/online/course/ebi-next-generation-sequencing-practical-course/what-you-will-learn/what-next-generation-dna
5. Applied Biological Materials Inc (abm) at www.abmgood.com
6. Kukurba KR, Montgomery SB (2015) RNA sequencing and analysis. Cold Spring Harb Protoc 2015(11):951–969. https://doi.org/10.1101/pdb.top084970
7. Piskol R, Ramaswami G, Li JB (2013) Reliable identification of genomic variants from RNA-seq data. Am J Hum Genet 93(4):641–651. https://doi.org/10.1016/j.ajhg.2013.08.008
8. Percharde M, Bulut-Karslioglu A, Ramalho-Santos M (2017) Hypertranscription in development, stem cells, and regeneration. Dev Cell 40(1):9–21
9. Vedi A, Santoro A, Dunant CF, Dick JE, Laurenti E (2016) Molecular landscapes of human hematopoietic stem cells in health and leukemia. Ann N Y Acad Sci 1370(1):5–14
10. Au KF, Sebastiano V (2014) The transcriptome of human pluripotent stem cells. Curr Opin Genet Dev 28:71–77
11. Kasoju N, Wang H, Zhang B, George J, Gao S, Triffitt JT, Cui Z, Ye H (2017) Transcriptomics of human multipotent mesenchymal stromal cells: retrospective analysis and future prospects. Biotechnol Adv 35(4):407–418
12. Godini R, Lafta HY, Fallahi H (2018) Epigenetic modifications in the embryonic and induced pluripotent stem cells. Gene Expr Patterns 29:1–9
13. Zhou D, Alver BM, Li S, Hlady RA, Thompson JJ, Schroeder MA, Lee JH, Qiu J, Schwartz PH, Sarkaria JN, Robertson KD (2018) Distinctive epigenomes characterize glioma stem cells and their response to differentiation cues. Genome Biol 19(1):43

14. Götz M, Sirko S, Beckers J, Irmler M (2015) Reactive astrocytes as neural stem or progenitor cells: in vivo lineage, in vitro potential, and genome-wide expression analysis. Glia 63 (8):1452–1468
15. Golding MC, Snyder M, Williamson GL et al (2015) Histone-lysine N-methyltransferase SETDB1 is required for development of the bovine blastocyst. Theriogenology 84 (8):1411–1422. https://doi.org/10.1016/j.theriogenology.2015.07.028
16. Kang YK (2015) SETDB in early embryos and embryonic stem cells. Curr Issues Mol Biol 17:1–10
17. Namba T, Huttner WB (2017) Neural progenitor cells and their role in the development and evolutionary expansion of the neocortex. Wiley Interdiscip Rev Dev Biol 6:1
18. Harbom LJ, Michel N, McConnell MJ (2018) Single-cell analysis of diversity in human stem cell-derived neurons. Cell Tissue Res 371(1):171–179
19. Donati G (2016) The niche in single-cell technologies. Immunol Cell Biol 94(3):250–255
20. Wang H, Zhang Q, Fang X (2014) Transcriptomics and proteomics in stem cell research. Front Med 8(4):433–444
21. Stanton LW, Bakre MM (2007) Genomic and proteomic characterization of embryonic stem cells. Curr Opin Chem Biol 11(4):399–404
22. Kabir MH, O'Connor MD (2019) Stems cells, big data and compendium-based analyses for identifying cell types, signalling pathways and gene regulatory networks. Biophys Rev 11 (1):41–50
23. Godoy P, Schmidt-Heck W, Hellwig B, Nell P, Feuerborn D, Rahnenführer J, Kattler K, Walter J, Blüthgen N, Hengstler JG (2018) Assessment of stem cell differentiation based on genome-wide expression profiles. Philos Trans R Soc Lond Ser B Biol Sci 5:373
24. Heitman N, Saxena N, Rendl M (2018) Advancing insights into stem cell niche complexities with next-generation technologies. Curr Opin Cell Biol 55:87–95
25. Cvejic A (2016) Mechanisms of fate decision and lineage commitment during haematopoiesis. Immunol Cell Biol 94(3):230–235
26. Wen L, Tang F (2016) Single-cell sequencing in stem cell biology. Genome Biol 17:71
27. Hume DA, Summers KM, Rehli M (2016) Transcriptional regulation and macrophage differentiation. Microbiol Spectr 4:3
28. Rasmussen TP (2014) The epigenetics of early development: inferences from stem cells. Mol Reprod Dev 81(2):194–201
29. Viale-Bouroncle S, Felthaus O, Schmalz G, Reichert TE, Morsczeck C (2013) Transcription factors for dental stem cell differentiation. Int J Oral Maxillofac Implants 28(6):e478–e486
30. Natarajan KN, Teichmann SA, Kolodziejczyk AA (2017) Single cell transcriptomics of pluripotent stem cells: reprogramming and differentiation. Curr Opin Genet Dev 46:66–76
31. Picelli S (2017) Single-cell RNA-sequencing: the future of genome biology is now. RNA Biol 14(5):637–650
32. Tirosh I, Suvà ML (2018) Dissecting human gliomas by single-cell RNA sequencing. Neuro-Oncology 20(1):37–43
33. Acosta JR, Joost S, Karlsson K, Ehrlund A, Li X, Aouadi M, Kasper M, Arner P, Rydén M, Laurencikiene J (2017) Single cell transcriptomics suggest that human adipocyte progenitor cells constitute a homogeneous cell population. Stem Cell Res Ther 8(1):250
34. Goldman SL, MacKay M, Afshinnekoo E, Melnick AM, Wu S, Mason CE (2019) The impact of heterogeneity on single-cell sequencing. Front Genet 10:8. https://doi.org/10.3389/fgene.2019.00008
35. McLeod CM, Mauck RL (2017) On the origin and impact of mesenchymal stem cell heterogeneity: new insights and emerging tools for single cell analysis. Eur Cell Mater 34:217–231
36. Marioni JC, Arendt D (2017) How single-cell genomics is changing evolutionary and developmental biology. Annu Rev Cell Dev Biol 33:537–553
37. Huang K, Maruyama T, Fan G (2014) The naive state of human pluripotent stem cells: a synthesis of stem cell and preimplantation embryo transcriptome analyses. Cell Stem Cell 15 (4):410–415

38. Van Hoof D, Heck AJ, Krijgsveld J, Mummery CL (2008) Proteomics and human embryonic stem cells. Stem Cell Res 1(3):169–182
39. van den Hurk M, Bardy C (2019) Single-cell multimodal transcriptomics to study neuronal diversity in human stem cell-derived brain tissue and organoid models. J Neurosci Methods 325:108350
40. Katoh M (2017) Canonical and non-canonical WNT signaling in cancer stem cells and their niches: cellular heterogeneity, omics reprogramming, targeted therapy and tumor plasticity (review). Int J Oncol 51(5):1357–1369
41. Otte J, Wruck W, Adjaye J (2017) New insights into human primordial germ cells and early embryonic development from single-cell analysis. FEBS Lett 591(15):2226–2240
42. Liu N, Liu L, Pan X (2014) Single-cell analysis of the transcriptome and its application in the characterization of stem cells and early embryos. Cell Mol Life Sci 71(14):2707–2715
43. van Bruggen D, Agirre E, Castelo-Branco G (2017) Single-cell transcriptomic analysis of oligodendrocyte lineage cells. Curr Opin Neurobiol 47:168–175

Check for updates

Nanoparticles and Stem Cells

9

Contents

Abstract

Nanoparticles are particles of varying in diameter spanning in the range of 1–100 nanometres (nm) with physical properties that have extensive and diverse application in the human day-to-day life. Their biocompatible nature, ability to be engineered, has enabled their different biological applications. They find applications in (1) delivery of stem cells, (2) monitoring of the stem cells by imaging thereby providing enormous advantages in stem cell-mediated application in regenerative medicine. But despite its advantageous the toxicity issues associated as still restricting their usage. We discuss in this chapter the promises and risks associated with nanoparticles in application in stem cell-mediated therapy.

Keywords

Nanoparticles · Prostrate cancer · Glioblastoma · Ovarian cancer · Graphene, toxicity · Carbon quantum dots · Liposomes · Regenerative medicine.

Abbreviations

AD	Alzheimer's disease
ANXA3	Annexin A3

S. Chowdhury, S. Ghosh, *Stem Cells*, https://doi.org/10.1007/978-981-16-1638-9_9

AuNPs	Gold nanoparticles
BC	Breast cancer
BCSCs	Breast cancer stem cells
BMP	Bone morphogenetic protein
CB	Charnoly body
CDQs	Carbon quantum dots
CNT	Carbon nanotubes
CoSCs	Colorectal cancer stem cells
CSC	Cancer stem cell
CT	Computed tomography
DBCO	Dibenzyl cyclooctyne
DOX	Doxorubicin
EGCG	Epigallocatechin-3-gallate
EPR	Enhanced permeability and retention
FA	Folic acid
Fbg	Fibrinogen
Fbn	Fibrin
FDA	Food and Drug Administration
FITC	Fluorescein isothiocyanate
GBM	Glioblastoma
GLI1	Glioma-associated oncogene 1
GSCs	Glioma stem cells
HA	Hyaluronic acid
HMSN	Hollow mesoporous silica nanoparticles
HPMA	N-(2-hydroxypropyl)-methacrylamide
IL-6	Interleukin-6
IONPs	Iron oxide nanoparticle
iPSCs	Induced pluripotent stem cells
IS	Ischemic stroke
ITT	Immature testicular tissue
IVD	Intervertebral disc degeneration
MAPK	Mitogen-activated protein kinase
MDR	Multidrug resistance
microRNA	miRNA
MIRET	Minimally invasive and regenerative therapeutics
MMSNs	Magnetic mesoporous silica nanoparticles
MNPs	Magnetic nanoparticles
MNPs@SiO_2(RITC)	Silica-coated magnetic nanoparticles
MRI	Magnetic resonance imaging
MS	Multiple sclerosis
MSC	Mesenchymal stem cell
MSN-HAs	Hyaluronic acid conjugated mesoporous silica nanoparticles
MT	Metallothioneins
MWNTs	Multi-walled carbon nanotubes

NF-κB	Nuclear factor kappa-light-chain-enhancer of activated B cells
NIR	Near infrared
Nos	Noscapine
NPs	Nanoparticles
NSC	Neural stem cell
OvCSCs	Ovarian cancer stem cells
PC	Prostrate cancer
PCEC	Prostate cancer epithelial cell
PCSCs	Pancreatic cancer stem cells
PD	Parkinson's disease
PDMAAm-coated γ-Fe_2O_3:	Poly(*N*,*N*-dimethylacrylamide)-coated maghemite
PEI	Polyethylene imine
PI3k	Phosphoinositide 3-kinase
PLGA NPs	Poly(lactic-co-glycolic acid) nanoparticles
PLGA-co-PEG	Poly[(D,L-lactide-co-glycolide)-co-PEG]
PLL	Poly-L-lysine
PSCA	Prostate stem cell antigen
PSMA	Specific membrane antigen
PSP@MB	Poly(ethylene glycol)-SS-polyethylenimine-loaded microbubbles
PTCH1	Human homologue of patched 1
PTEN	Phosphatase and tensin homologue
PTT	Photothermal therapy
PTX	Paclitaxel
QDs	Quantum dots
RBC	Red Blood cells
rGO-AgNPs	Reduced oxide-silver nanoparticles
ROS	Reactive oxygen species
SCI	Spinal cord injury
SHH	Sonic hedgehog
Si[PTX-NC]	Silica-coated paclitaxel (PTX) nanocrystals
siRNA	Small interfering RNA
SPIO@SiO_2-NH_2:	Amine-modified silica-coated polyhedral SPIO nanoparticles
SPIONS	Superparamagnetic iron oxide nanoparticles
SSCs	Spermatogonial stem cells
SWNTs	Single-walled carbon nanotubes
TNBC	Triple negative breast cancer
TSA	Trichostatin A
uPA	Urokinase plasminogen activator
uPAR	Urokinase plasminogen activator receptor
WBC	White blood cells

9.1 Introduction

Nanotechnology owes its origin and name from the Greek word "nanos" denoting dwarf. Nanotechnology encompasses functions of matters or operations in the range of a nanometre or 1/1,000,000,000. Today's world includes a vast range of applications of nanotechnology encompassing devices to applications in our daily lives, plant and biology of agriculture, medical, healthcare, and chemical applications. [1–10]. Nanotechnology has found diverse application in solving problems in study of Biological Sciences, in disease diagnosis, prognosis, and monitoring of Cancer therapy through different applications of molecular imaging, application of nanovectors and nanoimaging contrast agents, diagnosis, and targeting therapy with delivery of anticancer drugs. Nanowires and nanocantilever arrays find application in detection of precancerous and malignant lesion in biological fluid samples [1–10].

Nanoparticles have found application in the therapeutics, stem cell biology, bioimaging, microsurgery procedures with minimal invasive procedures in regenerative medicine thereby leading to the regeneration of different tissues in the body, including bone, cartilage, cardiac, liver, neural tissues, pancreas, skeletal muscle, and skin (Fig. 9.1).

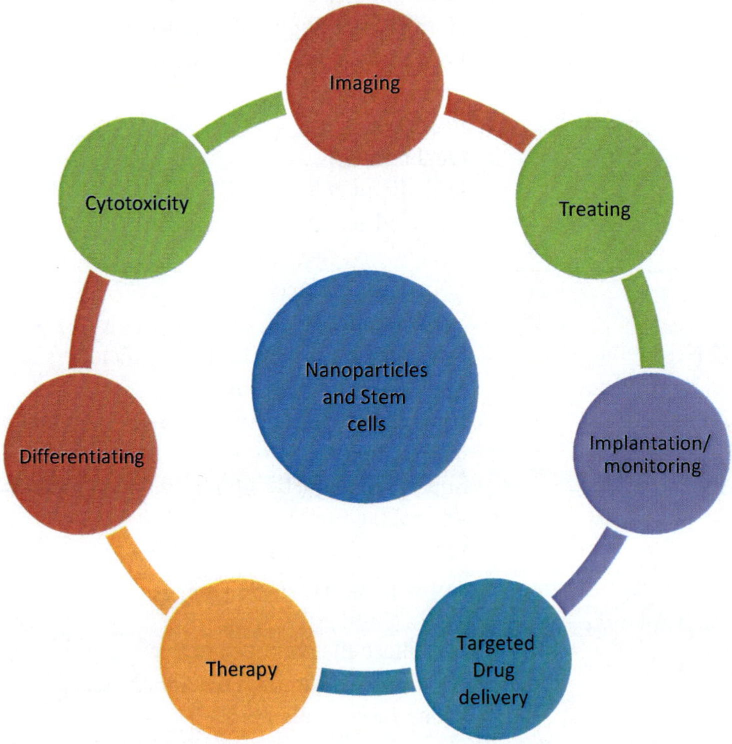

Fig. 9.1 Stem cells and nanoparticles

Table 9.1 Nanocarriers of drugs in targeting BCSCs (Adapted from He et al. [49] with permission)

Delivery systems	Therapeutic agents	Targeting moieties	Mechanisms of action
PEG-b-PLA nanopartioles	Bortezomib		Increased drug accumulation within cancer cells
PEG-b-PLA nanoparticles	Decitabine/ doxorubicin		Combination of drugs for sensitizing BCSCs to chemotherapy
PEG-b-PLA nanoparticles	All-*trans*-retinoic acid/doxorubicin		Co-delivery of drugs for inducing BCSCs to differentiate into non-CSCs and inhibiting tumor cell
PEG-b-PLA nanoparticles	Chloroquine/ doxorubicin/ docetaxel		Co-delivery of drugs for sensitizing BCSCs to chemotherapy
PLGA nanoparticles	Salinomycin/ paclitaxel	Hyaluronic acid	Co-delivery of drugs targeted toward BCSCs and bulk breast cancer cells
PLGA nanoparticles	Paclitaxel	Anti-CD133 monoclonal antibody	Selective inhibition of BCSCs
iTEP nanoparticles	Salinomycin		Increased drug accumulation within cancer cells
Chitosan modified Pluronic F127 nanoparticles	Doxorubicin	Chitosan	Selective inhibition of BCSCs
Cationic lipid-assisted polymeric nanoparticles	siRNA		Increased accumulation of siRNA within cancer cells
HA based self-assembled nanosystems	siRNA	Hyaluronic acid	Increased accumulation of siRNA within cancer cells
Gold nanoparticles	Doxorubicin		Increased accumulation of drug within BCSCs
Iron oxide magnetic nanoparticles	Gemcitabine derivatives	Anti-CD44 antibody	Selective inhibition of BCSCs
Hyaluronan modified/ non-modified mesoporous silica nanoparticles -supported lipid bilayers	8-hydroxyquinoline/ docetaxel	Hyaluronan	Combination therapy
Zinc sulfide nanoparticles			Selective inhibition of BCSCs
PEG-b-PCL micelles	Paclitaxel/ salinomycin	Octreotide	Combination therapy

(continued)

Table 9.1 (continued)

Delivery systems	Therapeutic agents	Targeting moieties	Mechanisms of action
PLGA-co-PEG micelles	Paclitaxel	Anti-CD44 antibody	Selective inhibition of BCSCs
PEG-PAC and PEG-PUC mixture micelles	Thioridazine/ doxorubicin		Co-delivery of drugs targeted toward BCSCs and bulk breast cancer cells
Phospholipid nanomicelles	Curcumin	Vasoactive intestinal peptide	Selective inhibition of BCSCs
Inorganic calcium and phosphate ions coated micelles	Curcumin/ paclitaxel	Hyaluronan oligosaccharides	Co-delivery of drugs targeted toward BCSCs
Cross-linked multilamellar liposomal vesicles	Salinomycin/ doxorubicin		Co-delivery of drugs targeted toward BCSCs and bulk breast cancer cells
Liposomes	Dexamethasone/ ESC8/NRP-1 shRNA-encoded plasmid		Co-delivery of drugs and shRNA-encoded plasmid
Stealth liposomes	All-*trans* retinoic acid/vinorelbine		Prolonged circulation, improved biodistribution and accumulation of drugs within tumors
Stealth liposomes	Vinorelbine/ parthenolide		Prolonged circulation, improved biodistribution and accumulation of drugs within cancer cells
Nanogels	Etoposide/ salinomycin/ curcumin	Hyaluronic acid	Increased accumulation of drug within cancer cells
Targeting moieties	*Targets*	*Therapeutic agents*	*Delivery systems*
Anti-CD44 antibody	CD44 receptor	Paclitaxel	(PLGA-co-PEG) polymeric micelles
Anti-CD44 antibody	CD44 receptor	Gemcitabine derivatives	Iron oxide magnetic nanoparticles
Hyaluronic acid	CD44 receptor	Etoposide/ salinomycin/ curcumin	Cholesteryl-hyalurcnic acid nanogel-drug conjugates
Hyaluronic acid	CD44 receptor	Salinomycin	PLGA nanoparticles
Hyaluronic acid	CD44 receptor	8-hydroxyquinaline	Mesoporous silica nanoparticles-supported ipid bilayers
Hyaluronic acid	CD44 receptor	siRNA	Hyaluronic acid based versatile self-assembling nanosystems

(continued)

Table 9.1 (continued)

Delivery systems	Therapeutic agents	Targeting moieties	Mechanisms of action
Oligosaccharides of hyaluronan	CD44 receptor	Curcumin/ paclitaxel	Inorganic calcium and phosphate ions coated hyaluronan oligosaccharides-histidine-menthone 1,2-glycerol ketal micelles
Chitosan	CD44 receptor	Doxorubicin	Pluronio F127-chitosan nanoparticles
Anti-CD133 antibody	CD133 receptor	Paclitaxel	PLGA nanoparticles
Vasoactive intestina peptide	Vasoactive intestinal peptide receptors	Curcumin	Sterically stabilized phospholipid nanomiceiies
Herceptin	Human epidermal growth factor receptor type 2	Salinomycin	PLGA nanoparticles

PLGA poly(lactic-co-glycolic acid), *PEG-b-PLA* poly(ethylene glycol)-block-poly(D,L-lactic acid), *(PEG-b-PCL) micelles*: poly(ethylene glycol)-block-poly(epsilon-caprolactone), *PLGA-co-PEG* PLGA-PEG-PLGA co-block polymer, *(PEG-b-PCL) micelles* Poly(ethylene glycol)-block-poly(epsilon-caprolactone), *PEG-PAC* polyethylene glycol-polyacrylate, PEG-PUC.

However, only delivery of stem cells or therapeutics is not enough and tracking, sensing, and monitoring of its effects on the targeted cells finds importance in dictating its biological functions in vivo.

Nanostructured particles or vehicles including lipid-based nanoparticles, mesoporous silica nanoparticles, and polymeric nanoparticles have been applied in delivery of regenerative therapy, diagnosis, medical devices designing, and visualization. Nano-biomaterials and soft bioelectronic devices have found application in biotracing of these therapaeutics or stem cells and this property of nanoparticles have led to the development of minimally invasive and regenerative therapeutics (MIRET) with high yield and associated low risks, reduced costs as compared to conventional methods [11].

Human induced pluripotent stem cells (iPSCs) are adult cells with genetical reprogramming generating an embryonic stem cell-like state. They find application in regenerative medicine, drug design and development, disease screening and disease modeling. Nanomedicine and iPSCs are being exploited for therapeutical applications [12].

Polymer nanocarriers of natural and synthetic origin have been applied to experiments on tissue regeneration, like regeneration of the blood vessels bone, cartilage, nerve, and skin tissue. Nanostructures of fibrinogen and fibrin based particles have been reported to interact with damaged tissues and cells. They have been designed in structures used to encapsulate and specifically deliver molecules including proteins, growth factors and stem cells and finds application in tissue regeneration [13].

9.2 Nanoparticles and Delivery of Stem Cells

Success in stem cell therapy encompasses effective delivery of molecules including nucleic acids, proteins, growth factors, cells, and small molecules for stem cell differentiation, and noninvasive, ways of tracking of transplanted stem cells. Magnetic, silica nanoparticles, quantum dots (QDs), and gold nanoparticles (AuNPs) find application in targeted biomolecule delivery due to their biocompatible nature [14].

We discuss in the different Cancers.

Cancer is a deadly malignant disease claiming millions of lives globally. The hallmark properties of cancer include relentless uncontrolled division of cells and malignancy. The Cancer stem cells (CSC) found in cancer tissues include cells that are resistant to chemotherapy and reveal multidrug resistance (MDR).

Several strategies have been tried to target the CSCs. Recently, microRNA (miRNA), metallothioneins (MTs), α-synuclein index, and Charnoly body (CB) have emerged as novel drug discovery biomarkers and selective CB agonists are being designed to target CSC thereby controlling MDR in malignancies [15]. Liposomes and carbon nanotubes/wires find application in gene delivery into stem cells. [16].

We discuss this section highlighting different Cancers.

9.2.1 Prostate Cancer (PC)

Prostate cancer (PC) is a major killer disease in males globally, overexpression of plasminogen activator (uPA) receptor (uPAR), prostate stem cell antigen (PSCA), and prostrate urokinase specific membrane antigen (PSMA) [17, 18] are known biomarkers in serum. Nanomaterials with characteristic optic, electronic, and magnetic properties have been designed for better and sensitive PSA detection. Gold nanoparticles (AuNPs), with high ratio of surface area to volume, have enabled improved PSA detection sensitivity in both serum [19] and urine samples [20]. PSMA, PF-4, interleukin-6 (IL-6), and annexin A3 (ANXA3) are other biomarkers in prostrate cancer (PC) that are being tested using nanomaterials [21, 22] to enable sensitive methods of detection of PC. Elevated blood level miRNA-141, reported from metastatic prostate cancer [23], is being detected by a polymer containing nanomaterial [24]. Spherical conjugates of gold nanoparticle with nucleic acids are being employed to design a miRNA array based detection system for detection of miRNA biomarkers in prostate cancer including miR −16, −20a, −21, −143, 200c, −205, −210. [25].

PSMA targeted anticancer drug - docetaxel is found to be more effective in targeting of PC cells. Superparamagnetic iron oxide nanoparticles (SPIONs) loaded with anticancer drug docetaxel, enabled delivery of docetaxol targeting PSCA ligands, and SPIONs acting as contrast agents for magnetic resonance imaging (MRI) enabled tracing of the delivery of the drug by imaging to understand the targeted efficiency.

Therapies encompassing administration of drugs like abiraterone acetate, cabazitaxel, docetaxel, enzalutamide, taxanes, which are inhibitors to androgen receptor pathway, find application in targeting prostate cancer. However, the major challenge is posed by the drug resistant prostrate cancer stem cells (PCSCs). Although application of bone targeting α-emitting radionuclide radium-223 chloride has revealed to improve survival of the prostrate cancer patients, the increasing resistance to chemotherapy by the PCSCs is challenging. Nanotechnology has enabled delivery of cancer drug delivery, targeting PCSCs, thereby increasing specificity and reduced toxicity [26]. Iron oxide nanoparticle (IONPs) delivering drug noscapine (Nos) to tumours targeting linked with uPA, is designed to target uPA receptor thereby inhibiting cancer cell growth and could be monitored by their MRI contrast effect [27].

Since an aberrant Wnt signalling is observed in prostrate cancer cells, targeting Wnt signalling is critical for PCSC treatment. Expression of mutated tomor suppressor, phosphatase and tensin homologue (PTEN) in chromosome 10, affecting the phosphoinositide 3-kinase (PI3K) signalling is a hallmark feature in prostrate cancer leading to activation of intracellular PI3K-Akt signalling pathway that promotes maintenance and self-renewal function of PCSC. Androgen signalling plays a significant role in controlling growth and progression of prostate cancer. Activated sonic hedgehog (SHH) pathway components, including overexpression of Glioma-associated oncogene 1 (GLI1), affected hedgehog receptor, human homologue of patched 1(PTCH1), and altered expression of SHH have been studied in tissues from prostate cancer in human and the three signalling including androgen signalling, PI3K-Akt, and Wnt signalling pathways are known to interact.

Green tea polyphenol epigallocatechin-3-gallate (EGCG) is known to induce apoptosis of prostate cancer cells (PCCs). A polylactic acid-polyethylene glycol encapsulated EGCG nanoparticle revealed efficient pro-apoptotic and angiogenesis-inhibitory effects [28].

RC-92a/hTERT cells are human prostate cancer epithelial cells designed to express human telomerase reverse transcriptase, overexpress CD133 [29]. N-(2-Hydroxypropyl)-methacrylamide or HPMA copolymer-docetaxel conjugate has been reported to reduce tumour load effectively [30, 31] (Fig. 9.2).

9.2.2 Breast Cancer

Breast cancer (BC) is major killer disease in females globally, [33] and the major challenges posed to conventional modes of therapy are provided by the drug resistant breast cancer stem cells (BCSCs) with expression of $CD44^{+}CD24^{-/low}$ cells [34], that can self-renew, differentiate, overexpress drug efflux transporters, and express antiapoptotic proteins and resistance to chemo and radiotherapeutic agents [34–36] and thus finds importance as therapeutic targets. Overexpressed CD133 (Prominin-1, 37), and aldehyde dehydrogenase ALDH that correlates with tumour size, recurrence, metastasis, progression, overall survival and drug resistance [37–41] act as marker of BCSCs.

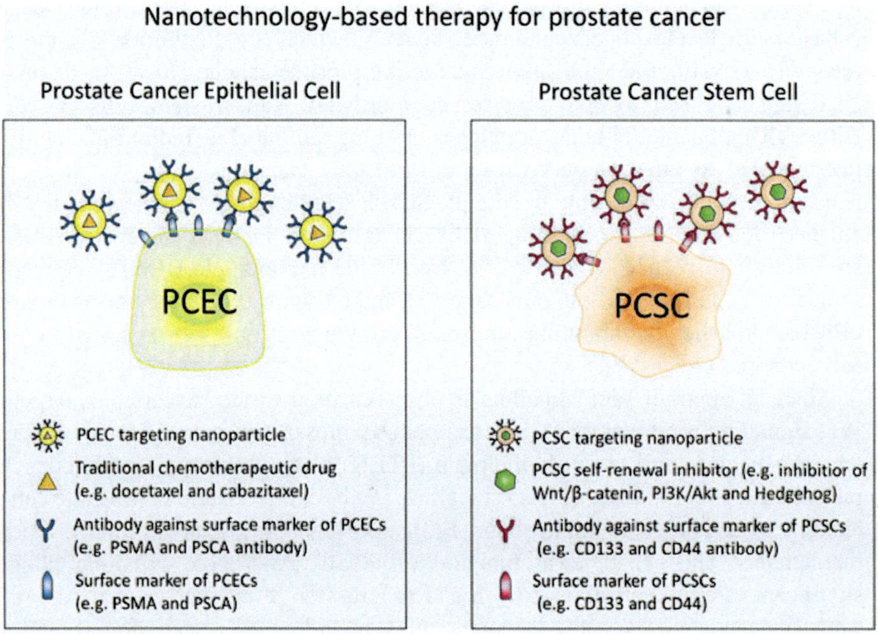

Fig. 9.2 Diagrammatic representation of experimental study of nanotechnology mediated therapy in prostate cancer. Experimental studies by targeting Prostate cancer epithelial cell (PCEC) with antibody against PCEC cell surface markers conjugated to nanoparticles carrying chemotherapeutic drugs and PCSC being targeted by antibodies against prostrate cancer stem cells (PCSC) surface markers conjugated with nanoparticles containing inhibitor molecules of self-renewal signalling pathway. Synergistic effects of both conventional and PCSC specific chemotherapy could cure prostrate cancer. Adapted as open access article under CC BY License [32]

Conventional drugs like salinomycin [42], disulfiram [43], chloroquine [44], curcumin [45], and siRNA [46] finding importance in targeting of breast cancer suffer from drawbacks of poor solubility, side effects, instability, short circulation half-life [47]. Therefore, specific therapy targeting the BCSCs finds importance. Nanoparticles of carbon nanotubes, liposomes, metal, polymer, polymeric micelles with properties of biocompatibility, enhanced permeability and retention (EPR) have enabled breast cancer detection, prevention, and treatment [48] (Fig. 9.3).

CD44 overexpressed in BCSCs is to target by nanosystems. Functionalized anti-CD44 antibodies and anticancer drug paclitaxel (PTX) loaded poly[(D,L-lactide-co-glycolide)-co-PEG] (PLGA-co-PEG, *PLGA-PEG-PLGA co*-block polymer) micelles have been reported to promote the internalization of PTX in BCSCs [50]. The nanoprobe combined MRI with FI modality enabled detection of BCSCs in solid tumour which is difficult by conventional methods [51].

Overexpressed Hyaluronic acid (HA) in breast cancer cells can bind to CD44 [52] leading to multiple drug resistance (MDR) [53]. Conjugates of cholesteryl-hyaluronic acid (CHA) nanogel drug has been reported to be efficient in treatment drug resistant breast cancer stem-like cells expressing CD44 [54]. Citosan-linked

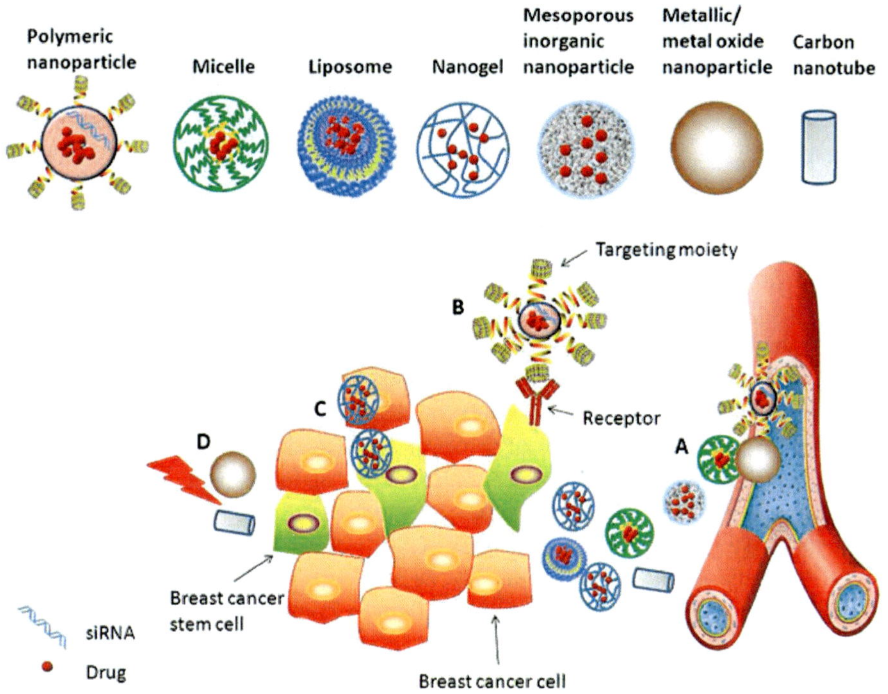

Fig. 9.3 Diagrammatic view of targeting BCSCs by nanomedicines by use of different nanoparticles [49] (Table 9.1)

and ***Doxorubicin*** (DOX)-carrying polymeric nanoparticle have been used in successful targeting of CD44 receptors of BCSCs [55].

Polymeric nanoparticle carrying anti cancer drug Paclitaxel conjugated to anti-CD133 antibody have been used to target CD-133 positive BCCs [56]. A gadolinium-rhodamine nanoparticle has been in labeling and tracking of cells by MRI based studies and optical imaging [57]. Nanoparticle including gold based, ion oxide and carbon nanotubes used in thermal therapy has revealed satisfactory results in targeting triple negative breast cancer cells by an infrared (NIR) light source, or alternating magnetic field. [58–62]. Injected gold nanoshells have been used to successfully target BCSCs [63] and bring about hyperthermia. Cell membrane coated nanoparticles (NPs) designed with NP coated with membrane extracted from cells like stem cells, red blood cells (RBC), white blood cells (WBC), platelets cancer cells, or bacterial cells find application in diagnostics, therapeutics, drug delivery, immune modulation, vaccination, and detoxification [64].

9.2.3 Ovarian Cancer

Different strategies have been employed to target ovarian cancer. Azide labelled mesenchymal stem cell (MSC) has been used to deliver high binding nanoparticles conjugated with dibenzyl cyclooctyne (DBCO) as targeting ligand has been reported to have promising effects in mice [65]. Poly(ethylene glycol)-SS-polyethylenimine-loaded microbubbles (PSP@MB) have been used in a gene delivery system [66]. Palladium NPs (PdNPs) could induce apoptosis in human ovarian cell cancer cell line SKOV3 [67]. Silica-coated paclitaxel (PTX) nanocrystals (Si[PTX-NC]) neural stem cells (NSCs) with PTX can kill ovarian tumours in vivo [68]. Small interfering RNA (siRNA) conjugated nanoparticles have been applied in ovarian cancer treatment [69]. Twist, is known to be a transcription factor with helix-loop-helix (HLH) structure and is known to play role in transition from epithelial to mesenchymal phenotype thereby promoting cancer metastasis. Overexpression of TWIST has been associated with stem-like features in cancer cells and cell express drug resistance. Thus targeting TWIST could be an effective strategy in treatment of cancer. Chemoresistant cancer stem-like cells (CSCs) in triple Negative Breast Cancer (TNBC) patients could be successfully targeted by delivery of nanoparticle based TWIST small interfering RNA (siRNAs) thereby combating cancer [70]. Phospholipid and polymer made hybrid nanoparticle with property of cold-triggered burst release of encapsulated drug at below ~12 °C could effectively target drug resistant ovarian cancer cells, CSCs, subcutaneous and orthotopic ovarian tumours [71]. Since CD 133 is a known marker in ovarian cancer, antiCD133 antibody conjugated with salinomycin-loaded poly(lactic-co-glycolic acid)-poly(ethylene glycol) nanoparticles (CD133-SAL-NP) has been experimentally used in targeting and elimination of $CD133^+$ ovarian CSCs [72].

TWIST-siRNA (siTWIST) delivered by hyaluronic acid conjugated mesoporous silica nanoparticles (MSN-HAs) finds importance in targeting cancer cells and overcoming chemoresistance [73]. Reduced graphene oxide-silver nanoparticle (rGO-Ag) conjugated with R-phycoerythrin (RPE) or RPE-rGO-Ag has been used to target $ALDH^+CD133^+$ cells and ovarian cancer stem cells (OvCSCs) [74]. siRNA targeting $p70^{S6K}$ in CSC delivered by dendrimer nanotechnology-based system has revealed promising result [75]. Reduced graphene oxide-silver nanoparticles (rGO-AgNPs) and trichostatin A (TSA) has been used to target human ovarian cancer cells with promising results in vitro [76]. Co-delivery of doxorubicin (DOX), NVP-AEW 541 (NVP) by hollow mesoporous silica nanoparticles (HMSN) in CD117 + CD44 + A2780 has been effective in targeting CSCs. Ovarian cancer stem-like cell (OCSCs) targeting IGF-1R could increase the apoptotic rates [77]. OCSCs has been targeted by the use of poly(lactic-co-glycolic acid) nanoparticles (PLGA NPs) delivering PTX together with folic acid (FA) as the ligand [78].

9.2.4 Colorectal Stem Cells

Targeting of colorectal Cancer Stem Cells (CoSCs) is a hard task by conventional therapies. Overexpressed chemokine receptor, CXCR4 is a marker for these cells. Engineered CXCR4-targeted toxin NPs have shown promising results in targeting $CXCR4^+$ human colon-CSCs in animal models [79]. X-ray irradiated gold nanoparticles (AuNPs) could kill HT29 CSLCs far more efficiently as compared to X-ray irradiation alone [80]. Curcumin-loaded CSO-SA micelles could target CD44 (+)/CD24(+) expression CSC cells in vitro and in vivo studies [81].

9.2.5 Glioblastoma

Glioblastoma (GBM) is most commonly occurring primary brain tumour affecting the central nervous system (CNS) has limited treatments and it is difficult to target GBM stem cell. Fluorescent AuNPs, formed by polyethylene imine (PEI), fluorescein isothiocyanate (FITC) or (AuPEI-FITC NPs) could transport DNA and drugs in GBM cells [82]. Glioma stem cells (GSCs) with expressed marker CD133 causes GBM resistance to Temozolomide TMZ delivered by magnetic mesoporous silica nanoparticles (MMSNs) tagged with αCT1 and CD133 antibody good biocompatibility, drug loading capacity and release properties could cause death of GSCs [83]. NFL-TBS.40–63 peptide has been reported to target tubulin binding site on neurofilaments, in experimental studies thereby halting the self-renewal property of GBSC [84].

9.2.6 Pancreatic Cancer

CD133 and CD44 are marker in pancreatic cancer stem cells (PCSCs) leading to chemoresistance and therefore bTiO2-Gd-CD133mAb finds importance in MRI based photothermal therapy (PTT) of pancreatic cancer [85]. Hyaluronic acid (HA) targeting CD44 receptors conjugated with copoly(styrene maleic acid) (HA-SMA) nanomicelles carrying 3,4-difluorobenzylidene curcumin (CDF) [86] which reveals anticancerous properties has been reported to target PSCSs effectively.

9.2.7 Neurodegenerative Diseases

Central nervous system (CNS) regenerative medicine finds application in increasing motor and memory functions in patients suffering from neurodegenerative disorders. NSCs find application in restoration of neurological function in such disorders. Nanostructured devices including scaffolds, drug delivery systems, and nanotechnology-based real-time imaging have largely improved the safety and the efficacy of neural stem cell-based treatments [87]. Multipotent MSCs, also with high self-renewal capability and broad tissue distribution, have potential in cell therapy but can have low survival rates posttransplantation, thereby reducing the therapeutic efficiency. Biomolecules and both NSC and MSC could be delivered by polymeric nanospheres and hydrogels that enable regeneration of the nerve and endogenous cell stimulation and provide regenerative therapies for the CNS [88].

9.2.8 Cardiovascular Diseases

Application of stem cells in cardiovascular diseases suffer from challenges of methods for their suitable monitoring when injected into the myocardium.

Stem cells has found applications in targeted imaging studies, drug delivery, hyperthermia, regenerative medicine and tissue repair studies. SPIONS has found importance in molecular and cellular imaging in myocardial infarctions [89]. Iron nanoparticles are used to track NSCs applied in cell therapies in Parkinson's disease [90]. Stem cell therapy is now revealing more and more successful attempts as a promising treatment strategy in repair of cardiac tissues in ischaemic heart diseases. [91]. SPIONS and magnetic nanoparticles are being used to magnetically label stem cells and monitor the viability of cells by MRI [92].

9.3 Other Disorders

Cryopreservation of immature testicular tissue (ITT) in prepuberty boys, undergoing chemotherapy is the sole option, to preserve spermatogonial stem cells (SSCs). Hydrogel based nanoparticles find application in the cryopreservation of the sperms [93]. Mesenchymal stromal cells finds application in regenerative biology in treatment of cardiac, neural, and musculoskeletal tissues or in the treatment of acute kidney injury and can be imaged by paramagnetic nanoparticle MRI approaches [94]. Cardiovascular disorders, brain injury, multiple sclerosis, urinary system diseases, cartilage lesions, and diabetes are being treated by stem cell transplantation. [95]. Intervertebral disc (IVD) degeneration causes low back pain due to loss of proteoglycans and stem cells enables regenerative medicinal application to IVD disease [96]. Stem cells and chondrocytes derived stem cell have been applied to repair and regeneration of joints in patients suffering from arthritis [97].

9.3.1 Different Nanoparticles and Stem Cells

1. *Carbon quantum dots* (CQDs) being chemically inert, water solubile, biocompatibile together properties of resistance to photobleaching and being optically superior in nature has found application in bioimaging, biosensing, drug delivery and catalysis [98].
2. *Magnetic nanoparticles* (MNPs) has been used label MSCs and has been reported to generate MSC with effective and long lasting retention at the target site which

when exposed to static magnetic fields could track their targets. However, the safety concerns to the human healthy tissues are still under research [99] and limit their immediate availability in medicine. Biocompatible magnetically responsive nanoparticles are being designed with enhanced retention in the desired treatment site and are used for targeted delivery of stem cells. This noninvasive treatment-localization strategy has shown promising results in regenerative medicine [100]. They are designed to label, target, pattern, and modify genes are engineered to form magnetic composite scaffolds. Magnetic fields applied to MNPs together with improved delivery of growth factor, drugs, gene transfections by the nano scaffolds are being used in engineering of bone tissue. MNP scaffolds used with magnetic fields and stem cells are being used in bone regeneration efficiently thereby increasing the osteogenic differentiation, angiogenesis by activating signalling pathways including mitogen-activated protein kinase (MAPK), integrin, bone morphogenetic protein (BMP), and nuclear factor kappa-light-chain-enhancer of activated B cells (NF-κB) significantly over conventional methods. They have also found important clinical applications in bone repair and regeneration, dental, craniofacial and in orthopedic treatments [101].

3. *Liposome* with the property of nanosize, biocompatibility, and biodegradability finds application as encapsulating agents and delivery of bioactive agents including pharmaceuticals [102]. Nanotechnology and stem cell preconditioning have found importance in efficient delivery of hormones in targeting cardiovascular diseases [103]..
4. *Iron oxide nanoparticles* (IONPs) with property of controlled responses under external magnetic fields have found applications in bone regeneration and reconstruction of bone tissue in disorders and defects caused by non-union, trauma, and bone tumour excision. IONPs have found application in delivery and control of stem cells for efficient treatment in stem cell-mediated therapy targeted towards repair of bone defect, repair therapy, and delivery of therapeutic agents, including chemical drug molecules, genetic materials, and growth factors playing role in regulation of signal transduction reactions thereby promoting osteogenesis, bone cell function, differentiation, in regeneration, repair, and growth of healthy bone tissue, reconstruction of damaged tissue and activation of membrane receptors [104].
5. *Graphene and graphene-derived particles* including functionalized graphene, graphene oxide, pristine graphene, and hybrid graphene particles by means of composites and coatings have profound applications in treatment of orthopaedic and neural tissue reconstruction [105]. Nanomaterials due to their unique properties encompassing their physical and chemical nature like design, morphology, shape, surface features, and chemical nature, and stem cell engineering interactions are complex and hypothesized to play role in triggering differentiation, immune reactions. Interactions of stem cells with their microenvironment is known to control stem cell fate, and also play role in regulating therapy [106].

9.4 Uses in Manipulation and Tacking of Stem Cells

Nanotechnologies enable quantitative measuring, understanding, study, and manipulation of stem cells. Magnetic nanoparticles and QDs have been used in stem cell labelling and in vivo tracking of cells in clinical medicine.

Nanoparticles, carbon nanotubes polyplexes, and engineered nanometre-scale scaffolds have been applied for the intracellular delivery of genetic materials, genes or oligonucleotides and protein or peptides, growth factors and used for stem cell differentiation and transplantation [107].

9.4.1 Nanoparticles in Stem Cell Imaging

MRI is being used to track stem cells as diagnostic tools for tracking of transplanted stem cell, detect their movement, migration and survival, and evaluate their efficacy in stem cell-mediated therapies [108].

Multipotent MSC has been used to monitor the delivery and biodistribution of transplanted stem cells by noninvasive methods [109]. MSCs finds application in treating ischemic cerebral and neurodegenerative disorders like Parkinson's disease (PD), stroke and amyotrophic lateral sclerosis (ALS). MRI of SPIONs and transplantation of SPIO-labeled MSCs in animal models enabled their detection and tracking [110].

Magnetic particle imaging (MPI) is a novel radiation-free tomographic imaging method providing a signal attenuation-free, background-free direct quantification of the spatial distribution of SPIONs with high temporal and spatial resolution, and high sensitivity [111]. Ferumoxytol (Feraheme™) is an IONP, approved by FDA (U.S. Food and Drug Administration) in the treatment of anaemia in patients as magnetic resonance (MR) contrast agent, and has been used to track stem cells, macrophages, and cell death mediators [112]. MSCs allow targeting, tracking, and local retention of cells at damaged tissue sites [113]. Fluorescent nanoparticles (NPs) and near-infrared (NIR) NPs finds importance with improved properties of sensitivity, property of capability of deep tissue penetration, temporal and spatial resolutions and has been applied in effectively tracking of transplanted stem cells in vivo [114].

QDs with fluorescence properties over traditional fluorescence probes find application in stem cell imaging technology [115].

Nanoparticles promotes stem cell proliferation and differentiation, imaging and tracking, vehicles for genes, drugs or growth factor delivery, and find therapeutic application in neurological diseases and neuronal repair and regeneration in diseases like Parkinson's disease (PD), Alzheimer's disease (AD) which are neurodegenerative disorders, ischaemic stroke (IS), spinal cord injury (SCI), multiple sclerosis (MS) which is an autoimmune disorder affecting the central nervous system, brain, spinal cord revealing inflammatory demyelinating *condition*, and gliomas which are tumours initiated in the glial cells of the brain.

Stem cells and NPs used as a contrast agent have been applied for imaging in glioma [116]. Computed tomography (CT) applied with gold nanoparticles (AuNPs) as contrast agents have found application as an imaging modality [117, 118]. However, they are not without potential cytotoxic effects on nerve cells [119]. Graphene has been used in live cell biosensing and stem cell therapy [120].

SPION labelled stem cells have been applied for diagnosis and tracking of cells in treatment of myocardial infarction by noninvasive methods [121]. SPIONs have been used in long-term clinical tracking of magnetically labelled stem cells after intracerebroventricular transplantation [122], and has been used as contrast agents for MRI in study and detection of inflammatory processes [123].

SPIONs are internalized by complex formation with agents of DNA transfection like poly-L-lysine (PLL) which helps in efficiently labeling cells. Ferumoxides as MRI contrast agents has been approved in detection of liver lesion commercially which are available as Feridex and Endorem and has applications as agents of cell labeling approved by the United Stated (U.S.) food and drug Administration (FDA). [124].

NSCs has been applied as vehicles in drug delivery, and nanotechnology-based real-time imaging studies has revealed their increased safety and the efficacy in neural stem cell-based treatments [125].

QDs find application as nanosensors in cancer diagnosis, for in vitro or in vivo stem cell imaging and has been applied in regenerative medicine [92]. MRI has been used in study of stroke and assessment of therapeutic role of stem cells [126].

Aptamers, single-stranded (ss) RNA or DNA oligonucleotides of 15–60 bases have been used to bind specifically to their cognate molecules, thereby undergoing conformational changes, transduced into a change in imaging contrast. This principle has been implemented in tracking of stem cells in the study of regenerative medicine, and tracking of stem cells in immune therapy [127].

Multifunctional nano-probes of Carbon nanotubes (CNTs), including single-walled carbon nanotubes (SWNTs) and multi-walled carbon nanotubes (MWNTs), have been used in biomedical imaging [128]. Nanostructures have been used in Regenerative dentistry [129].

Magnetoliposomes (MLs) has advantages of efficient functionalization, has enabled bimodal imaging by application of MRI and fluorescence and finds applications in theranostics. Magnetite cores with Gd(3+) -complexes attached onto the liposomal surface enable optimization of MLs for dynamic MRI [130].

Amine-modified silica-coated polyhedral SPIO nanoparticles (SPIO@SiO_2-NH_2) have been applied for MRI tracking of implanted MSCs in rabbit bone marrow-derived MSCs [131]. Multimodal, rhodamine B isothiocyanate (RITC)-labelled, silica-coated magnetic nanoparticles (MNPs@SiO_2(RITC)) labelled MSCs have been used to track the labelled cells transplanted into the subcutaneous tissue and liver of the experimental studies in animal models like mice and have been proved to be effective in labelling of human MSC and thereby enable cell tracking with multimodality imaging [131, 132].

MSCs labelled with fluidMAG IONPs have been used to image and study tumours with MRI [133]. Poly(*N*,*N*-dimethylacrylamide)-coated maghemite (PDMAAm-coated γ-Fe_2O_3) nanoparticles were used for labelling and tracking of MSCs were done by MRI [134].

9.4.2 Nanoparticles and Stem Cell Cytotoxicity

Metallic nanoparticles (MNPs) in different biological applications have been revealed to generate reactive oxygen species (ROS) that is known to cause cell toxicity, modulate cellular signalling in cell death, proliferation, and differentiation. Metal NP-induced ROS generation affects the stem cell delivery system also [135].

NP-based materials have breakthrough applications in engineering, communications, and textiles industries, food and environment, medicines, and cosmetics, but reveal toxic in pulmonary system [136]. Cytotoxicity of nanoparticles has been reported MSC in regenerative medicine [137]. Carbon-based nanomaterials are being tested for their toxic effects including overall toxicity carcinogenicity, cardiotoxicity, dermatotoxicity, epigenetic, genotoxicity, hepatotoxicity, immunotoxicity, neurotoxicity, and nephrotoxicity [138].

9.5 Discussions

NPs are being used widely in the field of biomedical sciences in the domain of antimicrobial drug delivery, cancer imaging, detection, therapy and drug delivery, and cell tracking. Nanomaterials with their unique ability of interaction with cell membranes and undergo internalization in the nucleus, thereby activating transcription factors and molecules thereby triggering the signalling pathway finds importance in communication with stem cell communication.

Extensive research is being carried out in the domain of nanotechnology and its application in stem cell research and therapy in disease. However, challenges remain encompassing the design of efficient methods to quantitate NP-stem cell interaction, calculation and design of optimal NP in stem cell interaction. Toxicity of NPs further is a challenge to large scale application of NPs in stem cell biology, calling in more research in this domain. More studies on the application and validation of different NPs need to be done.

References

1. Ghosh S (2018) Nanomaterials safety: toxicity and health hazards. DeGryter, Boston
2. Ghosh S, Ansar W (2014) Multidynamic liposomal nanomedicine: technology, biology, applications, and disease targeting. In: Nanoparticles' promises and risks—characterization, manipulation, and potential hazards to humanity and the environment. Springer, Berlin
3. Ghosh S, Saha S, Sur A (2015) Polymeric nanoparticles: application in cancer management. In: Encyclopedia of biomedical polymers and polymeric biomaterials. Taylor & Francis, London
4. Ghosh S (2018) Green synthesis of nanoparticle and fungal infection. In: Kumar A, Iravani SS (eds) Green synthesis, characterization and applications of nanoparticles. Springer, Berlin
5. Salata O (2004) Applications of nanoparticles in biology and medicine. J Nanobiotechnol 2:3. https://doi.org/10.1186/1477-3155-2-3
6. Khan I, Saeed K, Khan I (2019) Nanoparticles: properties, applications and toxicities. Arab J Chem 12(7):908–931

7. Wang EC, Wang AZ (2014) Nanoparticles and their applications in cell and molecular biology. Integr Biol (Camb) 6(1):9–26. https://doi.org/10.1039/c3ib40165k
8. McNamara K, Syed A, Tofail M (2017) Nanoparticles in biomedical applications. Adv Phys 2 (1):54–88
9. Dong L, Craig MM, Khang D, Chen C (2012) Applications of nanomaterials in biology and medicine. J Nanotechnol 2012:816184. https://doi.org/10.1155/2012/816184
10. Hofmann MC (2014) Stem cells and nanomaterials. Adv Exp Med Biol 811:255–275. https://doi.org/10.1007/978-94-017-8739-0_13
11. Ashammakhi N, Ahadian S, Darabi MA, El Tahchi M, Lee J, Suthiwanich K, Sheikhi A, Dokmeci MR, Oklu R, Khademhosseini A (2019) Minimally invasive and regenerative therapeutics. Adv Mater 31(1):e1804041
12. Jang SF, Liu WH, Song WS, Chiang KL, Ma HI, Kao CL, Chen MT (2014) Nanomedicine-based neuroprotective strategies in patient specific-iPSC and personalized medicine. Int J Mol Sci 15(3):3904–3925
13. Rajangam T (2013) An SS fibrinogen and fibrin based micro and nano scaffolds incorporated with drugs, proteins, cells and genes for therapeutic biomedical applications. Int J Nanomedicine 8:3641–3662
14. Patel S, Lee KB (2015) Probing stem cell behavior using nanoparticle-based approaches. Wiley Interdiscip Rev Nanomed Nanobiotechnol 7(6):759–778
15. Sharma S (2014) Nanotheranostics in evidence based personalized medicine. Curr Drug Targets 15(10):915–930
16. Asthana A, Lee JE, Rao CM (2014) Modulation of stem cell differentiation by the influence of nanobiomaterials/carriers. Curr Stem Cell Res Ther 9(6):458–468
17. Reiter RE, Gu Z, Watabe T, Thomas G, Szigeti K, Davis E, Wahl M, Nisitani S, Yamashiro J, Le Beau MM, Loda M, Witte ON (1998) Prostate stem cell antigen: a cell surface marker overexpressed in prostate cancer. Proc Natl Acad Sci U S A 95(4):1735–1740
18. Ghosh A, Heston WD (2004) Tumor target prostate specific membrane antigen (PSMA) and its regulation in prostate cancer. J Cell Biochem 91(3):528–539
19. Thaxton CS, Elghanian R, Thomas AD, Stoeva SI, Lee JS, Smith ND, Schaeffer AJ, Klocker H, Horninger W, Bartsch G, Mirkin CA (2009) Nanoparticle-based bio-barcode assay redefines “undetectable” PSA and biochemical recurrence after radical prostatectomy. Proc Natl Acad Sci U S A 106(44):18437–18442
20. Yuhi T, Nagatani N, Endo T, Kerman K, Takata M, Konaka H, Namiki M, Takamura Y, Tamiya E (2006) Gold nanoparticle based immunochromatography using a resin modified micropipette tip for rapid and simple detection of human chorionic gonadotropin hormone and prostate-specific antigen. Sci Technol Adv Mater 7(3):276–281
21. Chikkaveeraiah BV, Bhirde A, Malhotra R, Patel V, Gutkind JS, Rusling JF (2009) Single-wall carbon nanotube forest arrays for immunoelectrochemical measurement of four protein biomarkers for prostate cancer. Anal Chem 81:9129–9134
22. Kim YJ, Rahman MM, Lee J-J (2013) Ultrasensitive and label-free detection of annexin A3 based on quartz crystal microbalance. Sens Actuators B 177:172–177
23. Mitchell PS, Parkin RK, Kroh EM, Fritz BR, Wyman SK, Pogosova-Agadjanyan EL et al (2008) Circulating microRNAs as stable blood-based markers for cancer detection. Proc Natl Acad Sci U S A 105:10513
24. Tran HV, Piro B, Reisberg S, Tran LD, Duc HT, Pham MC (2013) Label-free and reagentless electrochemical detection of microRNAs using a conducting polymer nanostructured by carbon nanotubes: application to prostate cancer biomarker miR-141. Biosens Bioelectron 49:164–169
25. Alhasan AH, Kim DY, Daniel WL, Watson E, Meeks JJ, Thaxton CS et al (2012) Scanometric microRNA array profiling of prostate cancer markers using spherical nucleic acid-gold nanoparticle conjugates. Anal Chem 84:4153–4160
26. Del Vento F, Vermeulen M, de Michele F, Giudice MG, Poels J, des Rieux A, Wyns C (2018) Tissue engineering to improve immature testicular tissue and cell transplantation outcomes:

one step closer to fertility restoration for prepubertal boys exposed to gonadotoxic treatments. Int J Mol Sci 19(1):e286

27. Abdalla MO, Karna P, Sajja HK et al (2011) Enhanced noscapine delivery using uPAR-targeted optical-MR imaging trackable nanoparticles for prostate cancer therapy. J Control Release 149(3):314–322. https://doi.org/10.1016/j.jconrel.2010.10.030
28. Siddiqui IA, Adhami VM, Bharali DJ, Hafeez BB, Asim M, Khwaja SI et al (2009) Introducing nanochemoprevention as a novel approach for cancer control: proof of principle with green tea polyphenol epigallocatechin-3-gallate. Cancer Res 69:1712–1716
29. Miki J, Furusato B, Li H, Gu Y, Takahashi H, Egawa S et al (2007) Identification of putative stem cell markers, CD133 and CXCR4, in hTERT-immortalized primary nonmalignant and malignant tumor-derived human prostate epithelial cell lines and in prostate cancer specimens. Cancer Res 67:3153–3161
30. Zhao S, Fu J, Liu X, Wang T, Zhang J, Zhao Y (2012) Activation of Akt/GSK-3beta/beta-catenin signaling pathway is involved in survival of neurons after traumatic brain injury in rats. Neurol Res 34:400–407
31. Zhou Y, Kopecek J (2013) Biological rationale for the design of polymeric anti-cancer nanomedicines. J Drug Target 21:1–26
32. Qin W, Zheng Y, Qian BZ, Zhao M (2017) Prostate cancer stem cells and nanotechnology: a focus on Wnt signaling. Front Pharmacol 8:153
33. Torre LA, Bray F, Siegel RL, Ferlay J, Lortet-Tieulent J, Jemal A (2015) Global cancer statistics, 2012. CA Cancer J Clin 65(2):87–108. https://doi.org/10.3322/caac.21262. Epub 2015 Feb 4. PMID: 25651787
34. Al-Hajj M, Wicha MS, Benito-Hernandez A, Morrison SJ, Clarke MF (2003) Prospective identification of tumorigenic breast cancer cells. Proc Natl Acad Sci U S A 100:3983–3988
35. Geng R, Song F, Yang X, Sun P, Hu J, Zhu C, Zhu B, Fan W (2014) Association between cytotoxic T lymphocyte antigen-4 +49A/G, −1722T/C, and -1661A/G polymorphisms and cancer risk: a meta-analysis. Tumour Biol 35:3627–3639
36. Vinogradov S, Wei X (2012) Cancer stem cells and drug resistance: the potential of nanomedicine. Nanomedicine (Lond) 7(4):597–615
37. Mizrak D, Brittan M, Alison M (2008) CD133: molecule of the moment. J Pathol 214:3–9. https://doi.org/10.1002/path.2283
38. Vira D, Basak SK, Veena MS, Wang MB, Batra RK, Srivatsan ES (2012) Cancer stem cells, microRNAs, and therapeutic strategies including natural products. Cancer Metastasis Rev 31 (3–4):733–751. https://doi.org/10.1007/s10555-012-9382-8
39. Meyer MJ, Fleming JM, Lin AF, Hussnain SA, Ginsburg E, Vonderhaar BK (2010) CD44posCD49fhiCD133/2hi defines xenograft-initiating cells in estrogen receptornegative breast cancer. Cancer Res 70:4624–4633
40. Zhou W, Wang G, Guo S (1836) Regulation of angiogenesis via notch signaling in breast cancer and cancer stem cells. Biochim Biophys Acta 2013:304–320
41. Aomatsu N, Yashiro M, Kashiwagi S et al (2012) CD133 is a useful surrogate marker for predicting chemosensitivity to neoadjuvant chemotherapy in breast cancer. PLoS One 7(9): e45865
42. Muntimadugu E, Kumar R, Saladi S, Rafeeqi TA, Khan W (2016) CD44 targeted chemotherapy for co-eradication of breast cancer stem cells and cancer cells using polymeric nanoparticles of salinomycin and paclitaxel. Colloids Surf B Biointerfaces 143:532–546
43. Yip NC, Fombon IS, Liu P, Brown S, Kannappan V, Armesilla AL et al (2011) Disulfiram modulated ROS-MAPK and NFκB pathways and targeted breast cancer cells with cancer stem cell-like properties. Br J Cancer 104:1564–1574. https://doi.org/10.1038/bjc.2011.126
44. Liang DH, Choi DS, Ensor JE, Kaipparettu BA, Bass BL, Chang JC (2016) The autophagy inhibitor chloroquine targets cancer stem cells in triple negative breast cancer by inducing mitochondrial damage and impairing DNA break repair. Cancer Lett 376:249–258
45. Gülçür E, Thaqi M, Khaja F, Kuzmis A, Onyüksel H (2013) Curcumin in VIP-targeted sterically stabilized phospholipid nanomicelles: a novel therapeutic approach for breast cancer and breast cancer stem cells. Drug Deliv Transl Res 3:10. https://doi.org/10.1007/s13346-013-0167-6

46. Zuo ZQ, Chen KG, Yu XY, Zhao G, Shen S, Cao ZT et al (2016) Promoting tumor penetration of nanoparticles for cancer stem cell therapy by TGF-β signaling pathway inhibition. Biomaterials 82:48–59. https://doi.org/10.1016/j.biomaterials.2015.12.014
47. Hu K, Law JH, Fotovati A, Dunn SE (2012) Small interfering RNA library screen identified polo-like kinase-1 (PLK1) as a potential therapeutic target for breast cancer that uniquely eliminates tumor-initiating cells. Breast Cancer Res 14:R22. https://doi.org/10.1186/bcr3107
48. Shen S, Xia JX, Wang J (2016) Nanomedicine-mediated cancer stem cell therapy. Biomaterials 74:1–18. https://doi.org/10.1016/j.biomaterials.2015.09.037
49. He L, Gu J, Lim LY, Z-x Y, Mo J (2016) Nanomedicine-mediated therapies to target breast cancer stem cells. Front Pharmacol 7:313
50. Gener P, Gouveia LP, Sabat GR, de Sousa RDF, Fort NB, Arranja A et al (2015) Fluorescent CSC models evidence that targeted nanomedicines improve treatment sensitivity of breast and colon cancer stem cells. Nanomedicine 11:1883–1892. https://doi.org/10.1016/j.nano.2015.07.009
51. Chen H, Wang Y, Wang T, Shi D, Sun Z, Xia C, Wang B (2016) Application prospective of nanoprobes with MRI and FI dual-modality imaging on breast cancer stem cells in tumor. J Nanobiotechnol 14(1):52
52. Chen C, Zhao S, Karnad A et al (2018) The biology and role of CD44 in cancer progression: therapeutic implications. J Hematol Oncol 11:64. https://doi.org/10.1186/s13045-018-0605-5
53. Götte M, Yip GW (2006) Heparanase, hyaluronan, and CD44 in cancers: a breast carcinoma perspective. Cancer Res 66:10233–10237. https://doi.org/10.1158/0008-5472.CAN-06-1464
54. Rao W, Wang H, Han J, Zhao S, Dumbleton J, Agarwal P et al (2015) Chitosan-decorated doxorubicin-encapsulated nanoparticle targets and eliminates tumor reinitiating cancer stem-like cells. ACS Nano 9:5725–5740. https://doi.org/10.1021/nn506928p
55. Swaminathan SK, Roger E, Toti U, Niu L, Ohlfest JR, Panyam J (2013) CD133-targeted paclitaxel delivery inhibits local tumor recurrence in a mouse model of breast cancer. J Control Release 171:280–287. https://doi.org/10.1016/j.jconrel.2013.07.014
56. Park W, Yang HN, Ling D, Yim H, Kim KS, Hyeon T et al (2014) Multi-modal transfection agent based on monodisperse magnetic nanoparticles for stem cell gene delivery and tracking. Biomaterials 35:7239–7247. https://doi.org/10.1016/j.biomaterials.2014.05.010
57. Vuu K, Xie J, McDonald MA, Bernardo M, Hunter F, Zhang Y et al (2005) Gadolinium-rhodamine nanoparticles for cell labeling and tracking via magnetic resonance and optical imaging. Bioconjug Chem 16:995–999. https://doi.org/10.1021/bc050085z
58. Atkinson RL, Zhang M, Diagaradjane P, Peddibhotla S, Contreras A, Hilsenbeck SG et al (2010) Thermal enhancement with optically activated gold nanoshells sensitizes breast cancer stem cells to radiation therapy. Sci Transl Med 2:55ra79. https://doi.org/10.1126/scitranslmed.3001447
59. Burke AR, Singh RN, Carroll DL, Wood JC, D'Agostino RB Jr, Ajayan PM et al (2012) The resistance of breast cancer stem cells to conventional hyperthermia and their sensitivity to nanoparticle-mediated photothermal therapy. Biomaterials 33:2961–2970
60. Sadhukha T, Niu L, Wiedmann TS, Panyam J (2013) Effective elimination of cancer stem cells by magnetic hyperthermia. Mol Pharm 10:1432–1441
61. Xu Y, Wang J, Li X, Liu Y, Dai L, Wu X et al (2014) Selective inhibition of breast cancer stem cells by gold nanorods mediated plasmonic hyperthermia. Biomaterials 35:4667–4677
62. Paholak HJ, Stevers NO, Chen H, Burnett JP, He M, Korkaya H et al (2016) Elimination of epithelial-like and mesenchymal-like breast cancer stem cells to inhibit metastasis following nanoparticle-mediated photothermal therapy. Biomaterials 104:145–157
63. Atkinson RL, Zhang M, Diagaradjane P, Peddibhotla S, Contreras A, Hilsenbeck SG et al (2010) Thermal enhancement with optically activated gold nanoshells sensitizes breast cancer stem cells to radiation therapy. Sci Transl Med 2:55ra79. https://doi.org/10.1126/scitranslmed.3001447
64. Narain A, Asawa S, Chhabria V, Patil-Sen Y (2017) Cell membrane coated nanoparticles: next-generation therapeutics. Nanomedicine (Lond) 12(21):2677–2692

65. Layek B, Shetty M, Nethi SK, Sehgal D, Starr TK, Prabha S (2020) Mesenchymal stem cells as guideposts for nanoparticle-mediated targeted drug delivery in ovarian cancer. Cancers (Basel) 12(4):e965
66. Liufu C, Li Y, Tu J, Zhang H, Yu J, Wang Y, Huang P, Chen Z (2019) Echogenic PEGylated PEI-loaded microbubble as efficient gene delivery system. Int J Nanomedicine 14:8923–8941
67. Gurunathan S, Qasim M, Park CH, Arsalan Iqbal M, Yoo H, Hwang JH, Uhm SJ, Song H, Park C, Choi Y, Kim JH, Hong K (2019) Cytotoxicity and transcriptomic analyses of biogenic palladium nanoparticles in human ovarian cancer cells (SKOV3). Nanomaterials (Basel) 9(5): e787
68. Tiet P, Li J, Abidi W, Mooney R, Flores L, Aramburo S, Batalla-Covello J, Gonzaga J, Tsaturyan L, Kang Y, Cornejo YR, Dellinger T, Han E, Aboody KS, Berlin JM (2019) Silica coated paclitaxel nanocrystals enable neural stem cell loading for treatment of ovarian cancer. Bioconjug Chem 30(5):1415–1424
69. Halbur C, Choudhury N, Chen M, Kim JH, Chung EJ (2019) siRNA-conjugated nanoparticles to treat ovarian cancer. SLAS Technol 24(2):137–150
70. Glackin CA (2018) Nanoparticle delivery of TWIST small interfering RNA and anticancer drugs: a therapeutic approach for combating cancer. Enzyme 44:83–101
71. Wang H, Agarwal P, Zhao G, Ji G, Jewell CM, Fisher JP, Lu X, He X (2018) Overcoming ovarian cancer drug resistance with a cold responsive nanomaterial. ACS Cent Sci 4 (5):567–581
72. Mi Y, Huang Y, Deng J (2018) The enhanced delivery of salinomycin to CD133+ ovarian cancer stem cells through CD133 antibody conjugation with poly(lactic-co-glycolic acid)-poly (ethylene glycol) nanoparticles. Oncol Lett 15(5):6611–6621
73. Shahin SA, Wang R, Simargi SI, Contreras A, Parra Echavarria L, Qu L, Wen W, Dellinger T, Unternaehrer J, Tamanoi F, Zink JI, Glackin CA (2018) Hyaluronic acid conjugated nanoparticle delivery of siRNA against TWIST reduces tumor burden and enhances sensitivity to cisplatin in ovarian cancer. Nanomedicine 14(4):1381–1394
74. Choi YJ, Gurunathan S, Kim JH (2018) Graphene oxide-silver nanocomposite enhances cytotoxic and apoptotic potential of salinomycin in human ovarian cancer stem cells (OvCSCs): a novel approach for cancer therapy. Int J Mol Sci 19(3):e710
75. Ma J, Kala S, Yung S, Chan TM, Cao Y, Jiang Y, Liu X, Giorgio S, Peng L, Wong AST (2018) Blocking Stemness and metastatic properties of ovarian cancer cells by targeting p70S6K with dendrimer Nanovector-based siRNA delivery. Mol Ther 26(1):70–83
76. Zhang XF, Huang FH, Zhang GL, Bai DP, Massimo DF, Huang YF, Gurunathan S (2017) Novel biomolecule lycopene-reduced graphene oxide-silver nanoparticle enhances apoptotic potential of trichostatin a in human ovarian cancer cells (SKOV3). Int J Nanomedicine 12:7551–7575
77. Guo X, Guo N, Zhao J, Cai Y (2017) Active targeting co-delivery system based on hollow mesoporous silica nanoparticles for antitumor therapy in ovarian cancer stem-like cells. Oncol Rep 38(3):1442–1450
78. Abou-ElNaga A, Mutawa G, El-Sherbiny IM, Abd-ElGhaffar H, Allam AA, Ajarem J, Mousa SA (2017) Novel nano-therapeutic approach actively targets human ovarian cancer stem cells after xenograft into nude mice. Int J Mol Sci 18(4):e813
79. Serna N, Álamo P, Ramesh P, Vinokurova D, Sánchez-García L, Unzueta U, Gallardo A, Céspedes MV, Vázquez E, Villaverde A, Mangues R, Medema JP (2020) Nanostructured toxins for the selective destruction of drug-resistant human CXCR4+ colorectal cancer stem cells. J Control Release 320:96–104
80. Abbasian M, Baharlouei A, Arab-Bafrani Z, Lightfoot DA (2019) Combination of gold nanoparticles with low-LET irradiation: an approach to enhance DNA DSB induction in HT29 colorectal cancer stem-like cells. J Cancer Res Clin Oncol 145(1):97–107
81. Wang K, Zhang T, Liu L, Wang X, Wu P, Chen Z, Ni C, Zhang J, Hu F, Huang J (2012) Novel micelle formulation of curcumin for enhancing antitumor activity and inhibiting colorectal cancer stem cells. Int J Nanomedicine 7:4487–4497

82. Giesen B, Nickel AC, Garzón Manjón A, Vargas Toscano A, Scheu C, Kahlert UD, Janiak C (2020) Influence of synthesis methods on the internalization of fluorescent old nanoparticles into glioblastoma stem-like cells. J Inorg Biochem 203:110952
83. Xingyi J, Guonan C, Xin Z, Naijie L (2019) AbCD133 modified αCT1 loaded target magnetic mesoporous silica nano-drugcarriers can sensitizes glioma cancer stem cells to TMZ and have therapeutic potential on TMZ resistant glioblastoma. J Biomed Nanotechnol 15(7):1468–1481
84. Lépinoux-Chambaud C, Eyer J (2019) The NFL-TBS 40–63 peptide targets and kills glioblastoma stem cells derived from human patients and also targets nanocapsules into these cells. Int J Pharm 566:218–228
85. Wang S, Ren W, Wang J, Jiang Z, Saeed M, Zhang L, Li A, Wu A (2018) Black TiO2-based nanoprobes for T1-weighted MRI-guided photothermal therapy in CD133 high expressed pancreatic cancer stem-like cells. Biomater Sci 6(8):2209–2218
86. Kesharwani P, Banerjee S, Padhye S, Sarkar FH, Iyer AK (2015) Hyaluronic acid engineered Nanomicelles loaded with 3,4-Difluorobenzylidene curcumin for targeted killing of CD44+ stem-like pancreatic cancer cells. Biomacromolecules 16(9):3042–3053
87. Carradori D, Eyer J, Saulnier P, Préat V, des Rieux A (2017) The therapeutic contribution of nanomedicine to treat neurodegenerative diseases via neural stem cell differentiation. Biomaterials 123:77–91
88. Elliott Donaghue I, Tam R, Sefton MV, Shoichet MS (2014) Cell and biomolecule delivery for tissue repair and regeneration in the central nervous system. J Control Release 190:219–227
89. Santoso MR, Yang PC (2016) Magnetic nanoparticles for targeting and imaging of stem cells in myocardial infarction. Stem Cells Int 2016:4198790
90. Ramos-Gómez M, Martínez-Serrano A (2016) Tracking of iron-labeled human neural stem cells by magnetic resonance imaging in cell replacement therapy for Parkinson's disease. Neural Regen Res 11(1):49–52
91. Zhu K, Li J, Wang Y, Lai H, Wang C (2016) Nanoparticles-assisted stem cell therapy for ischemic heart disease. Stem Cells Int 2016:1384658
92. Santoso MR, Yang PC (2016) Magnetic nanoparticles for targeting and imaging of stem cells in myocardial infarction. Stem Cells Int 2016:4198790
93. Vermeulen M, Poels J, de Michele F, Des Rieux A, Wyns C (2017) Restoring fertility with cryopreserved prepubertal testicular tissue: perspectives with hydrogel encapsulation, nanotechnology, and bioengineered scaffolds. Ann Biomed Eng 45(7):1770–1781
94. Hossain MA, Chowdhury T, Bagul A (2015) Imaging modalities for the in vivo surveillance of mesenchymal stromal cells. J Tissue Eng Regen Med 9(11):1217–1224
95. Yahyapour R, Farhood B, Graily G, Rezaeyan A, Rezapoor S, Abdollahi H, Cheki M, Amini P, Fallah H, Najafi M, Motevaseli E (2018) Stem cell tracing through MR molecular imaging. Tissue Eng Regen Med 15(3):249–261
96. Handley C, Goldschlager T, Oehme D, Ghosh P, Jenkin G (2015) Mesenchymal stem cell tracking in the intervertebral disc. World J Stem Cells 7(1):65–74
97. Daldrup-Link HE, Nejadnik H (2014) MR imaging of stem cell transplants in arthritic joints. J Stem Cell Res Ther 4(2):165
98. Farshbaf M, Davaran S, Rahimi F, Annabi N, Salehi R, Akbarzadeh A (2018) Carbon quantum dots: recent progresses on synthesis, surface modification and applications. Artif Cells Nanomed Biotechnol 46(7):1331–1348
99. Silva LH, Cruz FF, Morales MM, Weiss DJ, Rocco PR (2017) Magnetic targeting as a strategy to enhance therapeutic effects of mesenchymal stromal cells. Stem Cell Res Ther 8(1):58
100. Cores J, Caranasos TG, Cheng K (2015) Magnetically targeted stem cell delivery for regenerative medicine. J Funct Biomater 6(3):526–546
101. Xia Y, Sun J, Zhao L, Zhang F, Liang XJ, Guo Y, Weir MD, Reynolds MA, Gu N, Xu HHK (2018) Magnetic field and nano-scaffolds with stem cells to enhance bone regeneration. Biomaterials 183:151–170

102. Panahi Y, Farshbaf M, Mohammadhosseini M, Mirahadi M, Khalilov R, Saghfi S, Akbarzadeh A (2017) Recent advances on liposomal nanoparticles: synthesis, characterization and biomedical applications. Artif Cells Nanomed Biotechnol 45(4):788–799
103. Acharya G, Hasan N, Yoo JW, Lee CH (2017) Hormone therapy and delivery strategies against cardiovascular diseases. Curr Pharm Biotechnol 18(4):285–302
104. Li Y, Ye D, Li M, Ma M, Gu N (2018) Adaptive materials based on iron oxide nanoparticles for bone regeneration. ChemPhysChem 19(16):1965–1979
105. Kumar S, Chatterjee K (2016) Comprehensive review on the use of graphene-based substrates for regenerative medicine and biomedical devices. ACS Appl Mater Interfaces 8 (40):26431–26457
106. Kerativitayanan P, Carrow JK, Gaharwar AK (2015) Nanomaterials for engineering stem cell responses. Adv Healthc Mater 4(11):1600–1627
107. Ferreira L, Karp JM, Nobre L, Langer R (2008) New opportunities: the use of nanotechnologies to manipulate and track stem cells. Cell Stem Cell 3(2):136–146
108. Darkazalli A, Levenson CW (2012) Tracking stem cell migration and survival in brain injury: current approaches and future prospects. Histol Histopathol 27(10):1255–1261
109. Liu X, Yang Z, Sun J, Ma T, Hua F, Shen Z (2019) A brief review of cytotoxicity of nanoparticles on mesenchymal stem cells in regenerative medicine. Int J Nanomedicine 14:3875–3892
110. Rosenberg JT, Yuan X, Grant S, Ma T (2016) Tracking mesenchymal stem cells using magnetic resonance imaging. Brain Circ 2(3):108–113
111. Meola A, Rao J, Chaudhary N, Song G, Zheng X, Chang SD (2019) Magnetic particle imaging in neurosurgery. World Neurosurg 125:261–270
112. Nejadnik H, Tseng J, Daldrup-Link H (2019) Magnetic resonance imaging of stem cell-macrophage interactions with ferumoxytol and ferumoxytol-derived nanoparticles. Wiley Interdiscip Rev Nanomed Nanobiotechnol 11(4):e1552
113. Kerans FFA, Lungaro L, Azfer A, Salter DM (2018) The potential of intrinsically magnetic mesenchymal stem cells for tissue engineering. Int J Mol Sci 14:19(10)
114. Chen G, Zhang Y, Li C, Huang D, Wang Q, Wang Q (2018) Recent advances in tracking the transplanted stem cells using near-infrared fluorescent Nanoprobes: turning from the first to the second near-infrared window. Adv Healthc Mater 7(20):e1800497
115. Yukawa H, Baba Y (2018) In vivo imaging technology of transplanted stem cells using quantum dots for regenerative medicine. Anal Sci 34(5):525–532
116. Deng SL, Li YQ, Zhao G (2018) Imaging gliomas with nanoparticle-labeled stem cells. Chin Med J (Engl) 131(6):721–730
117. Naseri N, Ajorlou E, Asghari F, Pilehvar-Soltanahmadi Y (2018) An update on nanoparticle-based contrast agents in medical imaging. Artif Cells Nanomed Biotechnol 46(6):1111–1121
118. Meir R, Popovtzer R (2018) Cell tracking using gold nanoparticles and computed tomography imaging. Wiley Interdiscip Rev Nanomed Nanobiotechnol 10:2
119. Zhang G, Khan AA, Wu H, Chen L, Gu Y, Gu N (2018) The application of nanomaterials in stem cell therapy for some neurological diseases. Curr Drug Targets 19(3):279–298
120. Kim TH, Lee D, Choi JW (2017) Live cell biosensing platforms using graphene-based hybrid nanomaterials. Biosens Bioelectron 94:485–499
121. Jasmin de Souza GT, Louzada RA, Rosado de Castro PH, Mendez-Otero R, Campos de Carvalho AC (2017) Tracking stem cells with superparamagnetic iron oxide nanoparticles: perspectives and considerations. Int J Nanomedicine 12:779–793
122. Janowski M, Walczak P, Kropiwnicki T, Jurkiewicz E, Domanska-Janik K, Bulte JW, Lukomska B, Roszkowski M (2014) Long-term MRI cell tracking after intraventricular delivery in a patient with global cerebral ischemia and prospects for magnetic navigation of stem cells within the CSF. PLoS One 9(2):e97631
123. Neuwelt A, Sidhu N, Hu CA, Mlady G, Eberhardt SC, Sillerud LO (2015) Iron-based superparamagnetic nanoparticle contrast agents for MRI of infection and inflammation. AJR Am J Roentgenol 204(3):W302–W313

124. Zhang H (2008) Iron oxide nanoparticles-poly-L-lysine complex. Molecular Imaging and Contrast Agent Database (MICAD) [Internet], [updated 2008 Mar 26]. National Center for Biotechnology Information, Bethesda
125. Carradori D, Eyer J, Saulnier P, Préat V, des Rieux A (2017) The therapeutic contribution of nanomedicine to treat neurodegenerative diseases via neural stem cell differentiation. Biomaterials 123:77–91
126. Aghayan HR, Soleimani M, Goodarzi P, Norouzi-Javidan A, Emami-Razavi SH, Larijani B, Arjmand B (2014) Magnetic resonance imaging of transplanted stem cell fate in stroke. J Res Med Sci 19(5):465–471
127. Wiraja C, Yeo D, Lio D, Labanieh L, Lu M, Zhao W, Xu C (2014) Aptamer technology for tracking cells' status & function. Mol Cell Ther 2:33
128. Gong H, Peng R, Liu Z (2013) Carbon nanotubes for biomedical imaging: the recent advances. Adv Drug Deliv Rev 65(15):1951–1963
129. Mitsiadis TA, Woloszyk A, Jiménez-Rojo L (2012) Nanodentistry: combining nanostructured materials and stem cells for dental tissue regeneration. Nanomedicine (Lond) 7 (11):1743–1753
130. Soenen SJ, Velde GV, Ketkar-Atre A, Himmelreich U, De Cuyper M (2011) Magnetoliposomes as magnetic resonance imaging contrast agents. Wiley Interdiscip Rev Nanomed Nanobiotechnol 3(2):197–211
131. Shan L (2008) Amine-modified silica-coated polyhedral superparamagnetic iron oxide nanoparticle–labeled rabbit bone marrow–derived mesenchymal stem cells. National Center for Biotechnology Information, Bethesda. Molecular Imaging and Contrast Agent Database (MICAD) [Internet]. [updated 2010 Jan 28]
132. Shan L (2008) Multimodal, rhodamine B isothiocyanate-incorporated, silica-coated magnetic nanoparticle–labeled human cord blood–derived mesenchymal stem cells for cell tracking. National Center for Biotechnology Information, Bethesda. Molecular Imaging and Contrast Agent Database (MICAD) [Internet]. [updated 2010 Jan 28]
133. Shan L (2008) FluidMAG iron nanoparticle-labeled mesenchymal stem cells for tracking cell homing to tumors. National Center for Biotechnology Information, Bethesda. Molecular Imaging and Contrast Agent Database (MICAD) [Internet]. [updated 2010 Jan 28]
134. Shan L (2008) Poly(N,N-dimethylacrylamide)-coated maghemite nanoparticles for labeling and tracking mesenchymal stem cells. National Center for Biotechnology Information, Bethesda. Molecular Imaging and Contrast Agent Database (MICAD) [Internet]. [updated 2010 Feb 16]
135. Abdal Dayem A, Hossain MK, Lee SB, Kim K, Saha SK, Yang GM, Choi HY, Cho SG (2017) The role of reactive oxygen species (ROS) in the biological activities of metallic nanoparticles. Int J Mol Sci 18(1):e120
136. Osman NM, Sexton DW, Saleem IY (2019) Toxicological assessment of nanoparticle interactions with the pulmonary system. Nanotoxicology 10:1–38
137. Liu X, Yang Z, Sun J, Ma T, Hua F, Shen Z (2019) A brief review of cytotoxicity of nanoparticles on mesenchymal stem cells in regenerative medicine. Int J Nanomedicine 14:3875–3892
138. Madannejad R, Shoaie N, Jahanpeyma F, Darvishi MH, Azimzadeh M, Javadi H (2019) Toxicity of carbon-based nanomaterials: reviewing recent reports in medical and biological systems. Chem Biol Interact 307:206–222

Stem Cells in Clinical Research and Therapy 10

Contents

Abstract

Regenerative medicine is an emerging promising wing of medical sciences involving restoration of tissue and organ function in chronic disorders and organ failures. Stem cells and their therapeutic application find importance in regenerative medicine and clinical research in different complex disorders of cancer, spinal cord injury (SPI), neurodegenerative disorders. We discuss in this chapter stem cells and their applications in regenerative medicine and clinical settings and in different cardiovascular disorders, diseases of the central nervous system (CNS), injury of the spinal cord, eye disorders, dental disorders, liver disorders, diabetes, autoimmune and infectious diseases and in the context of global pandemic COVID-19.

S. Chowdhury, S. Ghosh, *Stem Cells*, https://doi.org/10.1007/978-981-16-1638-9_10

Keywords

Spinal cord injury · Embryonic stem cells · Human pluripotent stem cells · Induced pluripotent stem cells · Regenerative medicine · Stem cell therapy

Abbreviations

AD	Alzheimer's disease
ADSCs	Adipose-derived stem cells
ALI	Acute lung injury
AS	Asherman syndrome
ASCs	Adult stem cells
BCI	Brain–computer interface
BM-MNCs	Bone marrow mononuclear cells
CNS	Central nervous system
CRISPR	Clustered regularly interspaced short palindromic repeats
CVD	Cardiovascular disease
DMD	Duchene muscular dystrophy
DPSCs	Dental pulp stem cells
DR	Diabetic retinopathy
EPSC	Epidermal stem cells
ES	Embryonic stem
ESCs	Embryonic stem cells
EVs	Extracellular vesicles
FLSCs	Foetal liver stem cells
GVD	Graft versus host disease
hISC	Human intestinal stem cells
hPSC	Human pluripotent stem cells
HSCs	Haematopoietic stem cells
IBD	Intestinal bowel disease
IM	Intra muscular
iPSCs	Induced pluripotent stem cells
IV	Intravenous
MenSCs	Menstrual blood-derived stem cells
MHC	Major histocompatibility complex
MI	Myocardial infarction
MN	Motor neurons
MNDs	Motor neuron diseases
MS	Multiple sclerosis
MSCs	Mesenchymal stem cells
NPC	Neural precursor cells
NSC	Neural stem cell
NSCs	Neural stem cells
OECs	olfactory-ensheathing cells
PD	Parkinson's disease

PSCs	Pluripotent stem cells
RA	Rheumatoid arthritis
RGCs	Injured retinal ganglion cells
RPE	Retinal pigment epithelial
SC	Subcutaneous
SCI	Spinal cord injury
T1D	Type 1 diabetes
T2DM	Type 2 diabetes mellitus
TALENS	Transcription activator-like effector nucleases
Tregs	Regulatory T cells
UCB	Umbilical cord blood
WHO	World Health Organization
WJ-MSCs	Wharton jelly MSC
ZNFs	Zinc finger nucleases

10.1 Introduction

Induced pluripotent stem cells (iPSCs) are type of pluripotent stem cells in an embryonic stem (ES) cell-like condition that are generated by genetic reprogramming of adult somatic cells. They found application in regenerative medicine. Patient-specific iPSCs can be designed by genetic engineering approaches using gene editing tools of clustered regularly interspaced short palindromic repeats (CRISPR)/Cas9, transcription activator-like effector nucleases (TALENs), zinc finger nucleases (ZFNs), and has found applications in cinical medicine. They can be differentiated into target cells and finds application in treatment of various genetic and degenerative disorders circumventing the problems associated with risk factors encompassing graft versus host disease (GVD) [1] and efficient drug discovery and treatment of intractable diseases [2]. Pluripotent or multipotent stem cell capable of differentiation into germ layers finds therapeutic application [3]. Stem cells derived from adipose also called as adipose-derived stem cells (ADSCs), mesenchymal stem cells (MSCs) find application in engineered tissue and therapy [4] and find application in wound healing promoting angiogenesis and/or vascularization [5] with application as regenerative medicine and reconstruction of soft tissue [5, 6] together with application in treatment of melanoma [7]. MSC extracellular vesicles (EVs) secreted by live cells reveal regenerative potential and properties of immunomodulation and find application in treatment of cancer and regenerative medicine [8]. MSCs find application in stem cell based regenerative therapy [9].

Menstrual blood-derived stem cells (MenSCs) are MSCs, that find applications in therapy [10] and treatment of diseases including Asherman syndrome (AS), neurodegenerative diseases like Alzheimer's disease (AD), acute lung injury (ALI), cardiovascular disease (CVD), cutaneous wound, autoimmune disorders like diabetes, Duchene muscular dystrophy (DMD), liver disease, endometriosis stroke, ovarian diseases, myocardial infarction (MI) [11].

Fig. 10.1 Major applications of stem cell therapy in different disorders

Human pluripotent stem cells (hPSC) engineered with gene editing tools have found applications in tissue regeneration and cancer treatment [12]. MSCs and human intestinal stem cells also termed as hiSC have been used in therapy for intestinal diseases like inflammatory bowel disease (IBD) [13, 14] (Fig. 10.1).

10.2 Applications in Disorders of the Nervous System (NS)

Neural stem cell (NSC) is being exploited for their potential in therapy in treatment of disorders including stroke, psychiatric illnesses, foetal alcohol spectrum disorders, and cancer with promising reports of enabling enhanced recovery [15] in patients. iPSCs have been used in regenerative medicine and is being tested to reverse psychiatric problems [16]. Stem cell therapy is undergoing clinical trials in therapeutic applications for neurodegenerative disorders like Alzheimer's disease (AD) [17]. Neural cells can be developed from embryonic stem cells (ESCs) and NSCs but suffer from challenges of setting an immunorejection as they are not autologous cells or risk of transplantation of ESCs lead to development of teratoma. Mesenchymal stem cells with overexpressed ESC gene, nanog, could differentiate into neural cells and may have a potent application in treating AD [17]. MSC have been reported of its potential in replacing lost functional dopaminergic neurons and thereby providing promises in regenerative therapy of Parkinson's disease (PD) [18].

10.3 Application in Spinal Cord Injury (SPI)

Spinal cord injury (SCI) affects the central nervous system (CNS) leading to long term disability and 3D scaffolds of stem cells and hPSCs including ESCs and iPSCs has found application in neural tissue engineering and repair of SCI [19, 20]. Transplantation of neural precursor cells and iPSC has been applied in treatment of SCI [20]. Dental pulp stem cells also termed as DPSCs are MSCs that have been applied for tissue regeneration in dental, systemic diseases [21], SCI [22, 23], PD, AD, cerebral ischemia, diabetes, eye diseases MI, MD, liver diseases, oral diseases and immune diseases. IPSC, oligodendrocyte precursor cells, and MSC in scaffolds, biogels, and growth factors have been used for stem cell survival, function, and engraftment [24].

SCI could mediate effective axonal re-growth and sprouting and neuronal regeneration [24]. Neural precursor cell (NPC) and iPSC developed from somatic cells by genetic reprogramming and induction of transcription factors, and are induced to differentiate into NPCs with CNS properties [24, 25]. Stem cells are being applied in regenerative therapy in traumatic SCI. Transplanting NSC and pluripotent stem cells by biomaterial scaffold increases cell survival and improves differentiation efficiency [25]. While hESCs have found importance in regenerative medicine for SCIs [26], iPScs have been applied in regeneration of stroke and brain tumours [27]. MSCs and olfactory-ensheathing cells (OECs) have been applied in SCI [28]. hESCs and hiPSCs have been used for retinal repair and SCI [28, 29]. hiPSCs have been used in targeting motor neuron diseases (MNDs) [30, 31]. iPSCs have been used in conferring neuroprotection and regeneration, that can divide into astrocytes, mesenchymal stromal cells, NSC, neural crest cells, neurons, and oligodendrocytes [32, 33]. Stem cells have been applied to spinal fusion surgery, SCI, and intervertebral disc degeneration [34–38]. Neurons and glia cells develop from stem cells with neuroprotective effect. MSCs with regenerative and neuroprotective properties find application into patients with neurological disorders of the spinal cord [39].

Autologous or allogeneic stem cells or progenitor cells from various sources are being used in SCI [19]. Research in the domain of SCI using brain and computer interface (BCI) technology is being designed to develop interfaces of neural networks and their interaction by artificial tools in order to gain back motor control and make the injured area [33] mobile. SCI after the transplantation of NSCs derived from foetus or from ESCs has been successful in stem cell transplantation therapy [38]. Degeneration of motor neurons due to diseases or injury can cause paralysis and prove to be fatal. ESCs have been applied to regeneration of degenerated motor neurons (MN) [40, 41] and prove to be a promising application in restoration of normalcy.

Motor neurons in mouse, monkey, and human have been experimentally differentiated from ES and iPS cells involving formation of embryoid bodies and stimulation with factors like retinoic acid and Sonic hedgehog [34, 38].

Bone marrow stem cells transplantation into spinal cord lesions in experimental animal studies increases chances of axonal regeneration, promoting its functional activity. Adult bone marrow stem cell includes hematopoietic stem cells (HSCs), and MSCs [33, 39] and adult mesenchymal progenitor stem cells have been reported to play significant role in spine surgery [33–39].

Non-ES cells of affected individuals located in the bone marrow, adipose tissue, and umbilical cord blood (UCB) find application as alternative to ES cells that can differentiate into the three germ layers including ectodermal, endodermal, and mesodermal germ lineages. NSC has found application in differentiation into neurons and glia by vitro and in vivo [32–42] experimental methods. Stem cells find application in acute traumatic SCI neuroprotective and neurodegenerative treatments and cellular transplantation strategies have been designed for SCI and translational neurobiology in post-SCI injury for restoration of motor and sensory functions.

10.4 Bone Disorders

MSCs revealing the property of self-renewal and differentiation into lineages including adipocytes chondrocytes and osteoblasts find application and importance in replacement and regeneration therapies for disorders affecting the musculoskeletal system including bone defects and fractures, focal chondral lesions, osteoarthritis, spinal cord related disorders and tendon injuries [43]. Adipose-derived stem cells (ASCs) have found applications in regenerative therapy for PD and AD, SCI, heart disease, and autoimmune diseases as rheumatoid arthritis (RA), or for replacing lost tissue from trauma or tumour removal [44].

10.5 Cardiovascular Diseases

iPSCs with self-renew and differentiation properties have been reported to differentiate into different cells including cardiac myocytes and have been used to detect drug responses, source cell for cardiac tissue regeneration in animal models [45]. Adult stem cells (ASCs) have the advantage over embryonic and iPSC that does not form teratoma or generate any immune response and therefore offer advantage in transplantation in CVD including myocardial infarction (MI), doxorubicin-induced cardiomyopathy, and heart failure [46]. Although pluripotent stem cells (PSCs) find importance in cardiovascular tissue regeneration and have been used to treat different cardiovascular disorders, but challenges remain with respect to matters related to graft survival, immunogenicity, post-transplant arrhythmia, maintenance of proper function, and formation of tumour [47]. Stem cell transplantation is now reported to be a promising method to repair damaged myocardial tissue in ischemic heart diseases [48].

10.6 Eye Diseases

Stem cell has been applied in the therapy for degenerative eye disease, related to restoration of lost neuron, neural circuits by inducing growth of new network connections. Retinal progenitor like cells could be induced from ESC and iPSCs and lost retinal pigment epithelial (RPE) cells and photoreceptors could be replaced thereby restoring vision in the eye with lost vision. Injured retinal ganglion cells (RGCs) could be treated by MSC [49]. Diabetic retinopathy (DR), causing blindness in diabetic individuals and ASC, MSC generated from bone marrow and adipose tissue, has found application in treatment of DR in experimental set up and is under clinical trials [50].

MSCs with their properties of immunomodulation and pro-angiogenicity and exosomes of MSCs are finding importance in regenerative medicine based on cell-based approaches in disorders including glaucoma, diseases of the ocular surface, uveitis, retinal disorders and reduce eye injury and inflammation. They also find importance in delivery of immunomodulatory and trophic factors to the inner retina promoting survival and neuritogenesis of injured retinal ganglion cells [51].

10.7 Liver Diseases

Acute liver failure, cirrhosis, and end-stage liver fibrosis are some of the liver disorders that reveal pathophysiology of inflammation, necrosis, and fibrosis, cause high mortality, and are a concern globally. Although liver transplantation remains, the only option to cure a damaged liver, liver transplantation suffers from the problems of lack or limited availability of donor organs, major histocompatibility complex (MHC) compatibility, complex surgical procedure, immunological rejection, and costly hospital charges MSCs have been thought to be important in targeting liver disease as they can be treated to differentiate into hepatocytes. This property enables their usage in reduction of liver inflammation, cause regeneration of liver, and secrete cytokines [52–56]. MSCs are being used to target chronic liver diseases and the signalling pathways involved are being studied and are under clinical trials [55]. Foetal liver stem cells (FLSCs) have been reported to provide best candidates for stem cell transplantation and find application in regenerative medicine for inherited metabolic or other hepatic disorders [54].

But the major concerns over usage of MSC in transplants arise from risk of rejection of transplanted cell, probability of developing an infusional toxicity and iatrogenic tumour. MSC-secreted exosomes on the other hand have however proved advantageous in such transplantation experiments as they are smaller and less in complexity as compared to the parent cell and easy to store, less potentially immunogenic with no risk of forming tumour and can restore tissue homeostasis.

Exosomes have thus recently been tested in different liver disorders and injury, like acute liver injury, liver fibrosis, and hepatocellular carcinoma [52].

10.8 Diabetes

World Health Organization (WHO) revealed in 2014 a rise in cases to 422 million from around 100 million cases in 1980 with an estimated death of 1.6 million people in 2016 globally and found to be the major cause behind blindness, kidney failure, heart attacks, stroke, and amputation of the lower limb claiming above 2.2 million lives due to adverse effects of high blood glucose in 2012 [57]. Type 2 diabetes mellitus (T2DM) also called as non-insulin-dependent diabetes, is a chronic disease with high glucose level in the body of the patient, arising from inability of body to use insulin which could be due to multifactorial reasons involving the genetics and environmental factors, overweight, obesity, inactive life style. Therefore, restoration of insulin independence and normal blood glucose levels forms the only mode of therapy. Whole-organ and islet transplantation are the promising modes of treatment but suffers from the lack of proper matched donor tissues, complex procedure of maintaining the cells and post-transplantation complicacy and immensely costly procedures [58]. Autologous bone marrow mononuclear cells or autologous or allogeneic MSC from Wharton's jelly MSC (WJ-MSCs), umbilical cord MSCs (UCSCs) have been used in transplantation with promising results in treating T2DM by intravenous (IV), intramuscular (IM), or through subcutaneous (SC) route [59].

Type 1 diabetes (T1D) or juvenile diabetes is an autoimmune disorder wherein the insulin producing pancreatic β cells of the islets of Langerhans are gradually destroyed leading to insulin deficiency, occurs at an early age, affecting 5–10% of total diabetic population. The conventional mode of therapy includes injection of insulin in calculated doses to maintain a normal glucose level but this cannot remove the pathophysiology of autoimmunity and to regenerate lost islets. Islet transplantation, to replace the damaged β cells is a solution but suffers from lack of proper matched human islets, immune reactions, costly procedures. MSCs with hypoimmunogenic and immunomodulatory effect can be used in targeting T1DM [60, 61].

10.9 Treatment in COVID-19

COVID-19 is a global pandemic caused by a fast replicating strain of β-corona virus SARS-CoV-2 with symptoms varying from being asymptomatic, severe acute respiratory infection leading to multiorgan and systemic failure, septic shock, and multiple organ dysfunction (MOD) that started in the Wuhan district of China and spread across all the countries of the world, claiming millions of life, disrupting our daily lives and lifestyle, education, recession, national emergency restructuring of our economies globally (Fig. 10.2).

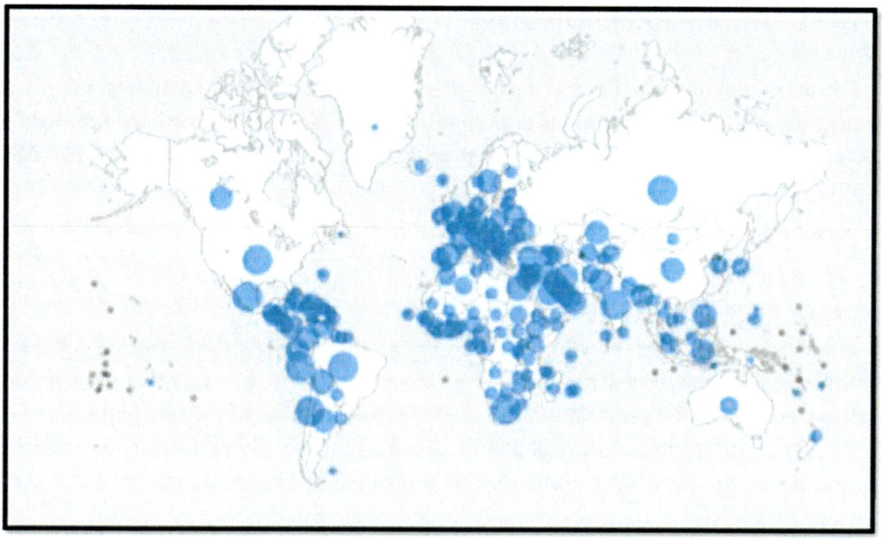

Fig. 10.2 https://covid19.who.int/WHOCoronavirusDisease (COVID-19) Dashboard Data last updated: 2020/7/22, 6:18 pm CEST [61] with permission

Remdesivir and hydroxychloroquine, plasma therapy were being administered by different countries [62] to combat the huge surge of infection. Although the scientist across the world toiled hard to come up with a solution of treatment, yet the disease was rampant and caused havoc in the lives of people. MSCs-based immunomodulation treatment by intravenous transplantation with accumulation in the lung was proposed to confer protection to alveolar epithelial cells [62]. Umbilical cord (UC)-MSCs have been employed in the treatment of COVID-19 infected patients in China in order to prevent the cytokine storm an immunological event observed in this disorder [63]. CD147, a host cell receptor was reported to play role in infection by SARS-CoV-2 mediated by viral spike protein thus disrupting the interaction between viral spike protein and CD147 or casing inhibition of expression of CD147 expression could possible halt viral invasion and infection process. Additionally the histology of COVID-19 revealed differentiation of resident lung stem cells to myofibroblasts during pulmonary fibrosis. Azithromycin possible interfering CD147-spike interaction, reducing activity of metalloproteinases could reduce viral load inducing anti-viral responses. Thus allogeneic MSCs from healthy donors were being tested towards endogenous lung repair and suppression of inflammation mechanism [64]. MSC, with its immuno-modulatory effect is known to suppress hyperactive immune responses and play role in tissue regeneration and repair and holds promise in treatment of patients with severely affected COVID 19 infection revealing a dysfunctional adaptive immune system [65].

10.10 Autoimmune Diseases

Autoimmune diseases, arising out of immune response generated against one's own antigen or self-antigens due to a malfunctioning immune system can lead to disorders affecting the body for which no cure exist. Transplantation of MSCs holds great promising therapeutic effects in targeting autoimmune disorders by its immunomodulatory properties [66]. HSCs find application in treating autoimmune disease T1DM, systemic sclerosis, multiple sclerosis (MS), and rheumatoid arthritis (RA) by regulatory T cells (Tregs) generation, inhibiting autoreactive T-cell thereby affecting the immune system [67].

MSC with properties of tissue repair, immunosuppression, and capability of differentiation has been reported as promising tool for treatment of autoimmune disorder Sjögren's syndrome [68]. hESCs and iPSCs have been experimentally reported to differentiate into mature β-like cells functionally producing and secreting insulin in vitro and have immense potential in transplantation therapy in diabetic individuals [69].

10.11 Skin Replacement in Burn Injury

Burn injuries and wounds which remain long time and difficult to manage with physiological and psychological impact are a major cause of global concern.

Therefore complete repair of the skin in case of injury wound or burn is aimed at reducing hypertrophic scarring and contracture and stem cells offer promises here in reconstructing the burnt area [70, 71]. Regenerative potential of stem cells has been exploited efficient and faster, healing of the burn reducing the inflammation, scar progression and pathophysiology of fibrosis and epidermal stem cells (EPSC) offer great promise [72].

10.12 Discussion

The last two decades have seen a great advancement in the study of stem cells, and their properties and potential in regenerative medicine and stem cell therapy are now being exploited enabling fast and efficient repair of the damaged tissue and many such applications are under clinical trials [1–80]. However questions pertaining to long-term applications related to safe applications, lifelong function, study on the use of light in control of nerve functions, gain or loss of function need to be carried out to understand and optimize dose of such cell therapy.

Although stem cells have been studied with great potential in replacement of the damaged or non-functional tissue and are a major step towards regenerative medicine, and it is being tested in clinical trials for several disorders, its immediate application suffers from limitations involving insufficient validation of action of donor cells and varies with the type of the transplanted cell type and the pathophysiology of host response to such transplantation. Therefore a real time study of post-transplantation events is necessary to confirm the therapeutic efficiency. Both basic and applied research need to be carried out against each application to understand the dynamics of pre- and post-transplantation events and the success of transplantation.

References

1. Al Abbar A, Ngai SC, Nograles N, Alhaji SY, Abdullah S (2020) Induced pluripotent stem cells: reprogramming platforms and applications in cell replacement therapy. Biores Open Access 9(1):121–136
2. Lee J, Bayarsaikhan D, Bayarsaikhan G, Kim JS, Schwarzbach E, Lee B (2020) Recent advances in genome editing of stem cells for drug discovery and therapeutic application. Pharmacol Ther 209:107501
3. Suman S, Domingues A, Ratajczak J, Ratajczak MZ (2019) Potential clinical applications of stem cells in regenerative medicine. Adv Exp Med Biol 1201:1–22
4. Shukla L, Yuan Y, Shayan R, Greening DW, Karnezis T (2020) Fat therapeutics: the clinical capacity of adipose-derived stem cells and exosomes for human disease and tissue regeneration. Front Pharmacol 11:158
5. Agrawal M, Alexander A, Khan J, Giri TK, Siddique S, Dubey SK, Ajazuddin PRJ, Gupta U, Saraf S, Saraf S (2019) Recent biomedical applications on stem cell therapy: a brief overview. Curr Stem Cell Res Ther 14(2):127–136
6. Zielins ER, Luan A, Brett EA, Longaker MT, Wan DC (2015) Therapeutic applications of human adipose-derived stromal cells for soft tissue reconstruction. Discov Med 19 (105):245–253
7. Mirzaei H et al (2016) Application of mesenchymal stem cells in melanoma: a potential therapeutic strategy for delivery of targeted agents. Curr Med Chem 23(5):455–463
8. Huang YC, Lai LC (2019) The potential roles of stem cell-derived extracellular vesicles as a therapeutic tool. Ann Transl Med 7(22):693
9. Yin JQ, Zhu J, Ankrum JA (2019) Manufacturing of primed mesenchymal stromal cells for therapy. Nat Biomed Eng 3(2):90–104
10. Chen L, Qu J, Cheng T, Chen X, Xiang C (2019) Menstrual blood-derived stem cells: toward therapeutic mechanisms, novel strategies, and future perspectives in the treatment of diseases. Stem Cell Res Ther 10(1):406
11. Chen L, Qu J, Xiang C (2019) The multi-functional roles of menstrual blood-derived stem cells in regenerative medicine. Stem Cell Res Ther 10(1):1
12. Allsopp TE, Ebneth A, Cabrera-Socorro A (2019) Deploying human pluripotent stem cells to treat central nervous system disorders: facts, challenges and realising the potential. Stem Cell Res 41:101581
13. Mao F, Tu Q, Wang L, Chu F, Li X, Li HS, Xu W (2017) Mesenchymal stem cells and their therapeutic applications in inflammatory bowel disease. Oncotarget 8(23):38008–38021
14. Holmberg FE, Seidelin JB, Yin X, Mead BE, Tong Z, Li Y, Karp JM, Nielsen OH (2017) Culturing human intestinal stem cells for regenerative applications in the treatment of inflammatory bowel disease. EMBO Mol Med 9(5):558–570
15. Tuazon JP, Castelli V, Lee JY, Desideri GB, Stuppia L, Cimini AM, Borlongan CV (2019) Neural stem cells. Adv Exp Med Biol 1201:79–91
16. Ratajczak MZ, Ciechanowicz AK, Kucharska-Mazur J, Samochowiec J (2018) Stem cells and their potential clinical applications in psychiatric disorders. Prog Neuropsychopharmacol Biol Psychiatry 80(Pt A):3–9
17. Duncan T, Valenzuela M (2017) Alzheimer's disease, dementia, and stem cell therapy. Stem Cell Res Ther 8(1):111
18. Sugaya K, Vaidya M (2018) Stem cell therapies for neurodegenerative diseases. Adv Exp Med Biol 1056:61–84

19. Venkatesh K, Sen D (2017) Mesenchymal stem cells as a source of dopaminergic neurons: a potential cell based therapy for Parkinson's disease. Curr Stem Cell Res Ther 12(4):326–347
20. Nagoshi N, Tsuji O, Nakamura M, Okano H (2019) Cell therapy for spinal cord injury using induced pluripotent stem cells. Regen Ther 11:75–80
21. Yamada Y, Nakamura-Yamada S, Kusano K, Baba S (2019) Clinical potential and current progress of dental pulp stem cells for various systemic diseases in regenerative medicine: a concise review. Int J Mol Sci 6:20(5)
22. Jin MC, Medress ZA, Azad TD, Doulames VM, Veeravagu A (2019) Stem cell therapies for acute spinal cord injury in humans: a review. Neurosurg Focus 46(3):E10
23. Chhabra HS, Sarda K (2017) Clinical translation of stem cell based interventions for spinal cord injury—are we there yet? Adv Drug Deliv Rev 120:41–49
24. Nagoshi N, Okano H (2017) Applications of induced pluripotent stem cell technologies in spinal cord injury. J Neurochem 141(6):848–860
25. Agbay A, Edgar JM, Robinson M, Styan T, Wilson K, Schroll J, Ko J, Khadem Mohtaram N, Jun MB, Willerth SM (2016) Biomaterial strategies for delivering stem cells as a treatment for spinal cord injury. Cells Tissues Organs 202(1–2):42–51
26. Zeb H, Khan IN, Munir I, Ramadan WS, Ahmad MA, Hussein D, Kamal MA, Al KS (2016) Updates on therapeutics in clinical trials for spinal cord injuries: key translational applications of human embryonic stem cells-derived neural progenitors. CNS Neurol Disord Drug Targets 15(10):1266–1278
27. Cefalo MG, Carai A, Miele E, Po A, Ferretti E, Mastronuzzi A, Germano IM (2016) Human IPSC for therapeutic approaches to the nervous system: present and future applications. Stem Cells Int 2016:4869071
28. Anna Z, Katarzyna JW, Joanna C, Barczewska M, Joanna W, Wojciech M (2017) Therapeutic potential of olfactory ensheathing cells and mesenchymal stem cells in spinal cord injuries. [published correction appears in Stem Cells Int. 2017;2017:9438310]. Stem Cells Int 2017:3978595. https://doi.org/10.1155/2017/3978595
29. Borooah S, Phillips MJ, Bilican B et al (2013) Using human induced pluripotent stem cells to treat retinal disease. Prog Retin Eye Res 37:163–181. https://doi.org/10.1016/j.preteyeres.2013.09.002
30. Goldschlager T, Oehme D, Ghosh P, Zannettino A, Rosenfeld JV, Jenkin G (2013) Current and future applications for stem cell therapies in spine surgery. Curr Stem Cell Res Ther 8 (5):381–393
31. Winner B, Marchetto MC, Winkler J, Gage FH (2014) Human-induced pluripotent stem cells pave the road for a better understanding of motor neuron disease. Hum Mol Genet 23:27–34. https://doi.org/10.1093/hmg/ddu205
32. Khazaei M, Siddiqui AM, Fehlings MG (2014) The potential for iPS-derived stem cells as a therapeutic strategy for spinal cord injury: opportunities and challenges. J Clin Med 4(1):37–65. https://doi.org/10.3390/jcm4010037
33. Gongora M, Peressutti C, Machado S, Teixeira S, Velasques B, Ribeiro P (2013) Progress and prospects in neurorehabilitation: clinical applications of stem cells and brain-computer interface for spinal cord lesions. Neurol Sci 34(4):427–433
34. Faravelli I, Bucchia M, Rinchetti P, Nizzardo M, Simone C, Frattini E, Corti S (2014) Motor neuron derivation from human embryonic and induced pluripotent stem cells: experimental approaches and clinical perspectives. Stem Cell Res Ther 5(4):87
35. Schroeder J, Kueper J, Leon K, Liebergall M (2015) Stem cells for spine surgery. World J Stem Cells 7(1):186–194
36. Aftab S, Chimutengwende-Gordon M, Malik A, Lee R (2013) The use of stem cells for the treatment of spinal surgical conditions. Curr Stem Cell Res Ther 8(6):456–463
37. Wei CC, Lin AB, Hung SC (2014) Mesenchymal stem cells in regenerative medicine for musculoskeletal diseases: bench, bedside, and industry. Cell Transplant 23(4–5):505–512
38. Forostyak S, Jendelova P, Sykova E (2013) The role of mesenchymal stromal cells in spinal cord injury, regenerative medicine and possible clinical applications. Biochimie 95(12):2257

39. López-González R, Velasco I (2012) Therapeutic potential of motor neurons differentiated from embryonic stem cells and induced pluripotent stem cells. Arch Med Res 43(1):1–10
40. Enomoto M, Wakabayashi Y, Qi ML, Shinomiya K (2004) Present situation and future aspects of spinal cord regeneration. Orthop Sci 9(1):108–112
41. Agbay A, Edgar JM, Robinson M et al (2016) Biomaterial strategies for delivering stem cells as a treatment for spinal cord injury. Cells Tissues Organs 202(1–2):42–51
42. Shah K, Zhao AG, Sumer H (2018) New approaches to treat osteoarthritis with mesenchymal stem cells. Stem Cells Int 2018:5373294. https://doi.org/10.1155/2018/5373294
43. Chu DT, Nguyen Thi Phuong T, Tien NLB et al (2019) Adipose tissue stem cells for therapy: an update on the progress of isolation, culture, storage, and clinical application. J Clin Med 8 (7):917. https://doi.org/10.3390/jcm8070917
44. Martins AM, Vunjak-Novakovic G, Reis RL (2014) The current status of iPS cells in cardiac research and their potential for tissue engineering and regenerative medicine. Stem Cell Rev Rep 10(2):177–190
45. Johnson TA, Singla DK (2017) Therapeutic application of adult stem cells in the heart. Methods Mol Biol 1553:249–264
46. Okano S, Shiba Y (2019) Therapeutic potential of pluripotent stem cells for cardiac repair after myocardial infarction. Biol Pharm Bull 42(4):524–530
47. Yu H, Lu K, Zhu J, Wang J (2017) Stem cell therapy for ischemic heart diseases. Br Med Bull 121(1):135–154
48. Mead B, Berry M, Logan A, Scott RA, Leadbeater W, Scheven BA (2015) Stem cell treatment of degenerative eye disease. Stem Cell Res 14(3):243–257
49. Gaddam S, Periasamy R, Gangaraju R (2019) Adult stem cell therapeutics in diabetic retinopathy. Int J Mol Sci 20(19):4876
50. Harrell CR, Simovic Markovic B, Fellabaum C et al (2018) Therapeutic potential of mesenchymal stem cell-derived exosomes in the treatment of eye diseases. Adv Exp Med Biol 1089:47–57
51. Lou G, Chen Z, Zheng M, Liu Y (2017) Mesenchymal stem cell-derived exosomes as a new therapeutic strategy for liver diseases. Exp Mol Med 49(6):e346
52. Eom YW, Shim KY, Baik SK (2015) Mesenchymal stem cell therapy for liver fibrosis. Korean J Intern Med 30(5):580–589
53. Yasen A, Tuxun T, Apaer S et al (2019) Fetal liver stem cell transplantation for liver diseases. Regen Med 14(7):703–714
54. Gazdic M, Arsenijevic A, Markovic BS et al (2017) Mesenchymal stem cell-dependent modulation of liver diseases. Int J Biol Sci 13(9):1109–1117
55. Boyd A, Newsome P, Lu WY (2019) The role of stem cells in liver injury and repair. Expert Rev Gastroenterol Hepatol 13(7):623–631
56. https://www.who.int/news-room/fact-sheets/detail/diabetes
57. Pawitan JA, Yang Z, Wu YN, Lee EH (2018) Towards standardized stem cell therapy in type 2 diabetes mellitus: a systematic review. Curr Stem Cell Res Ther 13(6):476–488
58. Sneddon JB, Tang Q, Stock P et al (2018) Stem cell therapies for treating diabetes: progress and remaining challenges. Cell Stem Cell 22(6):810–823
59. Wu H, Mahato RI (2014) Mesenchymal stem cell-based therapy for type 1 diabetes. Discov Med 17(93):139–143
60. Vanikar AV, Trivedi HL, Thakkar UG (2016) Stem cell therapy emerging as the key player in treating type 1 diabetes mellitus. Cytotherapy 18(9):1077–1086
61. https://covid19.who.int/
62. Golchin A, Seyedjafari E, Ardeshirylajimi A (2020) Mesenchymal stem cell therapy for COVID-19: present or future. Stem Cell Rev Rep 16(3):427–433
63. Atluri S, Manchikanti L, Hirsch JA (2020) Expanded umbilical cord mesenchymal stem cells (UC-MSCs) as a therapeutic strategy in managing critically ill COVID-19 patients: the case for compassionate use. Pain Physician 23(2):E71–E83

64. Ulrich H, Pillat MM (2020) CD147 as a target for COVID-19 treatment: suggested effects of azithromycin and stem cell engagement. Stem Cell Rev Rep 16(3):434–440
65. Rao Us V, Thakur S, Rao J et al (2020) Mesenchymal stem cells -bridge catalyst between innate and adaptive immunity in Covid 19. Med Hypotheses 143:109845
66. Chen Y, Yu Q, Hu Y, Shi Y (2019) Current research and use of mesenchymal stem cells in the therapy of autoimmune diseases. Curr Stem Cell Res Ther 14(7):579–582
67. Ben Nasr M, Bassi R, Usuelli V et al (2016) The use of hematopoietic stem cells in autoimmune diseases. Regen Med 11(4):395–405
68. Lu X, Wang X, Nian H, Yang D, Wei R (2017) Mesenchymal stem cells for treating autoimmune dacryoadenitis. Stem Cell Res Ther 8(1):126
69. Kondo Y, Toyoda T, Inagaki N, Osafune K (2018) iPSC technology-based regenerative therapy for diabetes. J Diabetes Investig 9(2):234–243
70. Cheng JZ, Farrokhi A, Ghahary A, Jalili RB (2018) Therapeutic use of stem cells in treatment of burn injuries. J Burn Care Res 39(2):175–182
71. Ghieh F, Jurjus R, Ibrahim A et al (2015) The use of stem cells in burn wound healing: a review. Biomed Res Int 2015:684084
72. Yang R, Liu F, Wang J, Chen X, Xie J, Xiong K (2019) Epidermal stem cells in wound healing and their clinical applications. Stem Cell Res Ther 10(1):229
73. Nakamura M, Tsuji O, Nori S, Toyama Y, Okano H (2012) Cell transplantation for spinal cord injury focusing on iPSCs. Expert Opin Biol Ther 12(7):811–821
74. Wright KT, El Masri W, Osman A, Chowdhury J, Johnson WE (2011) Concise review: bone marrow for the treatment of spinal cord injury: mechanisms and clinical applications. Stem Cells 29(2):169–178
75. Goldschlager T, Jenkin G, Ghosh P, Zannettino A, Rosenfeld JV (2010) Potential applications for using stem cells in spine surgery. Curr Stem Cell Res Ther 5(4):345–355
76. Harris DT (2008) Cord blood stem cells: a review of potential neurological applications. Stem Cell Rev 4(4):269–274
77. Shimazaki T (2003) Biology and clinical application of neural stem cells. Horm Res 60(Suppl 3):1–9
78. Bambakidis NC, Butler J, Horn EM, Wang X, Preul MC, Theodore N, Spetzler RF, Sonntag VK (2008) Stem cell biology and its therapeutic applications in the setting of spinal cord injury. Neurosurg Focus 24(3–4):E20
79. Reier PJ (2004) Cellular transplantation strategies for spinal cord injury and translational neurobiology. NeuroRx 1(4):424–451
80. Farzaneh M, Anbiyaiee A, Khoshnam SE (2020) Human pluripotent stem cells for spinal cord injury. Curr Stem Cell Res Ther 15(2):135–143

Check for updates

Plant Stem Cell Biology

11

Contents

Abstract

Plant stem cells (PSC) have been extensively studied by many research laboratories across the globe in the last decade highlighting their regenerative potential, cellular function, and regulatory mechanisms. PSC reveal the hallmark characteristic of self-renewal and developmental plasticity leading to the development of different cell identities and their subsequent differentiation. Studies have highlighted the molecules including hormones auxin and cytokinin, balancing cell division and differentiation in organogenesis, transcription factors, microRNA (miRNA) s, associated gene regulatory networks (GRNs), signaling by small peptides and their involved mechanism in regulation of stem cell maintenance, cell positioning and cell fate specification and the process of regeneration that play role in plant cell and organ regeneration. We discuss in this chapter the plant stem cells and differentiation.

Keywords

Stem cells · Cell fate · Gene regulatory networks (GRNs)

S. Chowdhury, S. Ghosh, *Stem Cells*, https://doi.org/10.1007/978-981-16-1638-9_11

Abbreviations

A thaliana	*Arabidopsis thaliana*
ACR4	ARABIDOPSIS CRINKLY4
AGO	Argonaute
ASC	Adult stem cells
BLJ	BLUEJAY
BR	Brassinosteroids
BRAVO	Brassinosteroids at vascular and organizing center
CK	Cytokinins
CLE:40	CLAVATA3-ESR RELATED 40
CLV-1	CLAVATA1
CUC	CUPSHAPED
EPF	Epidermal patterning factor
EPLF	EPIDERMAL PATTERNING FACTOR-like protein
ERF115	ETHYLENE RESPONSE FACTOR 115
ESR	Enhancer of shoot regeneration
GRNs	Gene regulatory networks
IAA	Indole acetic acid
JKD	JACKDAW
LBD16	Lateral organ boundaries domain 16
LRR	Leucine rich receptor
miRNA	microRNA
MZ	Maturation zone
OC	Organizing centre
PAN	PERIANTHIA
PSC	Plant stem cell
QC	Quiescent centre
RAM	Root apical meristem
RGF	Root growth factors
SAM	Shoot apical meristem
SCNs	Stem cell niches
STM	SHOOT MERISTEMLESS
TZ	Transition zone
WOX11	Wuschel-related Homeobox-11
WOX5	WUSCHEL-RELATED HOMEODOMAIN 5
WUS	WUSCHEL

11.1 Introduction

Regeneration is a characteristic of every species but the potential of regeneration reveals differences within the species and also within different organs of the individual. Plants, however, reveal high regenerative potential with differences in regenerative potential expressed by different plant organs indicative of specific complex interacting cells and molecules and regulations involved. Plants are sessile and are exposed to various biotic and abiotic stress factors that can affect their growth. The regenerative potential in plants makes them capable of coping up with these stress factors. Plant regeneration also reveals spatial and temporal restriction of this process correlating with position of stem cells, progenitor cells, meristematic cells. In the last decade considerable research has been involved in understanding of the plant stem cell (PSC) and its role in regeneration highlighting the important molecules and complex regulatory mechanisms involved in the process of regeneration.

11.2 PSC in Regeneration and Plant Development

The major difference of plant growth with that of animal growth is that while plants reveal continuous postembryonic development and growth throughout their life in forming branches, leaves, flowers, roots, increased body thickness, animal axis development is mostly complete at their embryonic state [1], animal growth ceases after they reach a genetically determined size that is species specific.

But both animal and plant tissues undergo wear and tear and loss to predators, disease, and injury. Both the plant and animal world combat these constant threats of agents, due to their different strategies and one of them by the function of plant adult stem cells (ASC) and inducing the stem cell like property of self-renewal and differentiation into diverse lineages in the differentiated cells. This biology of regeneration, repatterning involves genes and regulatory network and signalling pathways and is intensely complex and far from our complete understanding [1] (Fig. 11.1).

It is hypothesized in plants that both postembryonic development and regeneration share some common patterns, the events in post-embryogenesis involving differentiation of new cells from pre-existing ones also called as regeneration is an intriguing science.

Callus are a group of undifferentiated mass of cells capable of regeneration into a whole plant body [3]. Research reveals that every plant cell reveal totipotency or the property of a single cell to divide into different lineages of cells in an organism at any developmental stage leading to callus formation.

Cells of the thin layer of pericycle lying between the endodermis and the phloem expressing J0121 are reported to reprogram into callus by hormones or endogenous stimulation [3, 4]. Stem cells are known to play role in organ regeneration and development [3, 5–8].

PSC are found in areas within the meristem termed as stem cell niches (SCNs) that play role in proliferation and growth of the plant. Asymmetric nature of division of the stem cells gives rise to daughter stem cell with property of self-renewal, proliferation, growth, and differentiation [2] during postembryonic development.

In the plant model organism *Arabidopsis thaliana (A thaliana)*, meristems are located at both ends of seedlings. While the aerial part is termed as the shoot apical

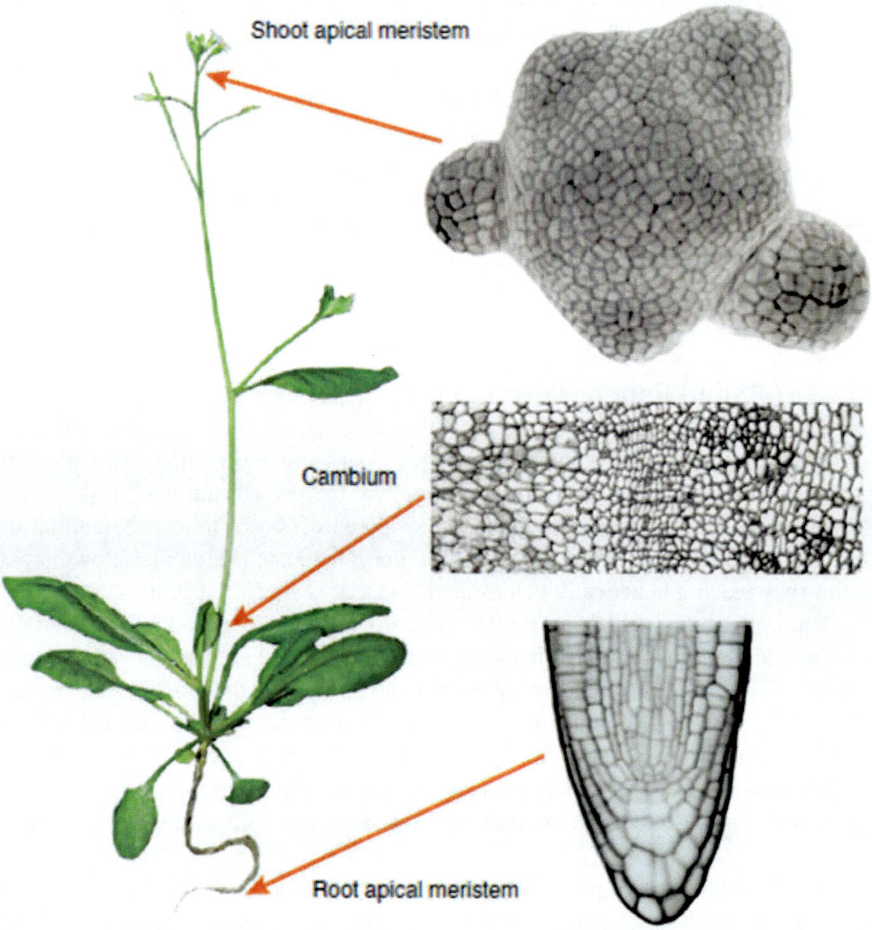

Fig. 11.1 In situ location and histological features of the three major stem cell systems in *Arabidopsis thaliana* (*A thaliana*). Image reproduced with permission from Greb and Lohmann [2]

meristem (SAM), the underground part is termed as the root apical meristem (RAM) [2, 9–11] and are developed during embryogenesis. In the postembryonic period, the underground part and lateral/vascular meristem further develops leading to thickening of the plant [12, 13]. Development of new meristems in the postembryonic growth leading to branches in the shoot and root system leads to the development of the adult plant body (Fig. 11.2).

11.3 Root Apical Meristem (RAM)

SCN housed in RAM both structurally and functionally is encompassed by quiescent centre (QC) surrounded by stem cells [2, 10, 11].

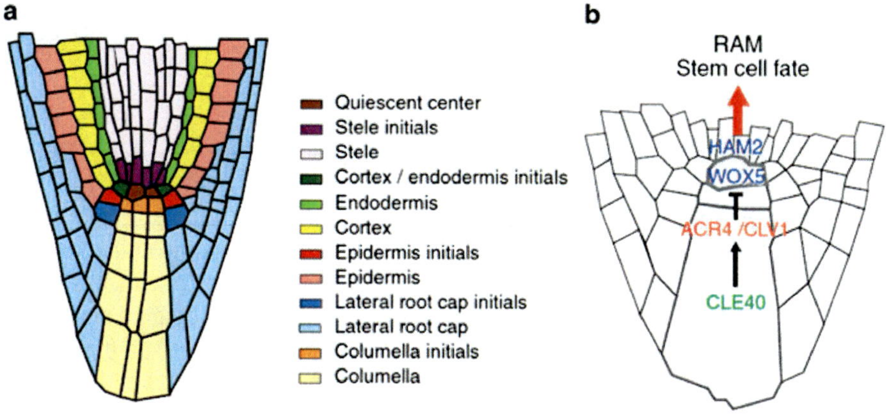

Fig. 11.2 The schematic representation of root apical meristem (RAM) with cell types (**a**), and regulations involved in maintaining the RAM activity (**b**). Image reproduced with permission from Greb and Lohmann [2]

Hormones are known to play role in plant growth regulation and auxin or indole acetic acid (IAA) and cytokinins (CKs) are plant hormones that regulate the maintenance and biological functions of RAM. Auxin reveals maximum auxin activity at the root tip QC cells that enables promoting SCN specification [14]. Cytokinins located towards the shoot is known to promote cell differentiation [15] together with a low auxin level promotes differentiation [16].

Brassinosteroids (BR) are polyhydroxylated steroidal plant hormones that is known to regulate the SCN activity, can cause plant cell elongation, growth, and development, and play role in providing responses to abiotic stress factors including extreme temperatures and drought. Inhibition of signalling function of BR has been reported to reduce cell division rate and renewal of distal stem cell, while increase in BR signalling in mutant varieties has been associated with events in early cell differentiation [17].

In RAM, QC is reported to express an R2-R3 MYB transcription factor termed as Brassinosteroids At Vascular And Organizing CENTER (BRAVO), that has been reported to play role in regulating BR signalling and is responsible for maintaining the low rate division or maintaining a more quiescent state [18].

Wuschel-Related Homeodomain 5 (WOX5) expressed by QC has been reported to regulate its maintenance and function [19] affecting the cell cycle regulators [20], and is regulated by the ligand–receptor interactions CLAVATA3-ESR RELATED 40 (CLE40)/ CLAVATA1 (CLV1) peptides and receptor kinase ARABIDOPSIS CRINKLY4 (ACR4) being a member of the family receptor-like kinases of CRINKLY4 (CR4) [21, 22].

CLE40 peptide expressed in the columella cells binds to CLV1, a leucine rich receptor (LRRs) or ACR4 which is its homologous expressed in columella stem cells restricting the expression and function of WOX5 to the QC. This enables the stem cell differentiation beyond SCN [21, 22].

Transcription factors, namely SCARECROW (SCR) and PERIANTHIA (PAN) and epigenetic modifications [23–25] are known to play role in WOX5 expression while its movement to adjacent cells in QC enables maintenance of their identity [24].

Transcriptional negative feedback loop of transcription factors SOMBRERO and FEZ, in the epidermal tissue, is known to play role in regulation of PSC [26].

Transcription factors, namely BIRD, BLUEJAY (BLJ), JACKDAW (JKD), and SHORT ROOT (SHR) enable asymmetric division of cells leading to the formation of cortex and endodermis tissue [14, 27]. Stem cell identity is maintained by transcription factors, namely BLJ, SCR, and JKD [14]. Wounds signal rapid regeneration of the SCN. Signalling by BR at the QC promotes synthesis of transcription factor ETHYLENE RESPONSE FACTOR 115 (ERF115) [28] and both transcription factors BRAVO and ERF115 play essential role in maintaining balanced cell divisions of QC.

Other genes including PLETHORA (PLT), SHR and SCR [29], BIRD, BLJ, and JKD are involved in root formation and in embryonic and postembryonic development and fine temporal and cell-type specific regulation [30].

11.4 Root Stem Cells (RSC) and Organ Regeneration

Root stem cells (RSC) are key cells in organ self-regeneration. An interplay of signalling by hormones, peptides, microRNA (miRNA), and transcription factors dominate the process (Fig. 11.3).

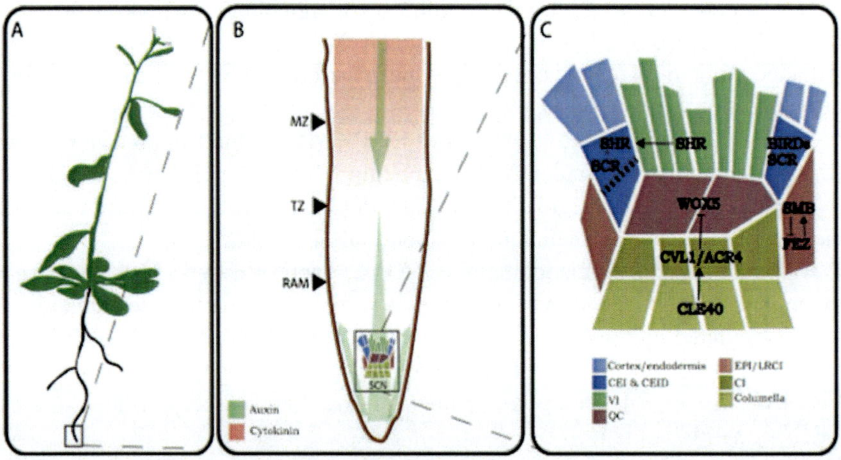

Fig. 11.3 Schematic representation of RAM in A thaliana. (**a**) adult plant primary root tip (**b**) Hormone flux in the root tip showing the Maturation zone (MZ), Root apical meristem (RAM), Stem cell niche (SCN), and Transition zone (TZ). (C) Stem cell regulators at the SCN of the RAM. Reproduced with permission from Perez-Garcia and Moreno-Risueno [30]

11.5 Shoot Stem Cells (SSC) in Self-Regeneration

Located apically in the stem tissue, the SAM, reveals properties of self regeneration and self renewal, generates aerial postembryonic organs involving the activity at the central zone, with stem cell layers including L1–3 and organizing center (OC) [2] expressing marker homeodomain transcription factor WUSCHEL (WUS) which can travel upwards to regulate stem cell activity [31, 32]. SAM cells reveal asymmetrical division in unspecified direction moving more towards the meristem periphery further away from the SCN. The plant hormone cytokinin (CK) expressed in SAM promotes maintenance of SCN while auxin hormone activates differentiation of the cells [33, 34].

CLV3 peptide are expressed under activity of WUS transcription factor and are secreted to the apoplast, is identified by CLV1, receptor leading to downregulated WUS expression in the OC, by negative feed-back regulation mechanism [35] thereby maintaining homeostasis. Since only some cells of SAM reveal regenerative capacities, it is suggested to have less regenerative capacity as compared to the stem [36] (Figs. 11.4 and 11.5).

11.6 SSC in Whole Organ Regeneration

Factors controlling pluripotency in PSC have been attributed to *PLT* genes including PLT3, PLT5, and PLT7 [37] and Wuschel-related Homeobox-11 (WOX11-LBD16) pathway [38] that is known to activate Lateral Organ Boundaries domain 16 (LBD16), transcription factor in *Arabidopsis thaliana*.

These newly developed shoot cells are further specialized by activation of different signalling pathways.

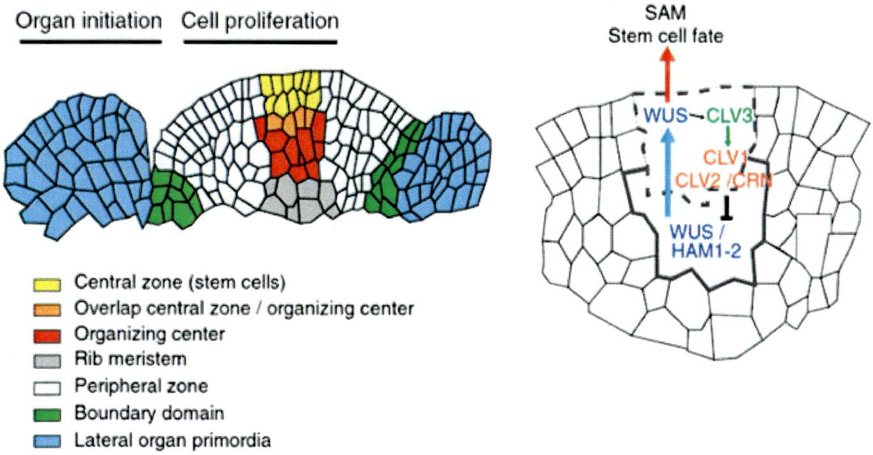

Fig. 11.4 Schematic representation of shoot apical meristem (SAM) (**a**). Central regulatory circuit in SAM (**b**). Image reproduced with permission from Greb and Lohmann [2]

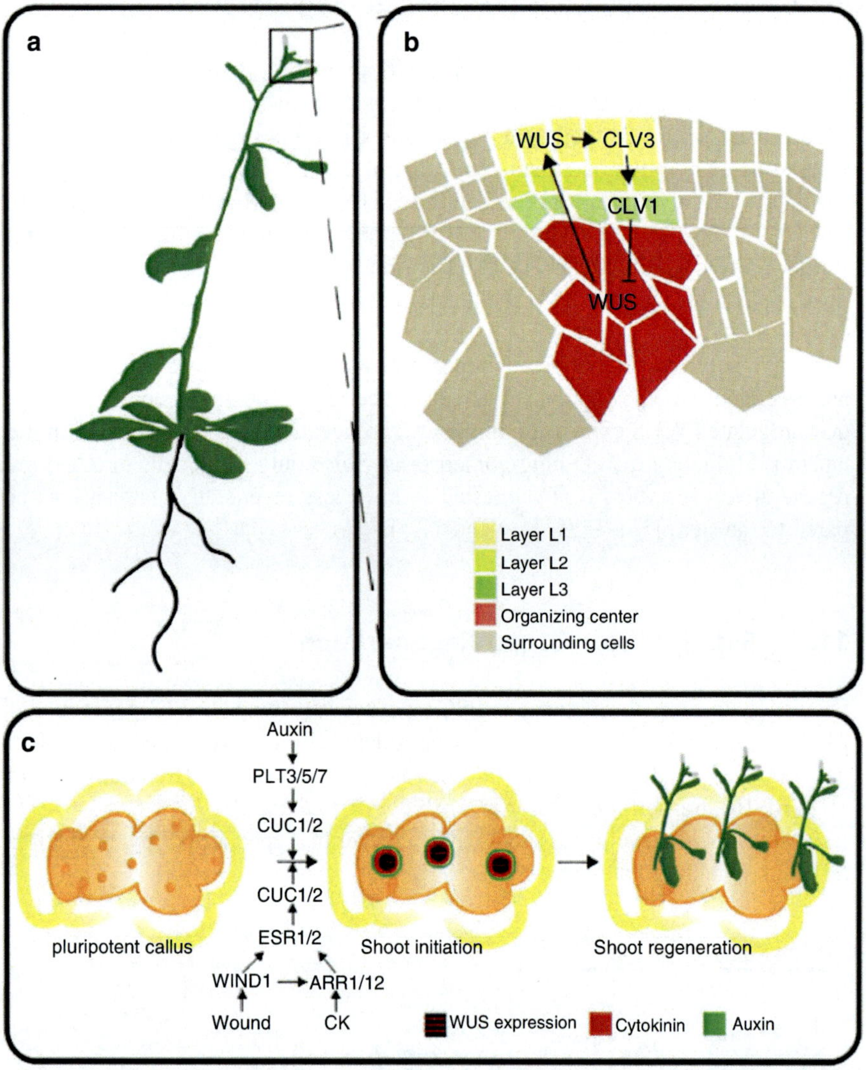

Fig. 11.5 Schematic representation of SSC regulators in regenerative processes. (**a**) Adult plant of *A thaliana* SAM, (**b**) Stem cell regulation at the SAM. (**c**) Shoot regeneration from Pluripotent callus. Reproduced with permission from Perez-Garcia and Moreno-Risueno [30]

PLT3, PLT5, and PLT7 mediate signalling of two NAC transcription factors including CUPSHAPED 1 (CUC1) and CUC2, regulate SAM initiation during embryogenesis [37, 39, 40]. Transcription factor, enhancer of Shoot Regeneration ESR1 and ESR2, is known to stimulate shoot regeneration promoting expression of CUC2, belonging to the NAC family of transcription factors [41].

SAM stem cell expresses transcription factor WOX2 [42] during embryogenesis but with limited role in shoot regeneration.

In vitro experiments have revealed that CK promotes maintenance of cellular identity and development of shoot tissue. Auxin can lead to regeneration of shoot from callus mediated by transcription factor PIN-FORMED1 (PIN1) even in presence of high levels of cytokines (CKs) [43–45]. Shoot can regenerate from roots.

Wounds have been reported to promote shoot regeneration from roots mediated by signalling by WIND1 in roots [46] together with the role of epidermis [47] and cell fate.

11.7 Stem Cell Differentiation and Regulation

Several factors play role in regulation events of cellular division process in plants including hormones, small peptides, miRNAs transcription factors that signal and mediate regulation of cell differentiation.

11.7.1 Hormones and Patterning

Plant hormone signalling, accumulation and function controls cell fates. Auxin (Au) and cytokinin (ck) plant hormones are known to play role in positioning and maintaining meristem activity, their balance regulates events of division and differentiation [48] and guides a cell towards gaining its identity. While in the root, auxin regulates stem cell divisions and cytokinin causes differentiation [49] the roles are exactly reversed in the stems. Auxin signal destabilizes signaling repressors and activates downstream transcription factors [50], while cytokinin signalling activate transcriptional regulators [51]. Both hormones reveal negative feedback regulation. Transcription factor *WUSCHEL-RELATED HOMEOBOX* (*WOX*) gene family play role in stem cell homeostasis in which *WUSCHEL* (*WUS*) and WOX5 play role in shoots and roots development, respectively. In SAM, *WUS* expression is governed by cytokinin [52]. In plants, expression of *WUS* in the organizing centre or OC controls fate of stem cell.

11.7.2 Role of Plasmodesmata

Movement of transcription factors, signalling molecules, and small RNAs through the selectively permeable plasmodesmata regulate development in plants [53]. With maturity the plasmodesmata can accommodate larger molecules [54].

11.7.3 Transcription Factors and miRNAs

In plants miRNAs are reported to play major role in transcriptional feedback mechanism. *miR165* and *miR166 that can form hairpin like structure play role* in regulation of processes of division and differentiation [55] in plants. While they promote differentiation in SAM, RAM, and other meristems, by repressing transcription factor family of HD-ZIP III [56], including members *PHAVOLUTA*, *PHABULOSA*, and *REVOLUTA*. Different argonaute (AGO) proteins in meristematic cells can promote differentiation in meristems by sequestering *miR165* and *miR166* away from targets.

11.7.4 Small Peptides

Small peptides including CLAVATA3/ESR-related (CLE) play role in signalling development and differentiation of plants [57] and CLE plays role in root development, lateral root, vascular tissue development. CLV3, CLE40, downregulates *WOX5* expression in stem cells encircling the root QC [58].

Epidermal patterning factor (EPF) peptides enable pattern formation [59] in leaf. EPF1 and EPF2 can promote division and restrict differentiation at different stages and play role in stomatal development [60]. EPIDERMAL PATTERNING FACTOR-like protein 2 EPFL2, together with auxin enables morphogenesis in plant leaf.

Small peptides function in maintenance of root endodermis [61]. Transcription factors CASPARIAN STRIP INTEGRITY FACTOR 1 (CIF1) and CIF2 also contribute to the complete development of Casparian strips.

11.7.5 Transcription Factors

Auxin accumulation can induce transcription of PLT [62] most in stem cells gradually declining shootwards [63] and PLTs are reported to regulate QC, stem cell maintenance and differentiation. Root growth factors (RGFs) play role in root meristem [64] maintenance. In *Arabidopsis, SHR* (short root) transcription factor enables root patterning [65]. SHOOT MERISTEMLESS (STM) enables SAM maintenance in *Arabidopsis* [66]. MUTE, immobile in *Arabidopsis* was reported to regulate cell divisions in leaf patterning [67, 68].

11.8 Discussion

The primary focus in plant developmental biology remains in the understanding of the underlying mechanisms that regulate maintenance of PSC and its transition into a functionally specialized fully differentiated cell.

Developmental pattern of position and identity leading to cellular lineage diversity and functional specialization are generated due to complex dynamic interactions and intercellular communications between genes and molecules including hormones like auxin and cytokinin, maintaining the balance in the events of processes of cell division and differentiation during organogenesis transcription factors, small RNAs, microRNA (miRNAs) and small peptides at cellular and molecular levels. Signalling encompasses dynamic interactions and feedback loops thereby specifying cell fate.

However, studies in this field are just at the tip of the iceberg. There are lots of answered questions related to how cells gain identity. One such important mind boggling question lies as to differential role of the same transcription factor in different cell types.

References

1. Birnbaum KD, Sánchez AA (2008) Slicing across kingdoms: regeneration in plants and animals. Cell 132(4):697–710
2. Greb T, Lohmann JU (2016) Plant stem cells. Curr Biol 26(17):R816–R821
3. Bustillo-Avendaño E, Ibáñez S, Sanz O, Barross JAS, Gude I, Perianez-Rodriguez J, Micol JL, del Pozo JC, Moreno-Risueno MA, Perez-Perez JM (2017) De novo root regeneration in Arabidopsis leaves. Plant Physiol 2017:00980
4. Sugimoto K, Jiao Y, Meyerowitz EM (2010) Arabidopsis regeneration from multiple tissues occurs via a root development pathway. Dev Cell 18(3):463–471. https://doi.org/10.1016/j.devcel.2010.02.004
5. Efroni I, Mello A, Nawy T et al (2016) Root regeneration triggers an embryo-like sequence guided by hormonal interactions. Cell 165(7):1721–1733
6. Iwase A, Mita K, Nonaka S et al (2015) WIND1-based acquisition of regeneration competency in Arabidopsis and rapeseed. J Plant Res 128(3):389–397
7. Kareem A, Durgaprasad K, Sugimoto K et al (2015) PLETHORA genes control regeneration by a two-step mechanism. Curr Biol 25(8):1017–1030
8. Perianez-Rodriguez J, Manzano C, Moreno-Risueno MA (2014) Post-embryonic organogenesis and plant regeneration from tissues: two sides of the same coin? Front Plant Sci 5:219. https://doi.org/10.3389/fpls.2014.00219
9. Gaillochet C, Daum G, Lohmann JU (2015) O cell, where art thou? The mechanisms of shoot meristem patterning. Curr Opin Plant Biol 23:91–97
10. Drisch RC, Stahl Y (2015) Function and regulation of transcription factors involved in root apical meristem and stem cell maintenance. Front Plant Sci 6:505
11. Fisher AP, Sozzani R (2016) Uncovering the networks involved in stem cell maintenance and asymmetric cell division in the Arabidopsis root. Curr Opin Plant Biol 29:38–43
12. DeRybel B, Vassileva V, Parizot B, Demeulenaere M, Grunewald W, Audenaert D, Van Campenhout J, Overvoorde P, Jansen L, Vanneste S, Möller B, Wilson M, Holman T, Van Isterdael G, Brunoud G, Vuylsteke M, Vernoux T, De Veylder L, Inzé D, Weijers D, Bennett MJ, Beeckman T (2010) A novel Aux/IAA28 signaling cascade activates GATA23-dependent specification of lateral root founder cell identity. Curr Biol 20:1697–1706
13. Ruonala R, Ko D, Helariutta Y (2017) Genetic networks in plant vascular development. Annu Rev Genet 51:335–359
14. Mähönen AP, Ten Tusscher K, Siligato R, Smetana O, Díaz-Triviño S, Salojärvi J, Wachsman G, Prasad K, Heidstra R, Scheres B (2014) PLETHORA gradient formation mechanism separates auxin responses. Nature 515:125
15. Ioio RD, Nakamura K, Moubayidin L, Perilli S, Taniguchi M, Morita MT, Aoyama T, Costantino P, Sabatini S (2008) A genetic framework for the control of cell division and differentiation in the root meristem. Science 322:1380–1384
16. Di Mambro R, De Ruvo M, Pacifici E, Salvi E, Sozzani R, Benfey PN, Busch W, Novak O, Ljung K, Di Paola L, Marée A, Costantino P, Grieneisen V, Sabatini S (2017) Auxin minimum triggers the developmental switch from cell division to cell differentiation in the Arabidopsis root. Proc Natl Acad Sci U S A 2017:05833

17. González-García M-P, Vilarrasa-Blasi J, Zhiponova M, Divol F, Mora-García S, Russinova E, Caño-Delgado AI (2011) Brassinosteroids control meristem size by promoting cell cycle progression in Arabidopsis roots. Development 138:849–859
18. Vilarrasa-Blasi M-P, González-García D, Frigola N, Fàbregas KG, Alexiou N, López-Bigas S, Rivas A, Jauneau JU, Lohmann PN, Benfey M, Ibañes A, Caño-Delgado (2014) Regulation of plant stem cell quiescence by a brassinosteroid signaling module. Dev Cell 30:36–47
19. Sarkar AK, Luijten M, Miyashima S, Lenhard M, Hashimoto T, Nakajima K, Scheres B, Heidstra R, Laux T (2007) Conserved factors regulate signalling in Arabidopsis thaliana shoot and root stem cell organizers. Nature 446:811
20. Forzani C, Aichinger E, Sornay E, Willemsen V, Laux T, Dewitte W, Murray JA (2014) WOX5 suppresses CYCLIN D activity to establish quiescence at the center of the root stem cell niche. Curr Biol 24:1939–1944
21. Stahl Y, Grabowski S, Bleckmann A, Kühnemuth R, Weidtkamp-Peters S, Pinto KG, Kirschner GK, Schmid JB, Wink RH, Hülsewede A, Felekyan S, Seidel C, Simon R (2013) Moderation of Arabidopsis root stemness by CLAVATA1 and ARABIDOPSIS CRINKLY4 receptor kinase complexes. Curr Biol 23:362–371
22. Stahl Y, Wink RH, Ingram GC, Simon R (2009) A signaling module controlling the stem cell niche in Arabidopsis root meristems. Curr Biol 19:909–914
23. de Luis Balaguer MA, Fisher AP, Clark NM, Fernandez-Espinosa MG, Möller BK, Weijers D, Lohmann JU, Williams C, Lorenzo O, R. (2017) Sozzani predicting gene regulatory networks by combining spatial and temporal gene expression data in Arabidopsis root stem cells. Proc Natl Acad Sci U S A 114:7632–7640
24. Pi L, Aichinger E, van der Graaff E, Llavata-Peris CI, Weijers D, Hennig L, Groot E, Laux T (2015) Organizer-derived WOX5 signal maintains root columella stem cells through chromatin-mediated repression of CDF4 expression. Dev Cell 33:576–588
25. Sabatini S, Heidstra R, Wildwater M, Scheres B (2003) SCARECROW is involved in positioning the stem cell niche in the Arabidopsis root meristem. Genes Dev 17:354–358
26. Willemsen V, Bauch M, Bennett T, Campilho A, Wolkenfelt H, Xu J, Haseloff J, Scheres B (2008) The NAC domain transcription factors FEZ and SOMBRERO control the orientation of cell division plane in Arabidopsis root stem cells. Dev Cell 15:913–922
27. Long Y, Smet W, Cruz-Ramírez A, Castelijns B, de Jonge W, Mähönen AP, Bouchet BP, Perez GS, Akhmanova A, Scheres B, Blilou I (2015) Arabidopsis BIRD zinc finger proteins jointly stabilize tissue boundaries by confining the cell fate regulator SHORT-ROOT and contributing to fate specification. Plant Cell 27:1185–1199
28. Heyman J, Cools T, Vandenbussche F, Heyndrickx KS, Van Leene J, Vercauteren I, Vanderauwera S, Vandepoele K, De Jaeger G, Van Der Straeten D, De Veylder L (2013) ERF115 controls root quiescent center cell division and stem cell replenishment. Science 342:860–863
29. Xu J, Hofhuis H, Heidstra R, Sauer M, Friml J, Scheres B (2006) A molecular framework for plant regeneration. Science 311:385–388
30. Perez-Garcia P, Moreno-Risueno MA (2018) Stem cells and plant regeneration. Dev Biol 442 (1):3–12
31. Daum G, Medzihradszky A, Suzaki T, Lohmann JU (2014) A mechanistic framework for noncell autonomous stem cell induction in Arabidopsis Proc. Natl Acad Sci USA 111:14619–14624
32. Mayer KF, Schoof H, Haecker A, Lenhard M, Jürgens G, Laux T (1998) Role of WUSCHEL in regulating stem cell fate in the Arabidopsis shoot meristem. Cell 95:805–815
33. Besnard F, Refahi Y, Morin V, Marteaux B, Brunoud G, Chambrier P, Rozier F, Mirabet V, Legrand J, Lainé S, Thévenon E, Farcot E, Cellier C, Das P, Bishopp A, Dumas R, Parcy F, Helariutta Y, Boudaoud A, Godin C, Traas J, Guédon Y, Vernoux T (2014) Cytokinin signalling inhibitory fields provide robustness to phyllotaxis. Nature 505:417
34. Zhao Z, Andersen SU, Ljung K, Dolezal K, Miotk A, Schultheiss SJ, Lohmann JU (2010) Hormonal control of the shoot stem-cell niche. Nature 465:1089

35. Fletcher JC, Brand U, Running MP, Simon R, Meyerowitz EM (1999) Signaling of cell fate decisions by CLAVATA3 in Arabidopsis shoot meristems. Science 283:1911–1914
36. Sena G, Wang X, Liu H-Y, Hofhuis H, Birnbaum KD (2009) Organ regeneration does not require a functional stem cell niche in plants. Nature 457:1150
37. Kareem A, Durgaprasad K, Sugimoto K, Du Y, Pulianmackal AJ, Trivedi ZB, Abhayadev PV, Pinon V, Meyerowitz EM, Scheres B, Prasad K (2015) PLETHORA genes control regeneration by a two-step mechanism. Curr Biol 25:1017–1030
38. Liu J, Hu X, Qin P, Prasad K, Hu Y, XuThe L (2018) WOX11–LBD16 pathway promotes pluripotency acquisition in callus cells during de novo shoot regeneration in tissue culture. Plant Cell Physiol 59:739–748
39. Aida M, Ishida T, Fukaki H, Fujisawa H, Tasaka M (1997) Genes involved in organ separation in Arabidopsis: an analysis of the cup-shaped cotyledon mutant. Plant Cell 9:841–857
40. Aida M, Ishida T, Tasaka M (1999) Shoot apical meristem and cotyledon formation during Arabidopsis embryogenesis: interaction among the CUP-SHAPED COTYLEDON and SHOOT MERISTEMLESS genes. Development 126:1563–1570
41. Ikeda Y, Banno H, Niu Q-W, Howell SH, Chua N-H (2006) The enhancer of shoot regeneration 2 gene in Arabidopsis regulates cup-shaped cotyledon 1 at the transcriptional level and controls cotyledon development. Plant Cell Physiol 47:1443–1456
42. Zhang Z, Tucker E, Hermann M, Laux T (2017) A molecular framework for the embryonic initiation of shoot meristem stem cells. Dev Cell 40:264–277
43. Atta R, Laurens L, Boucheron-Dubuisson E, Guivarc'h A, Carnero E, Giraudat-Pautot V, Rech P, Chriqui D (2009) Pluripotency of Arabidopsis xylem pericycle underlies shoot regeneration from root and hypocotyl explants grown in vitro. Plant J 57:626–644
44. Gordon SP, Heisler MG, Reddy GV, Ohno C, Das P, Meyerowitz EM (2007) Pattern formation during de novo assembly of the Arabidopsis shoot meristem. Development 2007 (134):3539–3548
45. Liu Z, Li J, Wang L, Li Q, Lu Q, Yu Y, Li S, Bai MY, Hu Y, Xiang F (2016) Repression of callus initiation by the mi RNA-directed interaction of auxin–cytokinin in Arabidopsis thaliana. Plant J 2016(87):391–402
46. Iwase A, Mita K, Nonaka S, Ikeuchi M, Koizuka C, Ohnuma M, Ezura H, Imamura J, Sugimoto K (2015) WIND1-based acquisition of regeneration competency in Arabidopsis and rapeseed. J Plant Res 128:389–397
47. Iwase A, Mitsuda N, Koyama T, Hiratsu K, Kojima M, Arai T, Inoue Y, Seki M, Sakakibara H, Sugimoto K, Ohme-Takagi M (2011) The AP2/ERF transcription factor WIND1 controls cell dedifferentiation in Arabidopsis. Curr Biol 21:508–514
48. Schaller GE, Bishopp A, Kieber JJ (2015) The yin-yang of hormones: cytokinin and auxin interactions in plant development. Plant Cell 27(1):44–63
49. Dello Ioio R, Linhares FS, Scacchi E, Casamitjana-Martinez E, Heidstra R et al (2007) Cytokinins determine *Arabidopsis* root-meristem size by controlling cell differentiation. Curr Biol 17(8):678–682
50. Weijers D, Wagner D (2016) Transcriptional responses to the auxin hormone. Annu Rev Plant Biol 67:539–574
51. Hwang I, Sheen J, Müller B (2012) Cytokinin signaling networks. Annu Rev Plant Biol 63:353–380
52. Chickarmane VS, Gordon SP, Tarr PT, Heisler MG, Meyerowitz EM (2012) Cytokinin signaling as a positional cue for patterning the apical-basal axis of the growing *Arabidopsis* shoot meristem. PNAS 109(10):4002–4007
53. Vatén A, Dettmer J, Wu S, Stierhof Y-D, Miyashima S et al (2011) Callose biosynthesis regulates symplastic trafficking during root development. Dev Cell 21(6):1144–1155
54. Nicolas WJ, Grison MS, Trépout S, Gaston A, Fouché M et al (2017) Architecture and permeability of post-cytokinesis plasmodesmata lacking cytoplasmic sleeves. Nat Plants 3:17082

55. Carlsbecker A, Lee J-Y, Roberts CJ, Dettmer J, Lehesranta S et al (2010) Cell signalling by microRNA165/6 directs gene dose-dependent root cell fate. Nature 465(7296):316–321
56. Williams L (2005) Regulation of *Arabidopsis* shoot apical meristem and lateral organ formation by microRNA *miR166g* and its *AtHD-ZIP* target genes. Development 132(16):3657–3668
57. Matsubayashi Y (2014) Posttranslationally modified small-peptide signals in plants. Annu. Rev. Plant Biol 65:385–413
58. Fernandez A, Drozdzecki A, Hoogewijs K, Vassileva V, Madder A et al (2015) The GLV6/RGF8/CLEL2 peptide regulates early pericycle divisions during lateral root initiation. J Exp Bot 66(17):5245–5256
59. Katsir L, Davies KA, Bergmann DC, Laux T (2011) Peptide signaling in plant development. Curr Biol 21(9):R356–R364
60. Pillitteri LJ, Torii KU (2012) Mechanisms of stomatal development. Annu Rev Plant Biol 63:591–614
61. Nakayama T, Shinohara H, Tanaka M, Baba K, Ogawa-Ohnishi M, Matsubayashi Y (2017) A peptide hormone required for Casparian strip diffusion barrier formation in *Arabidopsis* roots. Science 355(6322):284–286
62. Mähönen AP, ten Tusscher K, Siligato R, Smetana O, Díaz-Triviño S et al (2014) PLETHORA gradient formation mechanism separates auxin responses. Nature 515(7525):125–129
63. Galinha C, Hofhuis H, Luijten M, Willemsen V, Blilou I et al (2007) PLETHORA proteins as dose-dependent master regulators of *Arabidopsis* root development. Nature 449 (7165):1053–1057
64. Zhou W, Wei L, Xu J, Zhai Q, Jiang H et al (2010) *Arabidopsis* tyrosylprotein sulfotransferase acts in the auxin/PLETHORA pathway in regulating postembryonic maintenance of the root stem cell niche. Plant Cell 22(11):3692–3709
65. Hirsch S, Oldroyd GED (2009) GRAS-domain transcription factors that regulate plant development. Plant Signal Behav 4(8):698–700
66. Long JA, Moan EI, Medford JI, Barton MK (1996) A member of the KNOTTED class of homeodomain proteins encoded by the STM gene of *Arabidopsis*. Nature 379(6560):66–69
67. Pillitteri LJ, Bogenschutz NL, Torii KU (2008) The bHLH protein, MUTE, controls differentiation of stomata and the hydathode pore in *Arabidopsis*. Plant Cell Physiol 49(6):934–943
68. Pillitteri LJ, Sloan DB, Bogenschutz NL, Torii KU (2007) Termination of asymmetric cell division and differentiation of stomata. Nature 445(7127):501–505